AF902465

Springer Tracts in Civil Engineering

Series Editors

Sheng-Hong Chen, School of Water Resources and Hydropower Engineering, Wuhan University, Wuhan, China

Marco di Prisco, Politecnico di Milano, Milano, Italy

Ioannis Vayas, Institute of Steel Structures, National Technical University of Athens, Athens, Greece

Springer Tracts in Civil Engineering (STCE) publishes the latest developments in Civil Engineering - quickly, informally and in top quality. The series scope includes monographs, professional books, graduate textbooks and edited volumes, as well as outstanding PhD theses. Its goal is to cover all the main branches of civil engineering, both theoretical and applied, including:

- Construction and Structural Mechanics
- Building Materials
- Concrete, Steel and Timber Structures
- Geotechnical Engineering
- Earthquake Engineering
- Coastal Engineering; Ocean and Offshore Engineering
- Hydraulics, Hydrology and Water Resources Engineering
- Environmental Engineering and Sustainability
- Structural Health and Monitoring
- Surveying and Geographical Information Systems
- Heating, Ventilation and Air Conditioning (HVAC)
- Transportation and Traffic
- Risk Analysis
- Safety and Security

Indexed by Scopus

To submit a proposal or request further information, please contact:
Pierpaolo Riva at Pierpaolo.Riva@springer.com (Europe and Americas) Wayne Hu at wayne.hu@springer.com (China)

Ravi S. Jakka · Yogendra Singh · T. G. Sitharam ·
Bal Krishna Maheshwari
Editors

Earthquake Engineering and Disaster Mitigation

Contributions in the Honour of Late Professor D. K. Paul

Springer

Editors
Ravi S. Jakka
Department of Earthquake Engineering
Indian Institute of Technology Roorkee
Roorkee, Uttarakhand, India

Yogendra Singh
Department of Earthquake Engineering
Indian Institute of Technology Roorkee
Roorkee, Uttarakhand, India

T. G. Sitharam
Indian Institute of Technology Guwahati
Guwahati, Assam, India

Bal Krishna Maheshwari
Department of Earthquake Engineering
Indian Institute of Technology Roorkee
Roorkee, Uttarakhand, India

ISSN 2366-259X ISSN 2366-2603 (electronic)
Springer Tracts in Civil Engineering
ISBN 978-981-99-0080-0 ISBN 978-981-99-0081-7 (eBook)
https://doi.org/10.1007/978-981-99-0081-7

© Indian Society of Earthquake Technology 2023
This work is subject to copyright. All rights are solely and exclusively licensed by the Publisher, whether the whole or part of the material is concerned, specifically the rights of translation, reprinting, reuse of illustrations, recitation, broadcasting, reproduction on microfilms or in any other physical way, and transmission or information storage and retrieval, electronic adaptation, computer software, or by similar or dissimilar methodology now known or hereafter developed.
The use of general descriptive names, registered names, trademarks, service marks, etc. in this publication does not imply, even in the absence of a specific statement, that such names are exempt from the relevant protective laws and regulations and therefore free for general use.
The publisher, the authors, and the editors are safe to assume that the advice and information in this book are believed to be true and accurate at the date of publication. Neither the publisher nor the authors or the editors give a warranty, expressed or implied, with respect to the material contained herein or for any errors or omissions that may have been made. The publisher remains neutral with regard to jurisdictional claims in published maps and institutional affiliations.

This Springer imprint is published by the registered company Springer Nature Singapore Pte Ltd.
The registered company address is: 152 Beach Road, #21-01/04 Gateway East, Singapore 189721, Singapore

Preface I

Professor D. K. Paul obtained his B.Sc. Engineering from the University of Lucknow (1965), B.E. in Civil Engineering (1969), and M.E. in Earthquake Engineering (1971) from the erstwhile University of Roorkee. After completing his masters, he joined as a faculty member in the Department of Earthquake Engineering, University of Roorkee in 1972. In 1978, he went on to do his Ph.D. from the University of Wales, Swansea, UK, under a Commonwealth Scholarship and received it in 1982. He continued there as a postdoctoral researcher until 1986 before returning to the University of Roorkee. He served the University of Roorkee in many capacities such as Professor and Head, Department of Earthquake Engineering, Dean of Faculty Affairs, and Deputy Director. He was also the founding Head of the Centre of Excellence in Disaster Mitigation & Management and has actively participated in the various activities of the center. Professor Paul's contribution in postgraduate teaching in Earthquake Engineering and National Capacity Building Programs in Earthquake Engineering Education has helped greatly in Earthquake Disaster Reduction. As a part of the awareness program, he disseminated the basic knowledge on Earthquake Engineering and Earthquake Disaster Mitigation to the masses through simple lectures, guidelines, manuals, booklets, TV films, Symposia/Conferences, etc. These have made an immense impact on the Disaster Mitigation efforts in the country. After serving for 40 years, he retired as Professor/Deputy Director, IIT Roorkee. He continued to serve IIT Roorkee as an Emeritus Fellow.

Professor Paul was a leading authority in the field of Earthquake Engineering and Earthquake Disaster Mitigation. He was consulted for the assessment of the Seismic Safety of many Special Structures in the country requiring Earthquake Resistant Analysis and Design. He has made significant contributions in Earthquake Resistant Analysis and Design of Special Structures such as concrete gravity dams, earth and rockfill dams, bridges, nuclear power plants, multi-story buildings, etc., resulting in Earthquake Disaster Reduction in the country. He has served as Chairman and Member of several National Task Forces, Investigative Teams, National Expert Committees on Disaster Mitigation issues constituted by the DST, Ministry of Home Affairs, National Disaster Management Authority, Government of India, etc. He has also received several research awards and is a Fellow of several National Technical

Societies. Professor Paul served as the Chairman of the Earthquake Engineering Sectional Committee, CED 39 of the Bureau of Indian Standards. He was a Fellow of the Indian National Academy of Engineers (FNAE); the Indian Institute of Engineers (FIE); and the Indian Society of Earthquake Technology (FISET). He also served as the Vice President (1995–1997 & 1997–1999) and President (2007–2009 & 2009–2011) of the Indian Society of Earthquake Technology (ISET).

Professor Paul was the Chairman and Member of various national committees constituted by various ministries of the Government of India. He was Chairman of the Working Committee of Experts for Vulnerability Analysis and Risk Assessment constituted by NDMA, the National Steering Committee (NSC) on Seismic Microzonation of major cities constituted by the Ministry of Earth Sciences (MoES), the Working Group on Seismic Microzonation of Delhi Region constituted by the Department of Science and Technology (DST), and the Expert Group for the Preparation of a Comprehensive Proposal on Microzonation of Guwahati, constituted by the Ministry of Science and Technology (MoST). Also, he was a Member of the Expert Committee on 'Retrofitting of District Hospitals under National Earthquake Risk Mitigation Project (NERMP)' set up by the NDMA and the Environment Assessment Committee (MEAC) on River Valley Projects set up by the Ministry of Environment & Forest (MoE&F). The Ministry of Home Affairs (MHA) has appointed him Programme Coordinator of the National Programme for Capacity Building of Engineers in Earthquake Risk Management (NPCBEERM), Convener of the Committee on Model Building Byelaws and Review of City, Town and Country Planning Act and the Zoning Regulations, and Member of the National Core Group for Earthquake Disaster Mitigation. In view of the humongous contributions made by Professor Paul, the Indian Society of Earthquake Technology (ISET) had planned to bring out a Special Commemorative book volume entitled, "Earthquake Engineering and Disaster Mitigation—Contributions in the honour of Late Professor D. K. Paul".

Roorkee, India — Ravi S. Jakka
Roorkee, India — Yogendra Singh
Guwahati, India — T. G. Sitharam
Roorkee, India — Bal Krishna Maheshwari

Preface II

The editors are pleased to present this text to the readers on Earthquake Engineering and Disaster Mitigation—Contributions in the honour of Late Professor D. K. Paul. Extensive research study has been performed on the aspects of earthquake engineering and allied areas across the world, quite naturally because of its significance as well as relevance to the life safety of people. In the present edited book volume, the authors of various chapters have delineated a vast spectrum of research works carried out under several disciplines and sub-disciplines of earthquake engineering and technology. Brief coverage of various chapters is presented below.

Anbazhagan et al. (2023) presented seismic design criteria in the Indian Seismic Code IS1893 since its development, state-of-the-art procedure for the seismic hazard estimation, and the development of seismic design spectrum at Indian Rock Site from North India and South India seismic data separately. In this study, a modern smoothened and normalized way of developing the design spectrum using regional data is explained. Further, rock site seismic records from the southern and northern parts of India were collated and used to create the design spectrum. The derived design spectra presented are applicable at the rock sites for 5% damping based on inter and intraplate regions. This study has forecasted the fact that North and South Indian spectrums are different from the IS-1893 spectrum as the signature of each seismotectonic region is reflected in the proposed new spectral shape.

Narayan et al. (2023) presented relationships for predicting the amplification of SH-wave across two-dimensional deep basins at a fundamental frequency. In this study, numerical simulations and computation of spectral amplifications of the SH-wave across the considered rectangular and elliptical basins for different shape ratios, impedance contrast, sediment damping and development of relations to predict the amplification of SH-wave at fundamental frequency across 2D deep basins across the deep rectangular and elliptical basins have been presented. This study has developed relations for predicting the amplification at fundamental frequency across rectangular and elliptical deep basins in terms of amplification at center of respective basin, at an offset from the basin center as well as at half width of the basin. A comparison of numerically computed amplification at fundamental frequency across rectangular and elliptical deep basins with those obtained using developed relations reveals that the percentage differences are within the permissible limit.

Nath et al. (2023) performed earthquake-induced landslide hazard evaluation for seismic microzonation with emphasis to Garhwal Himalayas study area. Consideration of probabilistically generated peak ground acceleration for 10% exceedance probability in 50 years as a landslide triggering factor has been adopted in this study. This study recognized 464 landslides and 8 preparatory causative factors. The study depicted that around 40% of the study area falls under the zones of high and very high seismically induced landslide hazards.

The aspects of inherent randomness in the underlying geotechnical and geological formations make its characterization highly site-specific. Shreyasvi and Venkataramana (2023) outline the scope of a non-ergodic PSHA in the Indian scenario and the existing practices in capturing the uncertainties introduced by the site component. The authors believe that this manuscript can provide insight into improving the existing site-specific PSHA practices in the country.

Rangaswamy et al. (2023) presented a study on characterizing the dynamic properties of the ground soil profiles existing in southern Calicut city nearby the National Institute of Technology Calicut region. In this study, field MASW tests were carried out at 48 representative sites by using a geode seismograph. This study depicted that the average shear wave velocity (V_{s30}) of the uppermost 30m depth of soil strata is within the range of 273 to 615 m/s.

Kumar et al. (2023) in their study reported a comparative influence of two different types of loading mechanisms on the frequency-dependent dynamic properties of the cohesionless Brahmaputra sandy soil. In this study, both stress-controlled as well as strain-controlled experiments were performed on the aforementioned geomaterial which depicted that the loading frequency is found to influence the shear modulus and damping ratio with a contradicting response between the parameters. Further, this study also states that during strain-controlled loading, the frequency of loading is found to influence significantly the damping ratio only, whereas, for stress-controlled loading, both the damping ratio and shear modulus are found to be affected significantly by the loading frequency.

Padmanabhan and Maheshwari (2023) investigated the reliquefaction behavior of the Solani sand specimen by varying the acceleration amplitude and shaking duration subjected to repeated shaking events with two different shaking patterns. Experimental studies were performed with the saturated sand specimen at 25% relative density in a tank with repeated sinusoidal shaking. The experimental results demonstrated that the preshaking effect and shaking pattern are critical in influencing the reliquefaction potential of sand deposits and the pore pressure analysis depicts that the tendency of liquefaction increases even at lower acceleration with higher shaking duration.

Bharathi et al. (2023) investigated the behavior of batter piles under the influence of machine-induced vibrations. In this study, a series of dynamic lateral and vertical loading tests were conducted on a vertical piles and batter piles embedded into a layered silty sandy soil. Accelerometers were deployed on the pile cap in order to obtain the dynamic response of these piles. The results show that the resonant frequency of the soil–pile system decreases by 37–50% along the lateral direction and 43–50% along the vertical direction, with increasing force level. This study also depicted that in both lateral directions, the rotational stiffness of the soil–pile system decreases nonlinearly, whereas the damping ratio increases nonlinearly with an increase in lateral strain.

The catastrophic failure of pile foundations used for the support of bridges and other structures in liquefiable soils is still observed after every major earthquake. Basavanagowda et al. (2023) conducted three-dimensional finite difference analyses in order to evaluate the influence of non-liquefiable crust overlaid by a liquefiable crust with the pile embedded into non-liquefiable soil layer below the liquefiable soil in both level and sloping ground. Analysis was carried out for different cases of pile tip embedment into the liquefiable and non-liquefiable soil subjected to the 2001 Bhuj Earthquake motion. This study depicted that the maximum bending moments occurred at the interface of liquefiable and non-liquefiable layers.

Kumar and Takahashi (2023) investigated the efficacy of a hybrid foundation in order to mitigate liquefaction-induced effects under strong sequential ground motions. As per this study, the hybrid foundation is a combination of the gravel drainage system and friction piles having spiral blades devised under the footing as a hybrid mitigation technique against the liquefaction-induced effects on a shallow foundation. A series of dynamic centrifuge experiments were carried out to investigate the effectiveness of a hybrid foundation in the liquefiable ground under strong sequential ground motions. This study depicted that the second sequential ground motion is found to be diminishing the mitigation efficacy of the proposed hybrid foundation in comparison with the first sequential ground motion and the implications of strong sequential ground motions have also been elaborately presented.

Srivastav and Satyam (2023) investigated the contribution of vertical seismic coefficient in seismic analysis of hydro-tunnel in rock. This study provides a numerical analysis of the seismic response of a circular lined tunnel running through a jointed rock mass. The effect of tunnel depth, frequency, and Peak ground acceleration on axial force produced in the tunnel liner is also assessed in this study. This study concludes that incorporating the vertical seismic coefficient in the formulation results in a much higher maximum axial force during seismic loading and negating the vertical seismic coefficient might result in an underestimation of the earthquake effect.

Subramanya et al. (2023) conducted experimental studies to assess the dynamic response of a ten-story scaled building model supported on pile and piled raft foundations in soft clay. In this study, the response of the model structure was investigated for fixed and flexible base conditions considering two types of foundation systems, namely supported by pile groups and small piled raft foundations embedded in soft clay. This study delineated the fact that soil–pile structure system amplifies the lateral deflections and the story drift of the superstructure in comparison with the fixed base.

Sharma et al. (2023) investigated the seismic performance of masonry-infilled reinforced concrete frame building designed as per Indian codes. The current Indian design codes do not adequately cover the design of frame buildings with masonry infills. As a result, the designers continue to design frame buildings while ignoring the infills. This article reviews provisions of different national codes regarding the design of masonry infilled frames. Further, the effectiveness of the current design procedure being followed in the Indian design industry is also evaluated. Performance-based seismic engineering framework has been utilized in this study for quantifying the performance of frame designed using the mentioned design procedure.

It is indeed pleasing to see that a wide variety of topics have been dealt with in these chapters of the book. We feel that these contributed chapters in this book have elaboratively highlighted tenets of Earthquake Engineering and Disaster Mitigation aptly. Therefore, we believe that the latest developments in earthquake engineering and allied disciplines presented through these 13 chapters will prove to be highly informative to the readers and pave ways for further research. We thank all the staff of Springer for their full support and cooperation at all the stages of the publication of this book. We thank and acknowledge the service of authors and reviewers for their valuable time and efforts. We do hope that this book will be beneficial to students, researchers, and professionals working in the field of geotechnical earthquake engineering. The comments and suggestions from the readers and users of this book are most welcome.

Roorkee, India — Ravi S. Jakka
Roorkee, India — Yogendra Singh
Guwahati, India — T. G. Sitharam
Roorkee, India — Bal Krishna Maheshwari

Acknowledgements

We (editors) want to thank all the authors who have contributed to the book. We could bring this book out due to all the authors' timely contribution and cooperation. We would also like to thank and acknowledge the services of the following reviewers for their valuable time and efforts.

A Muralikrishna, Indian Institute of Technology Tirupati
Aditya Singh, Indian Institute of Technology Roorkee
Ashish Juneja, Indian Institute of Technology Bombay
Babloo Choudhury, National Institute of Technology Karnataka, Surathkal
Boominathan A, Indian Institute of Technology Madras
C R Parthasarathy, Sarathy Geotech Engineering Services
Deepankar Choudhury, Indian Institute of Technology Bombay
Dipti Ranjan Sahoo, Indian Institute of Technology Delhi
L Govindaraju, University Visvesvaraya College of Engineering, Karnataka
Madan Mohan Rout, Institute of Seismological Research
Madhavi G. Latha, Indian Institute of Science, Bangalore
Neelima Satyam, Indian Institute of Technology Indore
Paresh Nath Singha Roy, Indian Institute of Technology Kharagpur
Pradeep Kumar Dammala, Indian Institute of Technology Jodhpur
R Ayothiraman, Indian Institute of Technology Delhi
Robin Gee, GEM Foundation
S. K. Deb, Indian Institute of Technology Guwahati
S. K. Prasad, Vidyavardhaka College of Engineering, Mysore
Sanandam Bordoloi, The Hong Kong University of Science and Technology
Subhadeep Banerjee, Indian Institute of Technology Madras
Supriya Mohanty, Indian Institute of Technology BHU
Surendra Nadh Somala, Indian Institute of Technology Hyderabad
Vidya Bhushan Maji, Indian Institute of Technology Madras

Contents

About the Editors

Dr. Ravi S. Jakka is working as an Associate Professor in the Department of Earthquake Engineering, Indian Institute of Technology, Roorkee, and serving as Editor to the International Journal of Geotechnical Earthquake Engineering & ISET Journal of Earthquake Technology. He graduated in Civil Engineering from Andhra University Engineering College in the year 2001. He obtained master's and doctorate degrees from IIT Delhi in 2003 and 2007 respectively. His areas of interest are Site Characterization, Soil Liquefaction, Seismic Slope Stability of Dams, Landslides, Foundations & Seismic Hazard Assessment. He has published over 100 articles in reputed international journals and conferences. He has supervised over 40 Masters Dissertations and 8 Ph.D. thesis, while he is currently guiding 9 Ph.D. thesis. He has received prestigious DAAD and National Doctoral fellowships. He has obtained University Gold Medal from Andhra University. He also received 'Young Geotechnical Engineer Best Paper Award' from the Indian Geotechnical Society, and Professor TG Sitharam Mid-career Research Award from Indian Society of Earthquake Technology (ISET). He was instrumental in the development of Earthquake Early Warning System for northern India, a prestigious national project. He was also the Organizing Secretary to the 7th International Conference on Recent Advances in Geotechnical Earthquake Engineering and several other short-term courses/workshops/webinars. He also served as secretary of ISET for two terms.

Dr. Yogendra Singh is Railway Bridge Chair Professor, and former Head, Department of Earthquake Engineering, at Indian Institute of Technology Roorkee. He obtained his B.E. in Civil Engineering from IIT Roorkee (then University of Roorkee) in 1989 in First Division with Honours and M. Tech. and Ph.D. in Structural Engineering from IIT Delhi in 1990 and 1994, respectively. His has worked extensively on Performance Based Design of Buildings and Bridges; Seismic Vulnerability and Risk Evaluation; Non-Linear Modelling and Analysis; Seismic Evaluation and Retrofitting of Structures; and Earthquake Engineering in Hilly Regions. He has supervised 17 Ph.Ds., published more than 85 research articles in refereed journals and presented more than 100 papers in national and international conferences. He has been member of several expert teams for post-earthquake damage surveys and is convener of the BIS expert group on Performance Based Seismic Design of Structures. He is Fellow of Indian Society for Earthquake Technology, Life Member of Institution of Engineers (India), Member of American Society of Civil Engineers, and Earthquake Engineering Research Institute. He is recipient of IIT Roorkee Outstanding Teacher Award—2006, 2022, AICTE Career Award, Dr. Jai Krishna Prize and Medal, and Sir Arthur Cotton Memorial Prize.

Dr. T. G. Sitharam is the Director of Indian Institute of Technology Guwahati, Assam since July 2019. He is a member of the Science and Engineering Research Board (SERB), Established through an Act of Parliament: SERB Act 2008, Department of Science & Technology, Government of India. He is a Senior Professor in the Department of Civil Engineering, Indian Institute of Science (IISc), Bangalore, and served IISc for more than 27 years. He was Chairman of the Board of Governors at IIT Guwahati during 2019–2020. He was the former Chairman, Research Council, CSIR- CBRI (Central Building Research Institute, Roorkee). He is holding the position of Director (additional charge) of Central Institute of Technology, Kokrajhar, Assam (A Deemed to be University), since May 2021. Over the last 35 years, he has carried out research and development in the area of geotechnical and infrastructure engineering, seismic microzonation and soil dynamics, Geotechnical earthquake engineering and has developed innovative technologies in the area of earth sciences, leading to about 500 technical papers, 20 books with Google scholar H-index of 47 and I-10 index 137 with more than 7175 citations. He has guided 40 Ph.D. students and 35 Master's Students. He was listed in the world's top 2% of scientists for the most-cited research scientists in various disciplines by Stanford University in 2020. His name appeared again in the top 2% of scientists IN Elsevier by Stanford University in 2021. His broad area of research falls into Geotechnical Infrastructure engineering, earth sciences, hydrology, seismology, and natural hazards. He has carried out Seismic microzonation of many urban centers in India and he is an authority on seismic microzonation and site effects. Presently, he is the President of the Indian Society of Earthquake Technology and he was the chairman of the 7th International Conference on Recent Advances in Geotechnical Earthquake Engineering and Soil Dynamics held in July 2021. He is the chief editor of the International Journal of Geotechnical Earthquake Engineering, (IJGEE), PA, USA since 2010. He is the Editor-in-chief, Springer Transactions in Civil and Environmental Engg. series, Book Series, Singapore. He is the Fellow of ASCE, Fellow of ICE (UK), Diplomat of Geotechnical Engineering (D.Ge.) from Academy of Geo-Professionals, ASCE; Fellow of IGS, ISET, ISES, and many other Societies. Presently, he is Executive Council member of AICTE and Chairman, Eastern Zonal committee of AICTE. He is Chairman (Designate) of All India Council for Technical Education, Govt. of India. He is also a certified professional engineer and a chartered engineer from the Institution of Engineers.

Dr. Bal Krishna Maheshwari recently served as a Shamsher Prakash Chair Professor at Department of Earthquake Engineering, IIT Roorkee. He secured Ph.D. from Saitama University Japan, worked for about two years in industry in Tokyo,

Japan and then joined Washington University in St. Louis, Missouri, USA as a post-doctoral fellow. Since Dec. 2004, Dr. Maheshwari is a faculty at Department of Earthquake Eng., IIT Roorkee, India. Professor Maheshwari is working in the areas of Dynamic Soil-Structure Interaction, Liquefaction, Dynamic Soil Properties, Constitutive Modeling, Nonlinear Finite Element Analysis, Slope Stability, Landslides, Seismic Analysis of Tunnels and Disaster Mitigation & Management. He served as an Associate Editor of International Journal of Geomechanics, ASCE and as an Editor of ISET Journal of Earthquake Technology. Professor Maheshwari is a member of ASCE, EERI, ISSMGE, IACMAG, a life fellow of ISET, IGS, IEI and IWRS. He has been awarded "Excellent Research Contribution Award" in Kyoto, Japan in year 2014 and "John Booker Medal" in Torino, Italy in year 2022 of IACMAG. Professor Maheshwari served as a Head of Centre of Excellence in Disaster Mitigation and Management and as an Organizing Secretary for 16SEE. He is presently serving as a Vice-President of Indian Society of Earthquake Technology (ISET). He is elected as President, ISET for the term 2023–2025.

Chapter 1
Smoothed and Normalized Design Spectrum for Indian Rock Sites

P. Anbazhagan, Ketan Bajaj, and M. Shimna

Abstract Seismic resistance design requires the estimation of futuristic seismic force to the structure in terms of spectral acceleration/velocity/displacement at the corresponding natural period of the structure. These expected seismic forces are defined based on detailed seismic hazard analysis and design spectrums from recorded earthquakes in the region. In this study, we have presented seismic design criteria in the Indian Seismic Code IS 1893 since its development, state-of-the-art procedure for the seismic hazard estimation, and the development of seismic design spectrum at the Indian Rock Site from North India and South India seismic data separately. The first Indian seismic code of IS 1893 was released in 1962 based on the studies of the Geological Survey of India on past earthquakes. IS 1893 was frequently revised soon after major earthquakes in different parts of the country and the currently available version is IS 1893 (2016). The seismic zonation map of India is based on past earthquake intensities and not on systematic futuristic seismic hazard estimation accounting for probable location and size of earthquakes. The different natural period of structural design requires respective design spectral amplitude. The previous versions of IS 1893 have given seismic coefficients for seismic zones and spectral amplitude for the different periods based on earthquakes recorded in US at an epicentral distances of 50–70 km, with multiplication factors. A recent version of IS 1893 adopted a design spectrum from the Uniform Building Code, again without considering regional data. After discussing these points, a modern smoothened, and normalized way of developing the design spectrum using regional data is explained. Further, rock site seismic records from the southern and northern parts of India were collated and used to create the design spectrum. The derived design spectra presented are applicable at the rock sites for 5% damping based on inter- and intraplate regions.

P. Anbazhagan (✉) · M. Shimna
Department of Civil Engineering, Indian Institute of Science, Bengaluru, Karnataka, India
e-mail: anbazhagan@iisc.ac.in

K. Bajaj
Senior Earthquake Analyst, Swiss Re, Bangalore, Karnataka, India
e-mail: ketan_bajaj@swissre.com

M. Shimna
Department of Civil Engineering, BMS Institute of Technology and Management, Bengaluru, Karnataka, India

© Indian Society of Earthquake Technology 2023
R. S. Jakka et al. (eds.), *Earthquake Engineering and Disaster Mitigation*,
Springer Tracts in Civil Engineering, https://doi.org/10.1007/978-981-99-0081-7_1

Our study shows North and South Indian spectrums are different from the IS 1893 spectrum and the signature of each seismotectonic region is reflected in the proposed new spectral shape.

Keywords Seismicity · Seismic zone map · Seismic coefficient · Design spectrum · IS 1893

1.1 Introduction

India is rich in resources, culture, tradition, knowledge, and wisdom, which unfortunately are not truly reflected in the anti-seismic design and construction. Even today, the seismic code recommends seismic coefficient for the design of structure based on intensities and normalized spectrum arrived based on data recorded in US. On the other hand, the continuous tectonic strain buildup in the Himalayas causes several moderate and minor earthquakes, indicating the importance of Anti-Seismic Construction (ASC). Many ASC practices were traditionally adopted in several parts of India in the olden days and had slowly disappeared due to several reasons. The major reasons are an improper scientific explanation of those excellent practices, documentation, and lack of code of practice. Even now, simple rolling floor constructions above wooden beams in north Indian houses and sandbox techniques in temples in south India withstood several earthquakes. ASC might have been practiced from the experiences gained by our forefathers. These experiences might include two major aspects, one is past seismicity and expected future seismic force, and another is material and methods capable of handling expected seismic force. Later on, we can see via some rare current construction practices in the villages and age-old temples, but former ones are not available due to the unavailability of the historical scripts. At the same time, an increase in natural hazards, high population, and improper construction place India at high seismic risk and exposure at the global level [1]. Anti-seismic design and construction is a highly prioritized area to reduce seismic disasters. This is possible through proper estimation to provide reliable futuristic seismic forces for design in codes and make ASC a practice mandate. The former one is dealt with in this paper in detail to overcome some ambiguity in IS 1893-Indian Standard CRITERIA FOR EARTHQUAKE RESISTANT DESIGN OF STRUCTURES PART 1 GENERAL PROVISIONS AND BUILDINGS.

IS 1893 was first published in 1962 and revised soon after major earthquakes in the country, and the recent version was published in 2016. Seismic zones are marked based on past earthquake locations, zone factors are assigned based on past intensities, and design parameters are recommended based on the work in Western countries [2]. Indian seismic zonation maps and values are not based on a systematic estimation of potential hazards in each part of the country but are lumped values based on past known earthquakes [3]. The very first detailed seismic microzonation methodology was developed by the first author [4] by accounting geology, seismicity, seismotectonics, soil, site effects, and induced effects, a typical microzonation map of Bangalore was shown. But even today, we do not have a comprehensive futuristic seismic risk map of any city in India. Even though several seismic microzonation studies are carried out for Indian cities, the time taken for collating data and completing final maps makes these studies outdated. The seismic zonation map should be updated once in every 5 years or soon after a significant earthquake in the region, whichever is earlier. The current version of the IS 1893 seismic zonation map and design spectrum has several ambiguities and is not based on the state-of-the-art practice in the subject area; that could be the reason that the Sectional Committee mentions in every version of code that "there cannot be an entirely scientific basis for zoning in view of the scanty data available" and "Structures designed as per IS 1893 [5–11] are expected to sustain damage during strong earthquake ground shaking". So this study summarizes the development of seismic zone maps and seismic design coefficients in IS 1893 and highlights how to estimate futuristic seismic hazards at the bedrock level using rupture-based seismic hazard analysis developed at IISc [Indian Institute of Science].

This paper presents seismic records compiled by IISc for inter and intraplate regions of India, hereafter called South India and North India. Since both areas are entirely different regarding geology, seismicity, seismotectonic, and soil thickness and types, this is reflected in the seismic signatures, e.g., response spectrum. The complied acceleration time history data are separated based on region, and the cut-off periods for acceleration, velocity, and displacement-sensitive sections of the spectrum were estimated. Peak spectral acceleration, velocity, and displacement were estimated for the horizontal and vertical components for 5% damping at bedrock level. These results are further used to develop smoothened and normalized design spectrum for Peninsular India and North India. This is the first design spectrum of Peninsular India and North India using regional recorded earthquake acceleration time histories and state-of-the-art knowledge on the subject.

1.2 Indian Seismicity and Seismotectonic

India is rich in natural resources and aesthetic landscapes due to continuous seismotectonic and geological transformation, and these constant changes are non-uniform throughout the country. The degree and type of tectonic movement in different parts of India vary. Figure 1.1 shows the tectonic movements of India, which is part of Indo Australian Plate, moving with a speed of 26–36 mm/year in the Northeast direction and colliding with the Eurasian Plate, forming the Himalayan mountain ranges [12]. Indian landmass is predominately located on one side end of the Indo-Australian Plate [13]. Higher tectonic activities make this part to be called an Indian plate with unique understood seismotectonic activities. Shen [13] has highlighted that seismologists suspected from the 1980s that the Indo-Australian plate may be breaking up, ruptured four faults simultaneously within the Plate in April 2012 is a part of this breakup. One can narrow down the plate tectonic of India by taking effect of broadly distributed deformation of the northern Indian Ocean area within the composite India–Australia–Capricorn Plate. This area is well recognized as an Indian plate as a result of a reactivated fracture zone in the Indian Ocean Basin [13]. Jade et al. [14] showed that the "India plate borders the Eurasia plate on its northern and eastern boundary, Arabian plate on its western boundary; Somalia, Capricorn, and Australia plates to the south". Divergent boundary (pull apart) and associated deformation and activity are being documented with increased seismic activity in the Southern and Eastern boundary of the Indian plate. These south and southeastern and western boundaries may create large earthquakes resulting in Tsunamis affecting the Indian and Sri Lankan coastline and some associated moderate earthquakes in the landmass. So, special attention needs to be paid to understand these new seismic activities on the Indian Plate Southside as many nuclear power plants and harbors of India are located on the South Indian coastlines.

The major part of South India is located at mid of the Indian plate, which is a thin crust thickness. A major portion of it is called Peninsular India (PI) and is considered to be an intraplate region. The entire region is formed due to different geological transformations. Low plate thickness in the area causes rapid drifting towards the Himalayas in the northeastern direction with a high velocity of 5 cm per year [15]. Mohraz [16] interpreted that the earthquakes of the Indian plate interior are results of the periodic accumulation of stress/strain due to the shortening and release of accumulated strain along the same directions during the extension. This may be the reason that Central India has a fault plane at a depth of 5–38 km [17] and has caused significant earthquakes at Killari (Latur) and Jabalpur. Most of the intraplate earthquakes in PI are associated with unidentified local faults and weak zones Jade et al. [14] highlighted that the Indian plate interior is moving as a rigid block with a velocity similar to the Indian plate velocity and found no significant strain accumulation based on GPS measurement and the localized regional deformation specific to the active dislocations and faults in the region causes intraplate earthquakes. This is the main reason for isolated PI seismic events from each other, the movement along the regional dislocations and faults [14]. The northwestern part of the Indian plate covers the western part of India, i.e., edge of PI. The broad west boundary of India is a triple junction region where plates of India–Arabia–Eurasia meets. According to [18], Bhuj 2001 earthquake seems to be of the diffused Indian Plate western boundary rather than of intraplate tectonics. The western boundary of the Indian plate close to Kachchh is an active and transformed boundary and is the reorganization of plate velocities and directions [19], which induced a change in the Arabia–India–Somalia triple junction. Freeman [19] highlighted the chances of infrequent earthquakes of magnitude 7 and greater along the Arabia–India plate boundary unless deformation is in the form of aseismic creep. This scenario makes complex straining of western Gujarat and causes frequent moderate seismic events. Moving north; covering west and eastern parts of the north of Indian plate is the Himalayan Arc of 2500 km and characterized by several thrust faults that sole into the basal detachment of the Himalayan wedge or the main Himalayan Thrust. This entire region is a convergent boundary with a non-uniform slip rate and strain-locking zones. The

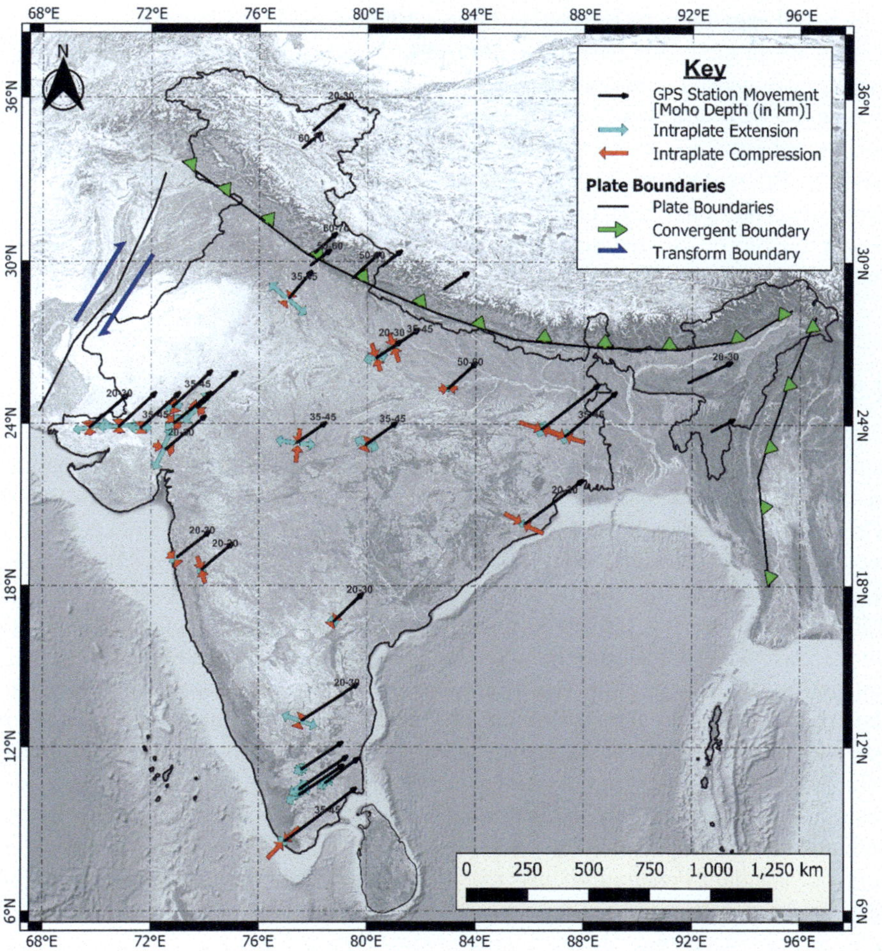

Fig. 1.1 Tectonic movements of India with Moho depth in km (marked in black long thin arrow) and stress direction of extension and compression (marked as a short thick arrow)

Himalayan Arc is seismically active due to the active under-thrusting of the Indian tectonic plate below the Eurasian plate and can be segmented from west to east into Kashmir, Ladakh, Gharwal, Kumaon, Nepal, Sikkim, Bhutan, Arunachal Himalaya, and Eastern Syntaxis [14]. We have recently estimated the futuristic seismic amplification of the Indo-Gangetic Plain, considering possible significant earthquakes due to the seismic gaps [20]. The eastern part of the Indian Plateau is much more complex, where three tectonic features of convergent and transformed boundaries and intraslab seismic activities take place. The northeastern side of the Indian plate, having transform motion with the Eurasian plate and the Eastern side with Burma–Sunda Plates. India–Burma convergence megathrust is currently accumulating strain and inactive/aseismic due to the lack of notable interplate instrumental earthquakes, which will eventually be released in future earthquakes [21]. There has been no big earthquake in the recent past in the northeastern part of India, but low to moderate events have caused extensive damages and liquefaction at several locations for a magnitude of 6 and less [22]. Overall, we can recognize that India has different seismic recurrences, seismotectonics, seismic sources, and depths. So, these may result in different seismic signatures and associated response spectrum even for the same site condition, i.e., layers with shear wave velocity (V_s) of more than 1500 m/s, which need to be incorporated in seismic design consideration in various parts of India.

1.3 Geology and Subsurface of India

Indian tectonic activities created a different type of surface and subsurface formation in India, where rock and soil layers are different in every kilometer grid of India. These variations are reflected in subsurface soil and soft rock type, thickness, and topography level. Subsurface layers causing Seismic Geo Hazards (SGH) of amplification, liquefaction, ground deformation, and landslide generally have V_s less than 1500 m/s and overlay hard rock, non-amplifying layer with V_s of 1500 m/s and above. Several earthquakes in India caused all types of SGH for a magnitude of 5.5 and above. But even now, there is no comprehensive SGH estimation using regional data and models. Researchers in India have made several attempts to estimate SGH and seismic microzonation maps since the work of [4]. But still, far away to estimate reliable SGH and microzonation maps using regional data and models. Here, we restrict our discussion only to the variation of surface and surface materials in the Indian landmass responsible for SGH. The shear strength of the subsurface layers in terms of standard Penetration Tests (SPT) N values or V_s values from geophysical tests are predominantly used for seismic site characterization and to estimate SGH at each place. Even though ample geotechnical data is generated as part of infrastructure projects, this data available for researchers is minimal. Even when the data is available, it is of little use since testing was not done as per the international standard requirement to use data for the estimation of SGH. In the last few years, shear wave velocity measurements have increased in different parts of India, and V_s is related to SPT N values [Uncorrected]. These correlations have different regression coefficients and goodness of fit within the region due to subsurface variation [23]. A couple of soil maps are published for India, but those are based on soil samples from very few centimeters with a concentration of geological classification. These surface-based soil maps may not help to arrive at a reliable SGH of any location. As per the author's knowledge, there is no comprehensive subsurface layer information required for SGH estimation.

In 2014, [2] reviewed geotechnical provisions in IS 1893 [10] and summarised soil type and its thickness in a different part of India using reliable data. Authors highlighted that "Geology and subsurface data collection show that India has diverse geology, soil and rock properties and site-specific variations in soil and rock properties must be accounted in seismic code similar to modern codes in foreign". It is worth mentioning that despite subsurface soil and rock variation in India, many researchers use SPT N or V_s seismic site classification developed based on American studies of NEHRP (National Earthquake Hazards Reduction Program) site classification [24]. NEHRP site classification is applicable for sites with rock depths 25–35 m and in shallow bedrock regions, it gives a higher site class and misunderstanding of amplification [25, 26]. At the same time, one should not forget that IGB has soil thickness up to 4–6 km deep with a very soft liquefiable surface soil deposit of up to 50 m. Systematic V_{SS} measurement up to a depth of 500 m and comparison with borelog by Anbazhagan and Ketan Bajaj [27] helped to understand the variation of amplification with depth in IGP. We found that amplification of subsurface layers several meters below the ground surface is much higher than that of surface layers, which needs to be accounted for in seismic design in those regions [20]. In principle, amplification correlations developed in other countries for peak ground acceleration/velocity and average spectral accelerations do not apply to India [25] and should not be used to site effect estimation. There is a need to understand the subsurface and surface geology and geotechnical properties and models for Indian soils at the micro-level and use them for reliable SGH estimation to reduce seismic risk due to SGH.

1.4 Regional Approach for Seismic Zonation Map

Several historic structures were designed for seismic forces and sustained several mega earthquakes in India. However, there is no evidence of a historical document explaining how it was done except for a few traditional practices in each state in the country. The seismic code initiative originated after a large-scale seismic disaster and destruction during Bihar–Nepal 1934 earthquake. The concept of seismic design was officiated only in 1962 in the IS 1893 seismic code. Buildings Sectional Committee [BSC] felt the need to rationalize the earthquake-resistant design of the structure to suit the Indian condition. BSC highlighted that IS 1893 [5] was based on accepted principles and practice in the field of earthquake-resistant design of structures before 1962. A number of important factors on the earthquake-resistant design of structures which are at the investigatory stage or not yet universally accepted were excluded from the IS 1893 [5] code and kept a scope for subsequent modification and revision. Many of the recommendations are primarily based on the research conducted abroad. Code clearly highlighted that it is not intended to lay down regulations such that no structure shall suffer any damage during earthquakes up all the magnitudes and the code, however, ensures that as far as possible, structures designed as per code are able to respond without structural damage to shocks of moderate intensities without total collapse to shocks of heavy intensities. Here, it is not clear to authors what moderate and heavy intensities of the different parts of India are. Only starting from IS 1893 [5] version, the earthquake-resistant design of normal structures and a detailed investigation were recommended for special and important structures. More or less above statements are repeated in most of the IS 1893 revised versions [6–11], and some of the statements are purposefully removed. Any seismic code recommendations can be broadly divided into three aspects; one is the recommendation of seismic hazard values at the bedrock level in the form of a seismic zonation map. Second is a recommendation of surface-level geohazard values based on different subsurface soil found in the region by considering site effects, liquefaction, and landslide. The third recommendation is building aspects such as configuration for earthquake resistance. This paper is limiting discussion only to the first and second recommendations in the IS 1893 code. The second one is not fully addressed in the code except few copied formulas and methods in the 2016 version without accounting for the testing practices and subsurface soil layers found in different parts of India.

Seismic zonation values are given in IS 1893 in the form of a map and the values in the table for each city for rock site conditions. IS 1893 [5] seismic zonation map was prepared using a rational approach based on the known magnitude and unknown epicenter. BSC assumed that all the other conditions were average and modified, such as average idealized isoseismal map in the light of tectonic, geology, and the maximum intensities as recorded from damage surveys, etc. The committee has also reviewed such maps in the light of past history and future possibilities and also attempted to draw a line demarcating the different zones to clear important towns, cities and industrial areas; after making a special examination of such cases, *the little modification in the zone demarcations may mean the considerable difference to the economics of the project in that area.* These points in IS 1893 [5] clearly show that the seismic zonation map was prepared based on past intensities and economic development of the area. The seismic zonation map of 1962 was modified in 1966, the number of seismic zone in the country kept similar, but the boundary of zones was modified. Figure 1.2 shows the comparison of the seismic zonation map released in 1962 and 1966 in IS 1893. A summary of seismic coefficients for cities with populations above 20 lakhs as per the 2011 census is given in Table 1.1. The seismic coefficient specified in the IS 1893 [5, 6] corresponds to the maximum acceleration that may be expected in any direction. At the same time, BSC said that seismic coefficient/factors are dependent on many variables and factors, and it is an extremely difficult task to determine the correct seismic acceleration at each location in the country. Hence, seismic coefficients are broadly adopted in different country zones, and rigorous analysis is recommended for important projects. These two codes give seismic coefficients [the ratio of the design acceleration due to the earthquake and the acceleration due to the gravity] for different subsurface layers broadly classified into three types. Table 1.2 shows subsurface layers of three types defined in IS 1893 by taking bearing capacity and SPT N value as a reference as per IS 2131. These subsurface layers should not settle appreciably due to the vibration loading for a few seconds. This means that IS 1893 design parameters are unsuitable for the site that undergoes displacement or settlement due to vibration loading. Figure 1.2 shows that few parts of the country are under seismic zone 0 since there are no intensities in that region. In Table 1.1, we can see that many south Indian cities have zero seismic coefficients as per IS 1893 [5, 6]. Unfortunately, several damaging earthquakes have occurred in the 0 zones of the country, leading to the removal of 0 zones and updating the 1966 zone in 1970.

The first time, BSC felt that no place in the country was free from the earthquake, so zero was removed, and zones VI and V were merged as zone V. So, in 1970, IS

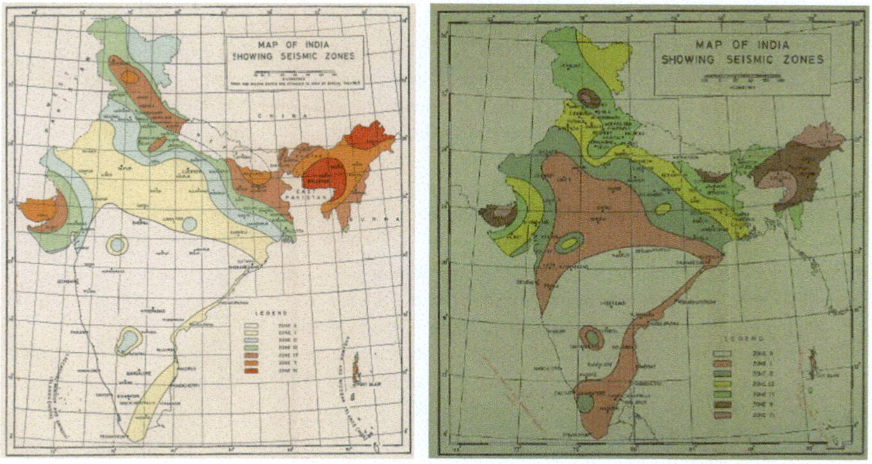

Fig. 1.2 IS 1893 seismic zonation maps published by Indian Standards Institution (IS 1893 [5, 6])

Table 1.1 Seismic coefficient/zone factor of cities with populations more than 20 lakhs as per 2011 census. The type of subsurface [T-I, T-II and T-III] is explained in Table 1.2

Sl no	Zonation year	1962			1966			1970/1975/1984		2002/2016
	Cities	T-I	T-II	T-III	T-I	T-II	T-III	T-I	Zone factor	Zone factor
1	Mumbai	0	0	0	0	0.01	0.02	0.04	0.2	0.16
2	Delhi	0.04	0.05	0.06	0.05	0.06	0.08	0.05	0.25	0.24
3	Bangalore	0	0	0	0	0	0	0.01	0.05	0.10
4	Hyderabad	0	0	0	0	0	0	0.01	0.05	0.10
5	Ahmedabad	–	–	–	0.04	0.05	0.06	0.04	0.2	0.16
6	Chennai	0	0.01	0.02	0	0.01	0.02	0.02	0.1	0.16
7	Kolkata	0.04	0.05	0.06	0.04	0.05	0.06	0.04	0.2	0.16
8	Surat	0.02	0.03	0.04	0.02	0.03	0.04	0.04	0.2	0.16
9	Pune	0	0	0	0	0.01	0.02	0.04	0.2	0.16
10	Jaipur	0	0.01	0.02	0	0.01	0.02	0.02	0.1	0.10
11	Lucknow	0	0.01	0.02	0.02	0.03	0.04	0.04	0.2	0.16
12	Kanpur	0	0.01	0.02	0.02	0.03	0.03	0.04	0.2	0.16
13	Nagpur	0	0	0	0	0	0	0.02	0.1	0.10

Table 1.2 Subsurface classification as per IS 1893 different versions [5–11]

Code year	Type I	Type II	Type III
1962	Hard soil having a bearing capacity greater than 45 tonnes/m^2	Average style having a bearing capacity greater than 20 tonnes/m^2 and equal or less than 45 tonnes/m^2	Soft soil having a bearing capacity greater than 10 tonnes/m^2 and equal or less than 20 tonnes/m^2
1966	Soil Type I having a bearing capacity greater than 45 tonnes/m^2	Soil Type II having a bearing capacity greater than 20 tonnes/m^2 and equal or less than 45 tonnes/m^2	Soil Type III having a bearing capacity greater than 10 tonnes/m^{2}and equal or less than 20 tonnes/m^2 provided that the standard penetration value (see IS2131-1963) is equal to or greater than 10
1970 1975 1984 2002	Type I rock or hard soil: well-graded gravel and sand gravel mixtures with or without clay binder, clayey sands, poorly graded, or sand–clay mixtures (GB, CW, SB, SW, and SC)* having N above 30	Type II medium soils—All soils with N between 10 and 30, and poorly graded sands or gravelly sands with little or no fines (SP*) with N > 15	Type III soft soils: all soil other than SP* with N < 10
2016	Type A rock or hard soil: well-graded gravel (GW) or well-graded sand (SW) both less than 5% passing 75 mm sieve (fines) Well-graded gravel—sand mixture with or without fines (GW-SW) Poorly-graded sand (SP) or Clayey sand (SC) all having N above 30 Stiff to hard clays having N above 30	Type B medium or stiff soils—Poorly graded sands or poorly graded sands with gravel (SP) with little or no fines having N between 10 and 30 Stiff to medium stiff fine-grained soils, like silts of low compressibility (ML) or clays of low compressibility (CL) having N between 10 and 30	Type C soft soils: all soft soils other than SP with N < 10. The various possible soils are: Silts of intermediate compressibility (MI); Silts of high compressibility (MH); Clays of intermediate compressibility (CI); Clays of high compressibility (CH); Silts and clays of intermediate to high compressibility (MI-MH or CI-CH); Silt with clay of intermediate compressibility (MI-CI); and Silt with clay of high compressibility (MH-CH)

Note N is the standard penetration value- Measured as per IS2131-1963/1981 method for standard penetration test for soil. *See IS1498-1959/1970 classification and identification of soil for general engineering purposes. In 2002 and 2016 versions of IS 1893 were given N as corrected N values code had only five seismic zones, and the seismic coefficient for different subsurface layers was given as a function of foundation type, i.e., "β for different soil foundation systems". For regular isolated footing, β is 1.0 for type I, 1.2 for type II, and 1.5 for Type III (see Table 1.2 for types of soil in the 1970 version). In the 1975 and 1984 versions, seismic coefficient method and response spectrum method factors are included. The latter factor is called a seismic zonation factor [Sa/g], and it is five times larger than the seismic coefficient as per IS 1893 [8, 9]. Not many changes were taken place in the later revision of IS 1893 [8] and 1984 when compared to the 1970 version. Three subsurface classifications given in the 1970, 1975, and 1984 versions of IS 1893 are almost similar and only desirable field N values are given with the updated version of IS 2131 and IS 1498. After sequence earthquakes at Latur (1993), Jabalpur (1997), and Bhuj (2001), IS 1893 was revised in 2002. Figure 1.3 shows the seismic zonation map of IS 1893 published in 1984 and 2016. The low seismic zone of I is completely removed, and several parts of the country are upgraded to higher zones. IS 1893 [10, 11] gives a zone factor for the Maximum Considered Earthquake [MCE], and a multiplication factor of ½ is suggested for Design Basis Earthquake to reduce MCE. Design seismic coefficient and zonation factor in seismic zonation maps from 1962 to the recent version of 2016 (see Table 1.1) are based on past earthquake locations and intensities reported in several parts of the country. The above discussion clearly shows that the seismic zonation map did not account for strain accumulation, the possible future seismicity, and systematic, rigorous seismic hazard analysis.

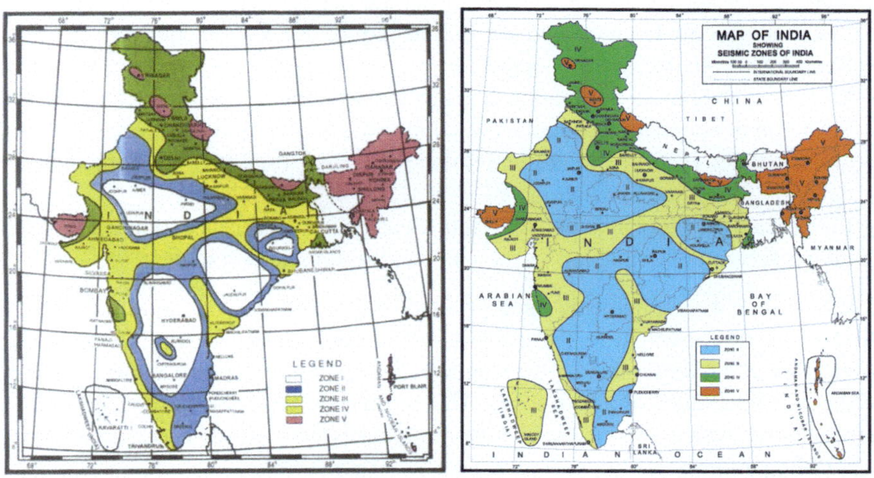

Fig. 1.3 IS 1893 seismic zonation maps published by Indian Standards Institution (IS 1893 [7, 11])

1.4.1 Regional Seismic Hazard Analysis

By knowing the seismotectonic and seismicity of India, uniform seismic zoning may lead to errors in arriving at the futuristic seismic zonation values. A study of IS 1893's different versions of codes clearly shows that the Indian seismic zonation map was not prepared based on the local variation of seismic aspects and state-of-the-art knowledge on the subject. Several seismic hazard analyses are carried out for cities in India or as a whole country, but the methodology and model adopted are almost similar. Some of the studies used outdated data and models. There are a lot of improvements have taken place in the seismic hazard analysis by accounting for regional seismicity and seismotectonic. These are yet to be systematically adopted by Indian researchers working in seismic hazard analysis. Even very critical structures and facilities in the country are designed using outdated analysis methods and results. As per the first author that many Indian researchers conduct seismic hazard analysis by adopting foreign procedures and models and are not doing any research to improve the same. Research carried out by the first author team reveals that a regional-specific approach is required to estimate reliably futuristic seismic factor estimation. Some interesting findings by accounting for regional seismic aspects in seismic hazard analysis can be found in [28]. We believe that seismic hazard analysis of the region should account for the following specific to the area.

- Identification of Seismic Study Area (SSA)—Area around the study area where earthquake occurrence in SSA and cause desirable ground motion at the study area.
- Location-specific Homogenization and magnitude conversion equation.
- Region-specific declustering approach.
- Maximum magnitude estimation considering regional rupture character [29].
- Selection of attenuation equations suitable to the region based on recorded earthquake data or intensity values and arrive ranks and weights of each attenuation equation for hazard estimation.
- Identification of probable potential future earthquake source locations in SSA by taking account of damaging earthquakes and strain released in the SSA.

Conventional deterministic and probabilistic-based seismic hazard analysis (SHA) give more weight to the location where earthquakes have occurred in the recent past. Also, they consider it a source in SHA, if the source has experienced an earthquake magnitude of 3.0 or above. This means conventional analyses give more seismic hazard values in a location where earthquakes have already occurred irrespective of the particular earthquake return period or repetition. The same concept was adopted in IS 1893 for assigning zone seismic coefficient/factor in all code versions. That may be the reason that seismic zones in the code are being modified soon after damaging earthquakes in the country. In order to overcome these issues, Anbazhagan et al. [30] developed rupture-based seismic hazard analysis by taking into account probable future earthquake locations. Anbazhagan and Silas Abraham [31] carried out Region-Specific Seismic Hazard Analysis of the Dam site by updating the procedure given by Anbazhagan et al. [30]. We believe that systematic region-specific seismic hazard analysis considering rupture character can help to arrive at reliable futurist seismic hazard values. The systematic region-specific hazard analysis steps are presented below:

(1) Selection of SSA based on Intensity/Damage Distribution/PGA interest to Structures (0.01 g) map.
(2) Identification of best magnitude conversion equations and Homogenization.
(3) Preparation of seismotectonic source map based on the regional data and seismic activity.
(4) Understand seismic distribution and delineation region for estimation of $M_{\max}$ and Recurrence relation.
(5) Estimation $M_{\max}$ for each source or region considering regional rapture character.
(6) Identification of Probable future rupture's location based on Anbazhagan and Silas Abraham [31].
(7) Characterize set of $M_{\max}$ and hypocentral distance based on regional seismotectonic considering damaging earthquakes in SSA in the last 50 years and Probable future rupture location as per (6).
(8) Selection of predictive relations, estimate weights, and ranks considering regional seismic data.
(9) Estimated PGA at the site for different combinations and identify controlling earthquake magnitude and distance resulting in higher PGA values. Deterministic RBSHA (RSHA-D).
(10) Probabilistic RBSHA (RSHA-P): Steps 1–8 remain the same. Some modifications in Deaggregation.
(11) Give more weightage to Probable Future Rupture Location by altering Deaggregation [20].
(12) Weights of the different models in the probabilistic logic tree are systemically estimated by considering the data support index in the RSHA-P.
(13) Map bedrock hazard and recommend design seismic coefficients or factors.

Considering differences in seismicity and seismotectonic throughout India, a region-specific approach could help to arrive at reliable futuristic seismic hazards in each section of India.

1.5 Soil Consideration and Design Spectrum

In general, seismic zones do not exclusively include site effects and induced effects of the earthquake, but emphasis on accounting for the same was spelt in the IS 1893 first version onwards. Site effects include wave modifications due to soil and topographic effects. The induced effects include ground deformation, liquefaction, and landslides. IS 1893 BSC stated that the intensity of shock greatly varies locally due to variations in the soil conditions. It is not possible to lay down the actual variations with present knowledge for different types of soils in quantitative terms. IS 1893 [5] and later version specifies that "design acceleration for the structure standing on the soil which will not settle appreciably due to the vibration lasting for a few seconds". Any soil subsurface that undergoes settlement during dynamic loading should be improved and, after achieving no settlement layer, then use code acceleration. Subsurface layers are divided into three types [hard, average, and soft soils] for estimating site effects. IS 1893 [5, 6] codes classified the soil based on the Bearing Capacity (BC) (see Table 1.2). Since no reference was given on how to estimate BC, we can reasonably assume that BC is based on soil strength of cohesion and angle of internal friction, as the settlement of these soil is considered to be negligible. Criteria on zero settlement were kept unchanged, and the definition of soil types was updated in the 1970–2002 version with errors in soil classification [2]. Error in soil classification was corrected in the 2016 version of IS 1893, and field SPT N value correction was introduced in the recent version. However, it can be noted here that SPT N values corrections are suggested without accounting for differences in SPT practices in different parts of the country and using foreign-developed correction factors. Diverse SPT equipment and operator practices in the country result in a change in SPT N values from 15 to 85% of measured field N values, which was not exclusively accounted for in the SPT code of IS 2131 and simply given in IS 1893 [11]. Table 1.2 shows that Type III/C or soft soils are defined as soils where SPT N values are less than 10 and are intermediate to highly compressible; these materials may undergo ground deformation and settlement during vibration shaking. We can see from IS 1893 that a few efforts have been made to address local site-specific effects in seismic design, but classification and associated recommendations are far from the modern knowledge and findings in the country, and most of them are copied from foreign literature. Unless we properly integrate region-specific hazard values and site-specific effects as per state-of-the-art knowledge, it may be difficult to achieve zero damage structure even for futuristic moderate seismic events similar to developed countries. The occurrence of earthquakes of magnitude 6.5 or less is just news in developed countries, but it will be a disaster in India.

According to this study, the maximum or average horizontal peak acceleration for 5% damping at rock site conditions from deterministic SHA or probabilistic SHA (for 10% or 2% probability of exceedance in period structures (in general 50 years)) is shown as contours values in the country seismic zonation map. Then design spectrum periods of a structure are suggested for different damping levels and various soil sites found in the region based on locally recorded earthquakes and site-specific soil models. But the Indian seismic zonation map gives seismic coefficient or zonation factor based on past earthquake intensities and response spectrum developed based on US data. IS 1893 has given different seismic coefficients for three types of the subsurface in the 1962 and 1966 versions, and a multiplication factor by taking foundation types in the 1970, 1975, and 1984 versions [see Table 1.1]. In all these versions, average acceleration spectrum curves developed by Dr. Housner and others from four California earthquakes with N [multiplication factor to get the proper values of spectrum quantities] were suggested throughout the country. It can be noted here that these earthquakes were recorded at the epicentral distance of 50–70 km and magnitude of 6.5–7.7. Average acceleration spectrum curves are given for damping values of 0, 2, 5, 10, 20, and 40%. During 1962 and 1966, the period of the spectrum was 0.2–2.8 s, zoomed Y-axis in 1970 with peak horizontal value from 0.2 to 0.4 s with all damping values. In 1975, 0 and 40% damping curves were removed, and the average acceleration spectrum was given up to 3.0 s, and the initial portion [0.1–0.3] curves are horizontal, the rest of them remain the same. Figure 1.4 shows the average acceleration spectrum given in IS 1893. In the 1984 version, the code suggested only zone factor and average acceleration spectrum as Sa/g as Y-axis and period of zero to 3 s in X-axis for damping values of 0, 2, 5, 10, and 20%. These curves initially increase in slope, then become horizontal and followed by curves (see Fig. 1.4). These curves for maximum horizontal components of ground motions and for vertical motions, the half value of these curves were recommended in IS 1893 [5–9] versions. IS 1893 [10, 11] has given a design spectrum [Sa/g versus period] for 5% damping and three curves for three subsurface materials. The seismic coefficient and multiplication factor for different soil types are removed. The difference in spectral acceleration coefficient is shown only after 0.4 s, and soil types II and III have more spectral coefficients than type I [see Table 1.2] after 0.67 s and up to 4.0 s. These versions are also given an equation to arrive Sa/g for the different periods and suggested 2/3 horizontal spectrum should be considered as a vertical spectrum. It can be noted that in the 2016 version of IS 1893, spectra were given up to 6 s, and constant values were suggested from 4 to 6 s. Also, the 2016 version of IS 1893 gives separate spectra for equivalent static method and response spectrum method in Sa/g variation up to 0.1 s, i.e., short period. The average acceleration/ response spectrum given in IS 1893 (a different version) is not directly comparable except for repeated ones. Figure 1.4 shows the comparison of all spectrums for rock site 5% damping. IS 1893, 1975 and 1984 versions are given lower S_a/g coefficient with a multiplication factor, which gives lower design values when compared to later versions. So we divided the curves by 0.2 values to get normalized value without multiplication factor replaced as IS 1893 (1975/1984)-C. This also confirms that S_a/g is much lower than the 2002 and 2016 versions of the same code. Further spectrum is given in different versions of IS 1893 based on acceleration time history and not account for velocity and displacement time of data, i.e., medium and longer period codal values are much lower than the actual value of the seismic event.

Code has clearly given the source of the average acceleration spectrum curves up to the 1984 version, but later design spectrum source information is not explicitly provided in IS 1893 [10, 11] version. The study by the authors reveals that the Design spectrum given in the 2002 and 2016 versions may be adopted from UBC Uniform Building Code [31]. In UBC, the effects of local soil conditions are accounted through foundation factor F and site factor S where. S is related to four subsurface layers with thickness. These were arrived based on the work of [33]. We can note here that most of the modern code gives peak ground acceleration (PGA) for rock sites and site amplification factors for different soil sites at all periods. In particular, zero spectral acceleration [ZSA] for rock and soil sites is different, which is not implemented in the Indian code spectrum; up to 0.4 s of the spectral period, rock and different soil

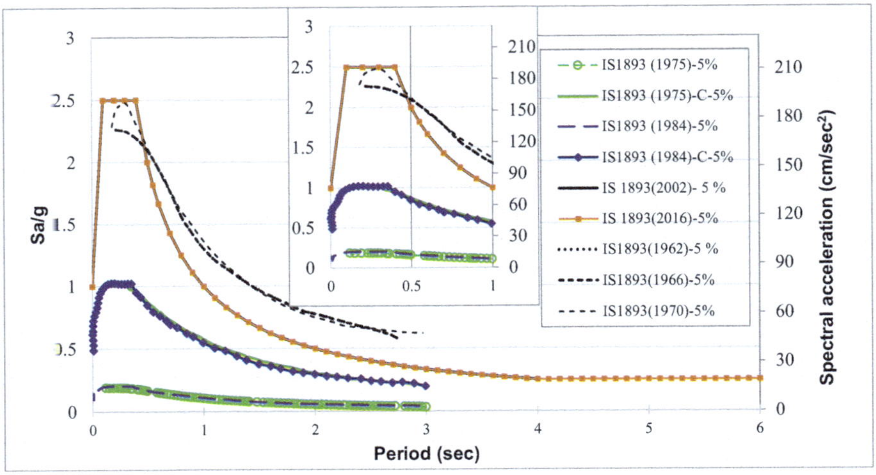

Fig. 1.4 Average acceleration spectrum curves and response spectra at rock site presented in IS 1893 from 1962 to 2016

sites have the same spectral amplitude. Several site response spectra were developed in India using measured data or site response analysis. Many of them are directly compared with the Indian code average response/ design spectrum.

1.6 Response and Design Spectrum

Any earthquake produces different types of seismic waves, which are recorded in seismometers in the form of velocity or acceleration with time. These time history data are recorded in two horizontal and one vertical components, and generally, one of the maximum horizontal components is used to arrive required time-domain parameters of peak ground acceleration/velocity/displacement (PGA/PGV/PGD) and durations of the events. Depending on the recording station, data can be classified as rock or different soil sites and further used for design parameter estimation. The vibration signal produces maximum acceleration or velocity or displacement and is depending on the frequency of the signal. As different stiffness and height of structure can have different natural periods/frequencies, if the natural frequency of the structure system matches with vibration maximum response frequency; then it can be subject to respective maximum acceleration or velocity or displacement. In general, short period [high frequency] is sensitive to acceleration, the intermediate period [medium frequency] is sensitive to velocity, and the long period [low frequency] is sensitive to displacement. So, it is necessary to characterize which frequency or period range that can produce maximum acceleration or velocity or displacement of recorded seismic signal in the region. These three regions are called acceleration-sensitive, velocity-sensitive, and displacement-sensitive regions of seismic events. Depending on the structure's natural period, respective acceleration or velocity or displacement should be considered in the design. So, any seismic design provision should reflect the same from regional seismic records.

A seismic zone map should give a variation of ZSA/PGA as a map in the region and give normalized and smoothed spectral shape by taking three sensitive regions and respective amplitude modification with the respective [PGA/PGV/PGD] normalized value. Rigorous and reliable SHA by accounting for all futuristic seismotectonic event possibilities can help to create a representative ZSA/PGA map. Seismic data recorded in the region at rock and different soil site stations can help to produce normalized and smoothed spectral shapes showing maximum amplitude and cut period of acceleration, velocity, and displacement in the region. Response spectrum is defined as the maximum relative linear response of a single degree of freedom system (SDOF) for excitation by a given strong earthquake seismic ground motion [34, 35]. Initially, Biot [36] introduced the concept of response spectra and proposed the standard spectral shape for earthquake-resistant design of the building. Housner [37, 38] averaged and smoothed the response spectra considering the four strong-motion records and proposed using average spectrum shape in earthquake engineering design. Newmark and Hall [39, 40] recommended a smooth response spectrum concentrating on three regions viz. acceleration (short period), velocity (medium period), and displacement (long period). The shape of Biot, Housner, Newmark, and Reg. Guide 1.60 spectra [41] were fixed by averaging the spectral shape with respect to the site conditions, distance, and earthquake size (magnitude or intensity). Various researchers (e.g., [42–44]) contributed towards the development of the Newmark–Hall spectrum. Mallick et al. [45] determined the amplification factors for acceleration, velocity, and displacement-sensitive regions of the spectrum for various damping values. Mallick et al. [45] showed that the displacement amplification factor is significantly different as compared to the previous studies. Malhotra [46] proposed a methodology to compute elastic response spectra for incompatible acceleration, velocity, and displacement time histories to account for all maximum possible responses. The procedure recommended by Malhotra [45, 46] for deriving the normalized response spectra is used here to develop the design spectrum of South India and North India.

1.6.1 Acceleration Time History to Design Spectrum

India is improving seismic instrumentation, and right now, 150 seismic stations are installed in different parts of the country and are being continuously monitored [47]. Here we have taken typical Indian data to explain how recorded acceleration data is converted as a response and design spectrum. An earthquake can have an acceleration time history along three components. Typical rock site record of NI earthquake data is shown in Fig. 1.5a. In this, the maximum acceleration time history is the NS component, which is used to compute velocity and displacement time history, as shown in Fig. 1.5b. These data are further used to arrive at a 5% damped acceleration, velocity, and displacement spectrum, as shown in Fig. 1.5c. These three responses are normalized with respect to maximum quantities, i.e., PGA, PGV, and PGD. Here, we can see that maximum acceleration response up to period of 0.92 s, maximum velocity response up to the period of 2.32 s, and the rest are maximum displacements. These values change with earthquake data, so compiling larger data will provide reliable values. Malhotra [46] and Mallick et al. [45] explained the procedure for deriving the normalized response spectra by solving a dynamic equilibrium for the single-degree-of-freedom equation. In-house MATLAB code has been used to solve Eq. 1.1 as per [45, 46] for a given acceleration, velocity, and displacement time history.

The spectral displacement (*SD*) is the maximum displacement of the SDOF system at any time, likewise the spectral velocity and spectral acceleration.

$$SD = |u|_{max} \tag{1.1}$$

Using the above equation, the spectral acceleration (*SA*) approaches to PGA at a short period, and spectral displacement approaches to PGD. The equation that relates *SA*, *SD*, and spectral velocity (*SV*) is as follows:

$$SA\left(\frac{T}{2\pi}\right)^2 = SD = SV\left(\frac{T}{2\pi}\right) \tag{1.2}$$

Further, the *SD* obtained from Eq. 1.1, has been converted to *SA* to obtain the tripartite response spectrum. The tripartite response spectra have been normalized as follows: firstly the central period (T_{cg}) of the seismic ground motion is calculated as

$$T_{cg} = 2\pi\sqrt{\frac{PGD}{PGA}} \tag{1.3}$$

This T_{cg} causes a horizontal shift in the response spectra, PGA and PGD change to $PGA \times T_{cg}/2\pi$ and $PGD \times 2\pi/T_{cg}$ respectively, however, PGV remains constant. Further, PGV and SV are normalized with respect to $\sqrt{PGA.PGD}$, this makes PGA and PGD unity and PGV and SV to make the following non-dimensional form

$$\overline{PGV} = \frac{PGV}{\sqrt{PGA.PGD}} \tag{1.4}$$

$$\overline{SV} = \frac{SV}{\sqrt{PGA.PGD}} \tag{1.5}$$

Further, the normalized spectrum has been smoothened considering the least-squares fitting of straight-line segments through the median curve. A typical smoothened median response spectrum is given in Fig. 1.6. The amplification factors above unity corresponding to acceleration, velocity, and displacement are denoted as α_A, α_V, and α_D. T_1, T_2, T_3, T_4, T_5, and T_6 in Fig. 1.6 are control periods (where

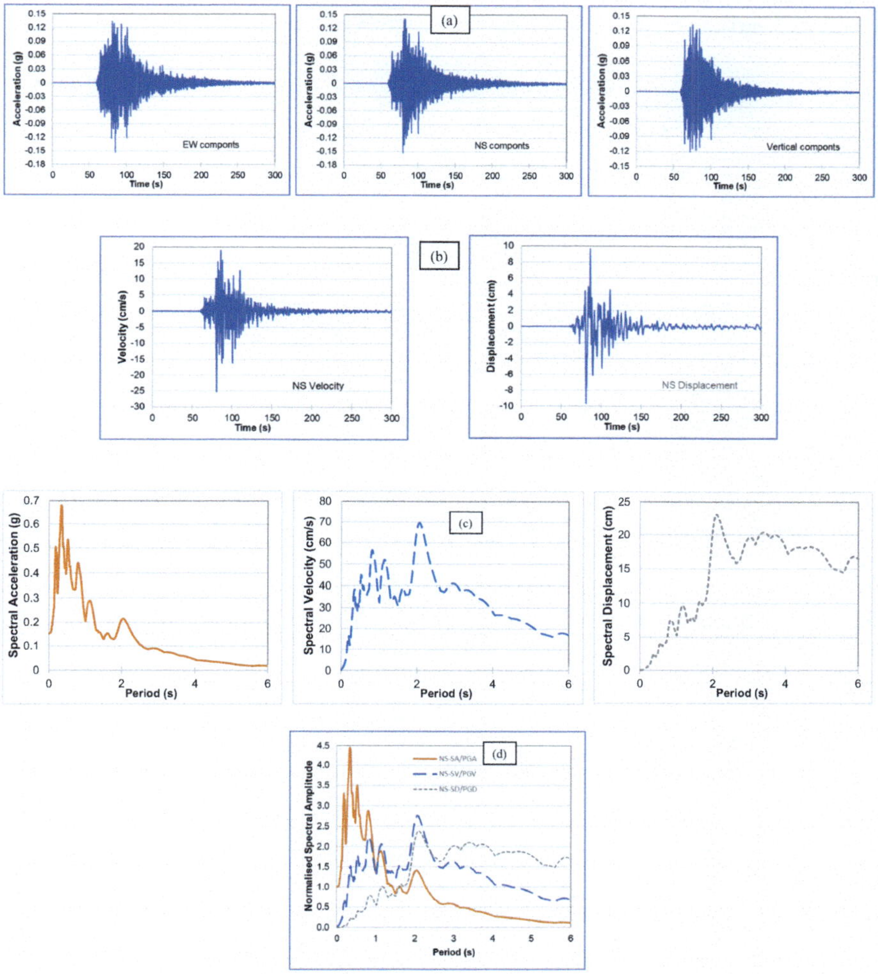

Fig. 1.5 Typical rock site earthquake time history and response spectrum **a** Acceleration time history of EW, NS, and Vertical record, **b** Maximum (NS) Velocity and Displacement time history data, **c** Response spectrum for acceleration, velocity, and displacement data and **d** Normalised spectral amplitude of acceleration, velocity, and displacement with respective peak values

the straight-line segments meet). A typical example of obtaining a normalized tripartite plot of 2011, Sikkim earthquake (6.9 M_w) recorded at a hypocentral distance of 102 km is given in Fig. 1.7. The recorded PGA, PGV, and PGD are 0.131 g, 0.025 m/s, and 0.002 m, respectively. Acceleration, velocity, and displacement time history are shown in Fig. 1.7a used for deriving the SD using MATLAB. Further, SD is converted to SV and a tripartite plot of 5% damping response spectrum is given in Fig. 1.7b. Using Eqs. 1.3, 1.4 and 1.5, the developed response spectrum is normalized and given as Fig. 1.7c. The whole procedure is further used for deriving the normalized site response spectrum for the deep and shallow bedrock sites of IGB and SI.

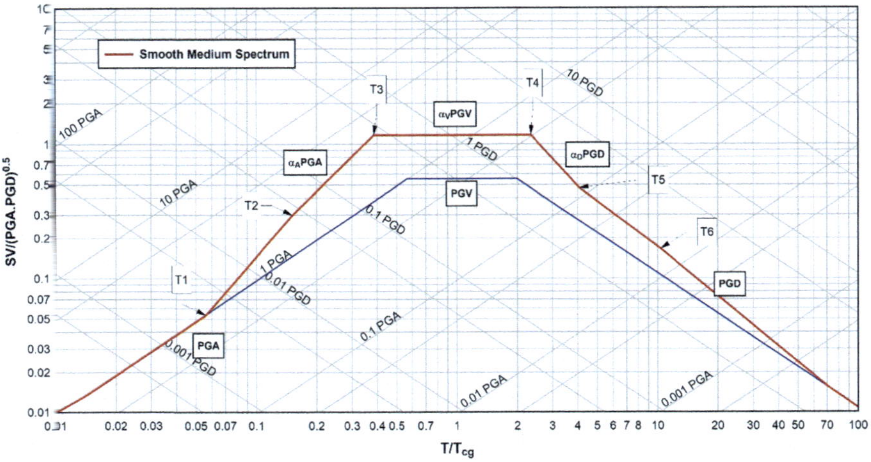

Fig. 1.6 Typical smooth medium response-spectrum considering PGA, PGV and PGD

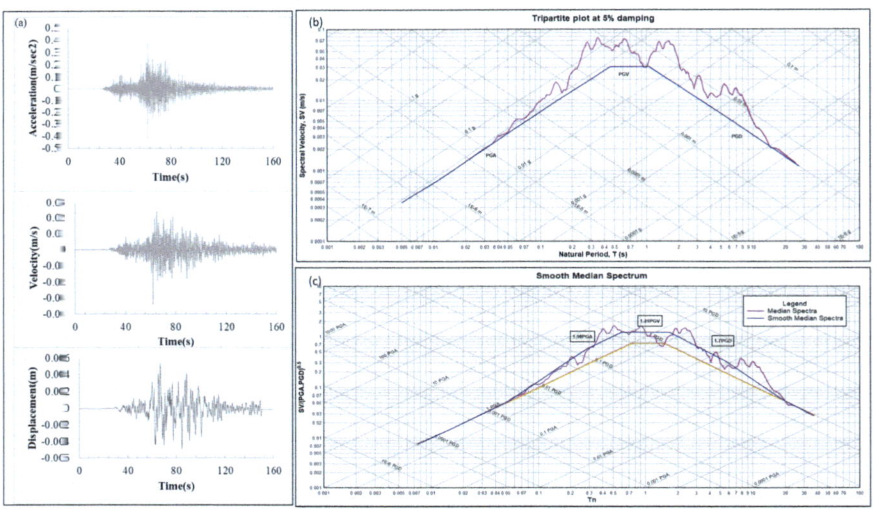

Fig. 1.7 Typical acceleration, velocity, and displacement time history on the left side (**a**), respective 5% damping response spectrum (**b**), and final smooth normalized response spectrum (**c**)

1.6.2 Cut-Off Period for Acceleration, Velocity, and Displacement

Cut-off period of acceleration, velocity, and displacement, a part of the design spectrum where the maximum response of respective quantities are represented, is the function of source, path, and site characters. These values in IS 1893 were adopted from UBC irrespective of Indian seismicity, path attenuation, and site geology in India. Even in modern codes, these cut-off regions are not well defined using a tripartite plot of seismic data from the region. Hall et al. [48], Malhotra [45] and Mohraz [42] defined the different period of the design spectrum that is sensitive to PGA, PGV, and PGD. Hall et al. [48] and Mohraz [42] assumed that SAs for periods 0.33 s, 0.33, and 3.33 s, and more than 3.33 s are sensitive to PGA, PGV, and PGD. Malhotra [45] showed that SAs for a period up to 0.62 s, 0.62–2.6 s, and the rest correlated well with PGA, PGV, and PGD for the same data. Malhotra [45] concluded that cut-off periods could change for different sets of seismic ground motions and need to be incorporated into the design spectrum. Hence, in this study, the PGA-, PGV-, and PGD-sensitive regions for NI and SI, correlation of SA at various periods with PGA, PGV, and PGD has been plotted. The recorded bedrock seismic ground motion data for the Himalayan region and Stable continental region has been used to determine the cut-off periods for PGA-, PGV-, and PGD-sensitive regions.

Figure 1.8 shows the correlation of SA with PGA, PGV, and PGD for NI and SI for rock sites with 5% damping. From Fig. 1.8, it can be noted that SA for the period up to 0.38 s and 0.30 s is correlating well with PGA for NI and SI, respectively. Further, for the period between 0.38 and 2.28 s, SA correlates well with PGV, and the rest correlates best with PGD for NI rock sites. Similarly, in the case of SI, SA correlates best with PGV for the period between 0.30 s and 1.55 s and above 1.55 s with PGD for rock sites. Bureau of Indian Standard (IS:1893 [11]) defined the cut-off period for rock sites as 0.1 and 0.4 s, which is significantly different from the present study. Similarly, it is noted here that the cut-off period calculated in this study is considerably different from [48, 42]. Hence, the methodology used in this study does not make any prior assumption in determining the cut-off period for three sections of the design spectrum.

1.7 Smoothed and Normalized Spectrum from India

Even though recorded time history data availability is very limited in India for engineers to understand earthquake effects, the first author continuously uses available data to bring Indian-specific seismic studies required for the seismic design in India. The first author and team are continuously trying to solve some of India's engineering

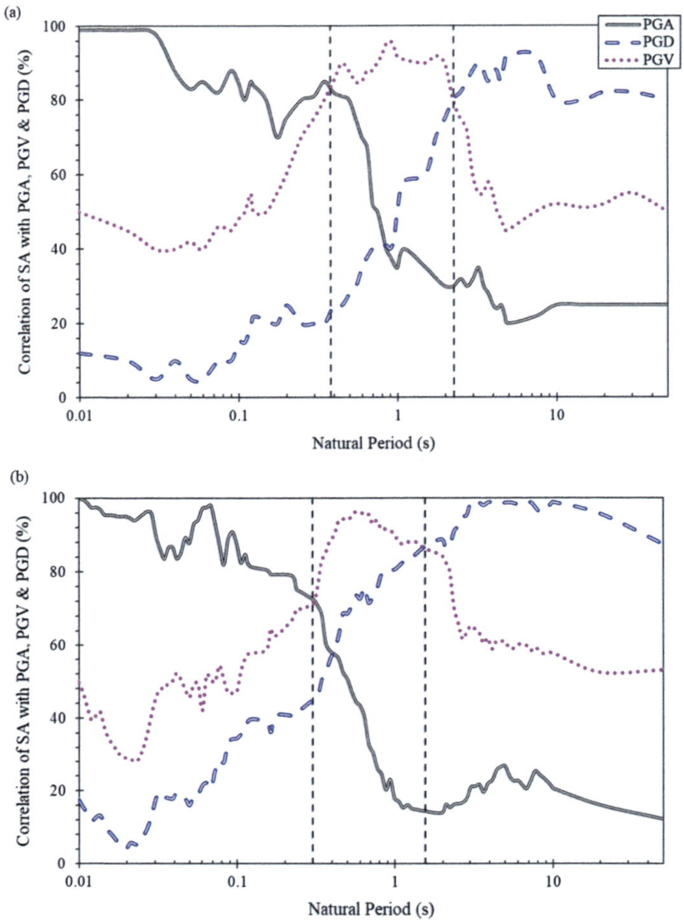

Fig. 1.3 Correlation of SA at various periods with PGA, PGV, and PGD for **a** Indo Gangetic Basin and **b** Stable continental region for bedrock

seismology and earthquake geotechnical problems through research and respective publications found at http://civil.iisc.ernet.in/~anbazhagan/List%20of%20Publications.html. Here, acceleration data complied from a different source and presented in the previous publications for rock sites are used for developing a 5% damped design spectrum for the horizontal and vertical components of ground motions. Since seismicity, seismotectonic, and subsurface information of North India (Himalayan region) are completely different from South India, a separate approach has been made to develop a design spectrum using the same methodology but corresponding data. Rock site acceleration time histories from NI and SI are used separately to develop the design spectra. Since the available rock sites' data are very little for south India (intraplate region), data from other intraplate areas of the world are added to the south Indian dataset.

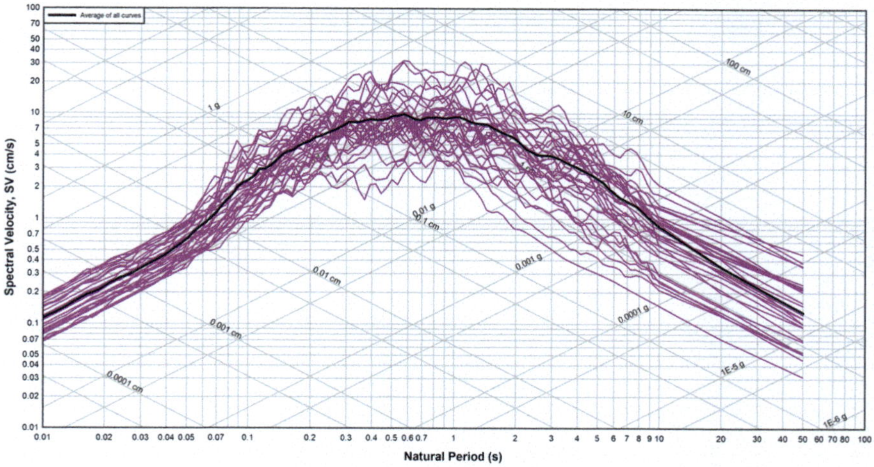

Fig. 1.9 Tripartite plot of horizontal seismic ground motions recorded at bedrock sites in North India

The tripartite plot of horizontal ground motions was recorded at bedrock sites at NI, as shown in Fig. 1.9. The procedure to calculate smoothed and normalized design spectrum from acceleration ground motion record is explained in the previous section. The median normalized spectrum versus normalized spectral period is arrived for NI and is shown in Fig. 1.10a. The shaded area in Fig. 1.10a corresponding to ± 1 standard deviation about the median. The smooth response spectrum has been obtained afterwards by using the least-squares fitting of straight-line segments through the median curve shown in Fig. 1.10b. Figures 1.10b show the smooth medium spectrum of north India covering the Himalaya, northeast and northwest, and Indo-Gangetic Basin. The factors above unit PGA, PGV, and PGD, i.e., α_A, α_V, and α_D for the Himalayan region are 2.29, 1.97, and 2.05, respectively. The control periods for acceleration, velocity, and displacement section of the spectrum, i.e., T_2, T_3, and T_4 found to be 0.15, 0.38, and 2.33 s, respectively.

The tripartite plot of horizontal ground motions of bedrock sites from the stable continental region of the world i.e., SI is shown in Fig. 1.11. A similar procedure discussed above has been used to arrive median normalized spectrum versus normalized spectral period is given in Fig. 1.12a, for stable continental data. The shaded area in Fig. 1.12a corresponds to $\pm$ 1 standard deviation about the median. Figure 1.12b shows the smooth response spectrum obtained by using the least-squares fitting of straight-line segments through the median curve. The factors of α_A, α_V, and α_D for the stable continental region is 2.18, 2.04, and 1.55, respectively. Here, we can notice that there is a slight difference that can be observed for both the region factors at rock sites. The control periods of T_2, T_3, and T_4 found to be 0.08, 0.28, and 1.52 s for SI, which is considerably different from NI. These observations are for 5% damping design spectrum factors using available data, and adding more regional data may refine findings.

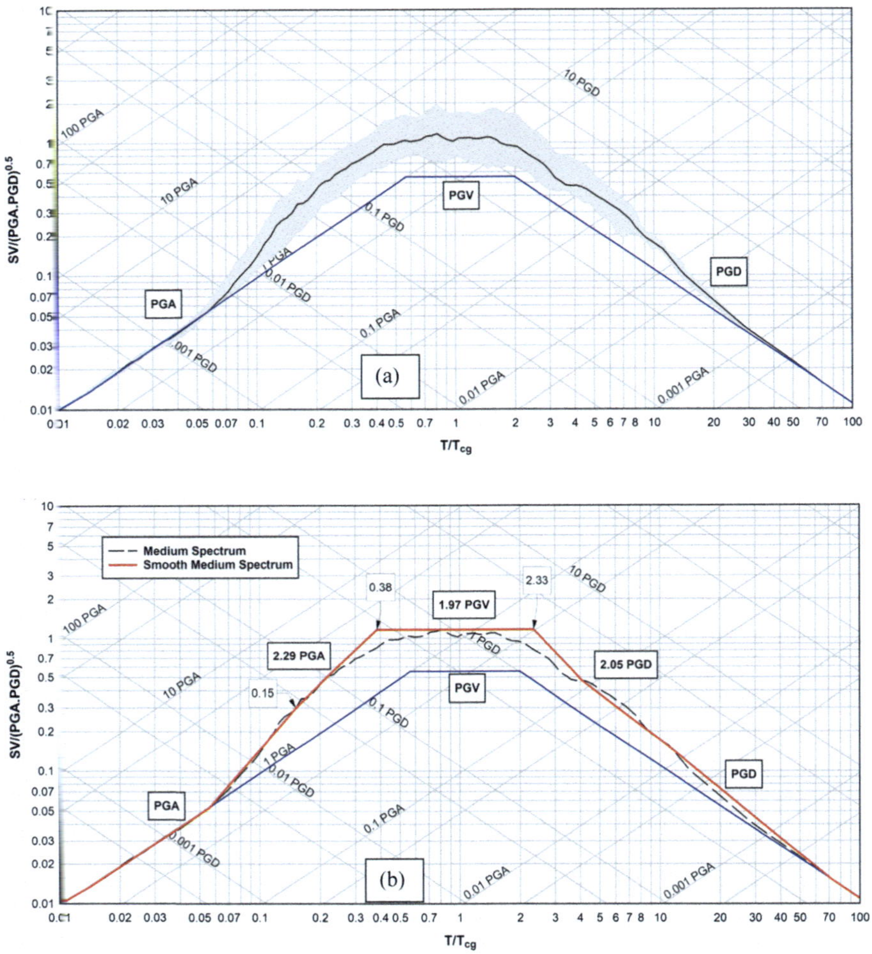

Fig. 1.10 North Indian normalized 5% damping median spectrum of horizontal seismic ground motions recorded at bedrock site (**a**) and smooth medium spectrum (**b**)

1.8 Code Design Spectrum

Many seismic codes have included the significant influence of the local site effect in their recent provisions. After the 1994 Northridge earthquake, seismic codes such as the National Earthquake Hazard Reduction Program [49], the Uniform Building Code [31], the International Building Code [50], and Eurocode 8 (EC8) [51] accounted for the site effects using the elastic design response spectra. In most modern seismic designs, the estimation of a seismic force on a typical structure is based on the 5% damped design response spectrum of recorded data in the region. In India, limited attempts have been made to develop the design spectrum with different damping

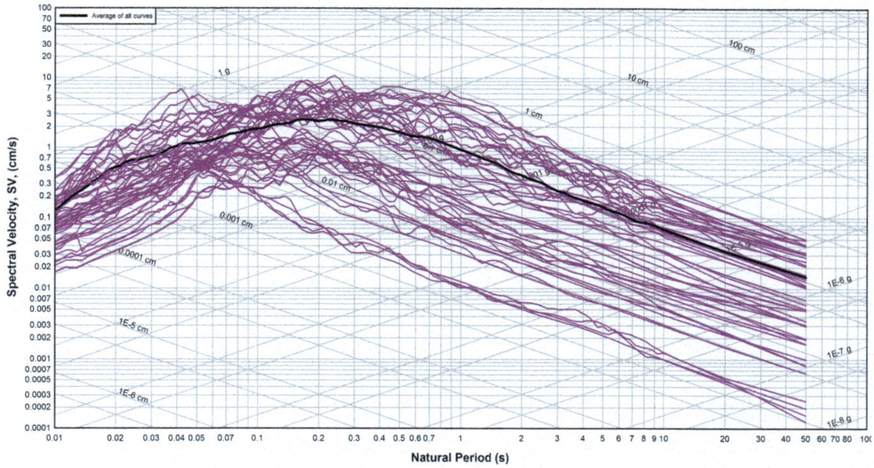

Fig. 1.11 Tripartite plot of horizontal seismic ground motions recorded at bedrock sites in the stable continental region, i.e., South India

level considering regional recorded data; however, some attempt has been made by Anbazhagan et al. [23]. Generally, the design spectra of a given site are obtained by modifying the uniform hazard spectrum by considering site factors corresponding to a particular seismic area. Conventionally, the design force is specified via response spectrum amplitude. However, with the increase in complexity of modern structures, to understand the seismic performance of the structure, it is now essential to define the amplitude and shape of the design spectra. EC8 [51] defined the normalized elastic design response spectra based on effective ground acceleration at the rock site, and the shape has been defined using the three corner periods. Similarly, the BIS, IS:1893 [11] defined the normalized elastic response spectra based on the effective ground acceleration at rock sites; whereas the recent international seismic codes (i.e., NEHRP, BSSC [52]) refined the elastic response spectra using the two parameters, namely spectral acceleration at short period and at a period of 1 s. The soil amplification has been accounted for using site amplification factors.

The work regarding the development of building codes began in Italy in 1908, after the Messina disaster; in Japan following the 1923 Tokyo earthquake; in California after the 1925 Santa Barbara earthquake [53], and in India after the 1934 Bihar–Nepal earthquake. Most of the building codes around the world have adopted the Newmark–Hall spectrum with some modifications [45]. New elastic design response spectra for corresponding rock sites with 5% damping ratio can be proposed in two ways. One is similar to EC8 [51], i.e., normalized elastic design response spectra which is based on one parameter (effective ground acceleration at rock). The other one, i.e., elastic design response spectra based on two input parameters that are spectral acceleration at a short period and at a period of 1 s, which is similar to NEHRP.

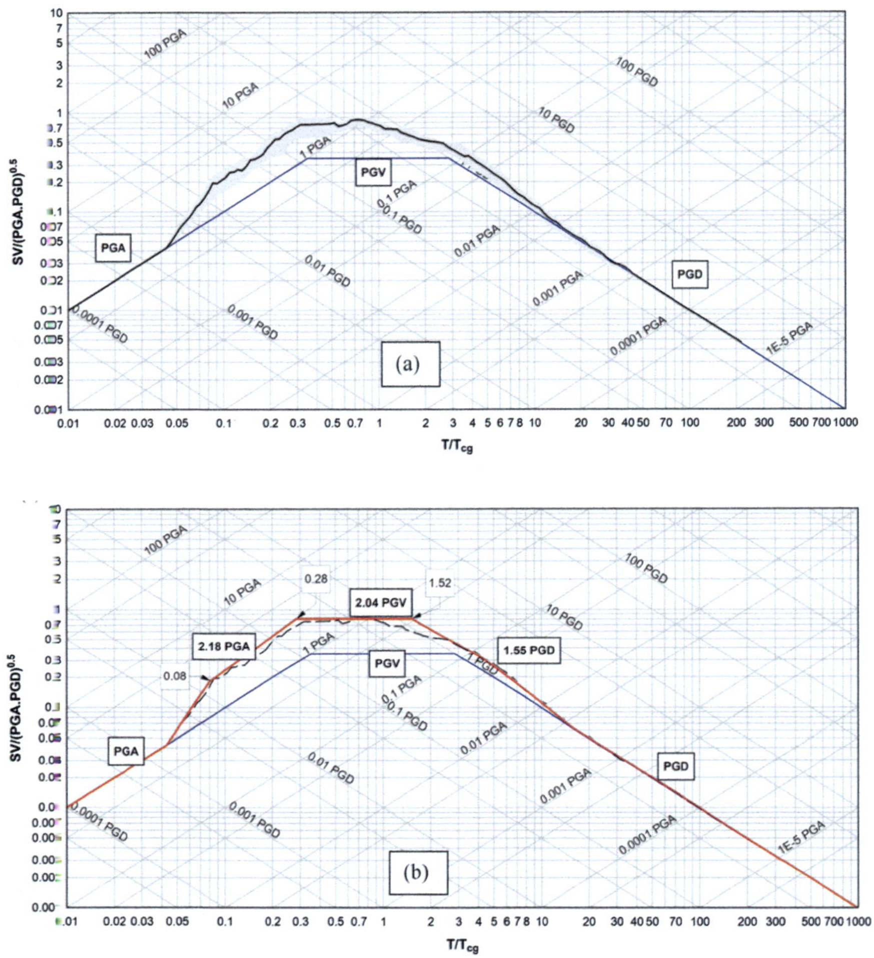

Fig. 1.12 South Indian normalized 5% damping median spectrum of horizontal seismic ground motions recorded at bedrock sites (**a**) and smooth medium spectrum (**b**)

1.8.1 Elastic Design Response Spectra for Single Input

The current Eurocode 8 (EC8, CEN 2005) [54] defined the two standard shapes for the linear response spectrum, viz. Type 1 and Type 2. Type 1 corresponds to more energy, and Type 2 corresponds to less energy in the long period motions. The former is used for the high seismicity area and the latter for the low seismicity area. The factors that determine the shape of the spectra depend on the ground acceleration at rock conditions and soil amplification factor, i.e., S, which accounts for the local soil and site effect. Further, the corner periods are defined as T_B, T_C, and T_D. The peak of the spectral amplitude is defined as 2.5 ηS, where η is the damping ratio, i.e., 5%. The typical plot for EC8 is shown in Fig. 1.13a. In the sixth revision of BIS:1893 [11], two methods, viz. equivalent static method and response spectra methods, are proposed to construct the acceleration design spectra for the different zone. However, unlike EC8, the SA coefficients are capped at 2.5 by classifying soil into three categories based on SPT-N value.

EC8 has proposed the generalized equation for constructing the normalized elastic response spectrum. The same equations are used in the present study to derive the elastic response spectra for the NI and SI. The general form of equations for the elastic response spectra for 5% damping is as

$$0 \leq T \leq T_B : \frac{S_a(T)}{PGA_{\text{rock}}} = s.\left[1 + \frac{T}{T_B}.(\beta - 1)\right] \tag{1.6}$$

$$T_B \leq T \leq T_C : \frac{S_a(T)}{PGA_{\text{rock}}} = s.\beta \tag{1.7}$$

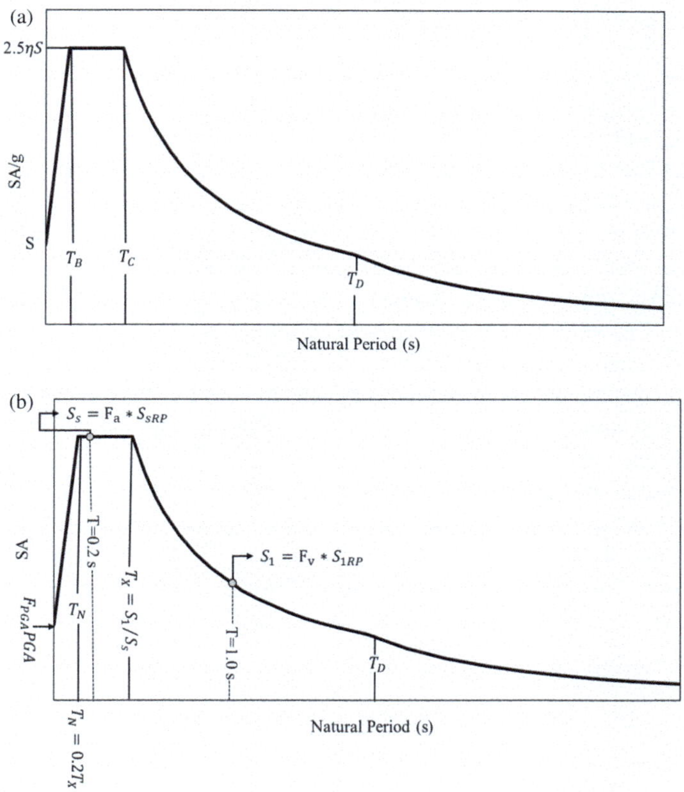

Fig. 1.13 Typical seismic design response spectrum proposed by EUROCODE (EC8). Typical seismic design response spectrum proposed by NEHRP/ASCE [55]/AASHTO [56]

$$T_C \leq T \leq T_D\colon \frac{S_a(T)}{PGA_{\text{rock}}} = s.\beta \frac{T_C}{T} \tag{1.8}$$

$$T_D \leq T\colon \frac{S_a(T)}{PGA_{\text{rock}}} = s.\beta.T_C \frac{T_D}{T^2} \tag{1.9}$$

Here, PGA_{rock} is the design ground acceleration at rock-site conditions, S and β are the soil amplification and spectral amplification factors. T_B and T_C are the limits of constant acceleration branch and T_D is the beginning of the constant displacement range of the spectrum.

1.8.2 Elastic Design Response Spectra for Two Inputs

In 1997, the Uniform Building code [31] used C_a to construct the acceleration region and C_v to construct the velocity region. The 1997 NEHRP [49] site coefficients for a short period i.e., F_a and long period i.e., F_v were defined during the 1992 national workshop. F_a and F_v are defined as spectral acceleration ratios averaged over period ranges of 0.1–0.5 s and 0.4–2.0 s respectively; whereas the site factors at a short period or 0.2 s (F_a) and long period or 1.0 s (F_v) are recommended by the American Society of Civil Engineers Standard ASCE 7-10 [55], the International Building Code [50], and the AASHTO guide [56] that was first defined in the 1994 NEHRP provisions [49]. F_a defined in the IBC is the average value and F_v is approximately the average $+1\sigma$ amplification values [57]. In IBC, F_a is estimated for the short-period band 0.1–0.5 s, whereas, F_v is defined over the long-period band 0.4–2.0 s [57]. These F_a and F_v values are used for constructing the acceleration response spectra for different seismic site classes based on NEHRP. The design response spectra are constructed from 5% damping at 0.2 s (S_S) and at 1 s (S_1) that are calculated from probabilistic seismic hazard analysis. The control period that defines the transition between the acceleration and velocity-sensitive regions is computed as $T_X = S_1/S_s$ and the period at the beginning of the flat acceleration-sensitive region is defined as $T_N = 0.2T_X$. A typical design response spectrum is given in Fig. 1.13b.

T_N T_X, and T_Y are the control periods. T_N and T_X are the limits of constant acceleration branch and T_Y is the beginning of the constant displacement range of the spectrum.

$$T_X = S_1 T_1 / S_s \tag{1.10}$$

$$T_N = T_X / \kappa \tag{1.11}$$

These control periods in this case depend on S_s and S_1. T_Y is analogous to T_4 (see Fig. 1.6) and equal to T_D and the values can be determined from Table 1.3 ($T_Y = T_D$). It can be noted that constant periods T_N and T_X are analogues to T_2 and T_3 in Fig. 1.6

Table 1.3 Parameters of the proposed new design response spectrum at rock site for 5% damping using Indian earthquake data

Parameters	North India (interplate)	South India (Intraplate)
T_B	0.15	0.08
T_C	0.38	0.28
T_D/T_Y	2.33	1.52
S	1.00	1.00
β	2.29	2.18

[45]. κ which is the ratio of the T_X/T_N is defined based on T_2 and T_3 for rock sites of NI & SI. κ value of 2.5 is derived by Malhotra [45] based on 63 rock sites recorded seismic ground motion and 4 as determined by NEHRP, ASCE and AASTHO. We believe that κ value for rock sites of NI and SI can be different. This may be because the design spectrum is conservative and it is a composite of several seismic events.

Further, S_s and S_1 are defined as

$$S_s = \mathrm{F_a} * S_{sRP} \tag{1.12}$$

$$S_1 = \mathrm{F}_v * S_{1RP} \tag{1.13}$$

Here, S_{sRP} and S_{1RP} are the reference maximum spectral acceleration corresponding to the constant acceleration branch and at a period of 1 s of the horizontal 5% damped elastic response spectra on bedrock. This method required input of PGA, average spectral factors [F_a and F_v] and spectral acceleration at constant acceleration region and 1 s [S_{sRP} and S_{1RP}]. As the author believes, that is no systematic futuristic hazard analysis was developed for NI and SI and providing this spectrum may lead to error, hence not exclusively included here. In this study, we developed a design spectrum by the first approach i.e., similar to Eurocode, which is also the current Indian code of practice. Then developed spectra are further compared with the Indian seismic code spectra at different revisions.

1.9 New Design Spectrum

1.9.1 Horizontal Motion Design Spectrum

The parameters S, β, T_B, T_C, and T_D depend on site/soil class and seismicity, as explained above. These periods are determined as per the [45] procedure in this study for both region separately. T_B, T_C, and T_D are analogues to T_2, T_3, and T_4 in Fig. 1.6 and the parameters β, T_B, T_C, and T_D have been determined using the normalized spectrum shown in Figs. 1.10b for NI and 1.12b for SI. S is unity for rock site (in this study) and more than unity for other soil sites depending on the local site effects. In the present study, these parameters are derived based on the

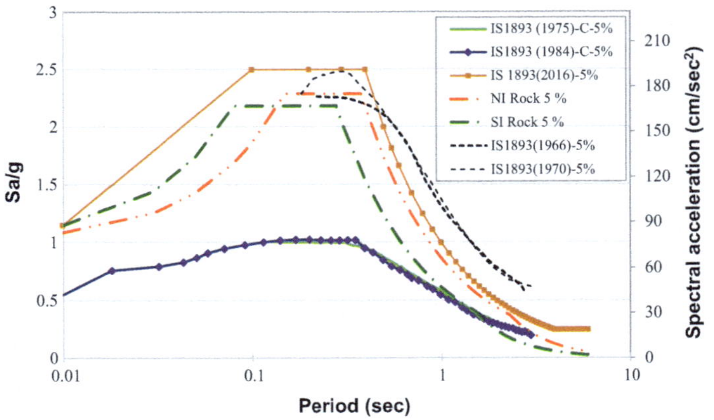

Fig. 1.14 Newly developed horizontal spectrum using Indian data and 5% acceleration, velocity, and displacement spectrum at rock sites for North India (NI) and South (SI) and also comparison with different versions of IS 1893 spectrum. IS 1893 [6, 7] spectrum with secondary Y-axis

interplate (North) and intraplate (South) region of India. Considering the spectra in the two regions separately helps better representation of the shape of the response spectra. β, T_B, T_C, and T_D are determined based on the shape of the normalized response spectra at rock site conditions. These factors are the result of fitting the smooth spectrum to the median normalized spectrum. The values are given in Table 1.3. Figure 1.14 shows elastic design response spectra for single input similar to EC8 for NI and SI. We can notice that data based newly derived spectrum is lower than most of IS 1893 spectra. Further cut-off period for acceleration, velocity and displacement and respective amplitudes are lower than the current codal provision of IS 1893 (2016). This spectrum is valid only for 5% damping and other damping reader can refer [23] or our future work. We are also working on the development of two input based spectrum similar to NEHRP and ASCE with bedrock level seismic hazard maps showing PGA, S_s, S_1 and respective factors.

1.9.2 Vertical Motion Design Spectrum

Generally, all code design spectrum developed are based on horizontal spectrum and have considered some reduction factor for vertical motions without studying the data. Indian code, IS 1893, suggested that the vertical seismic coefficient, where applicable, should be taken as half of the horizontal seismic coefficient given in 1962, 1966, 1970, 1975 and 1984 version values [in Fig. 1.4]. In 2002 and 2016, versions of IS 1893 recommended that the design acceleration spectrum for vertical motions may be taken as two-thirds of the design horizontal acceleration spectrum. Generally, it is assumed that the vertical spectrum is 2/3 of the horizontal spectrum is a recent code recommendation. This indicates that frequency distribution in horizontal spectra is the same as vertical spectra for any region. However, many studies [45] have proved that the frequency distribution of both spectra is different. This has been observed in various Indian subcontinent earthquakes (e.g., 2015 Nepal Earthquake, 2001 Bhuj Earthquake etc.). Hence, an attempt is made to determine the difference in the horizontal and vertical spectrum for the Indian data. Similar to [45], the vertical to horizontal ratio is calculated for 50 rock ground motions for North India data. The vertical to horizontal ground motion ratio is calculated for the different periods. The median vertical to horizontal ratio is derived for both rock sites. Figure 1.15a shows the vertical spectrum derived from the actual data along with the horizontal spectrum from data and 2/3 horizontal spectrum as a vertical spectrum. We can see a considerable difference between vertical horizontal and follow 2/3 based western countries recommendation is also not appropriate for North India. In comparison, it can be observed that control periods and amplification are considerably different for the horizontal spectrum. Further proposed vertical design response spectra are compared with the rock site of IS 1893 [11] horizontal and vertical spectrum and shown in Fig. 1.15b. Figure 1.15b shows that the actual North India earthquake data-based vertical spectrum is considerably different from the IS 1893 recommended American earthquake based spectrum for the horizontal and vertical components. This study developed the first time smoothed and normalised design spectrum considering acceleration, velocity, and displacement response of Indian earthquakes. These spectra can be further reviewed and modified by adding more and more data from rock sites in respective regions.

1.10 Summary and Conclusion

Several researchers summarised and reviewed Indian seismicity, seismotectonic, hazard analysis, and IS 1893 code in the literature. But most of them highlighted issues, but a minimal attempt was made to provide solutions based on the state-of-the-art knowledge and also using regional recorded earthquake data. Even though considerable effort has been made to record digital data of local earthquakes since 1997 and digitalization of analogue records of old significant earthquakes, very limited data is available to engineers. IISc complied earthquake data from different sources with a non-sharing agreement and used it to arrive at more realistic regional seismotectonic parameters, source parameters, and ground motion simulation and models. These models and methods are highly useful to arrive a reliable futuristic seismic zonation map of India. This paper summarized the diversity of Indian seismicity, seismotectonic and subsurface layers. It then highlighted the importance of a region-specific approach for arriving seismic hazard values at bedrock level based on region-specific rupture-based seismic hazard approach developed at IISc. A study of IS 1893 [5–11] codes indicates that the design spectrum of the given code was adopted from American data and not valid for Indian conditions due to diversity in seismicity and soil type. Collected time history data at the bedrock level has been used to arrive at smoothed and normalized design spectrum parameters for north (interplate) and

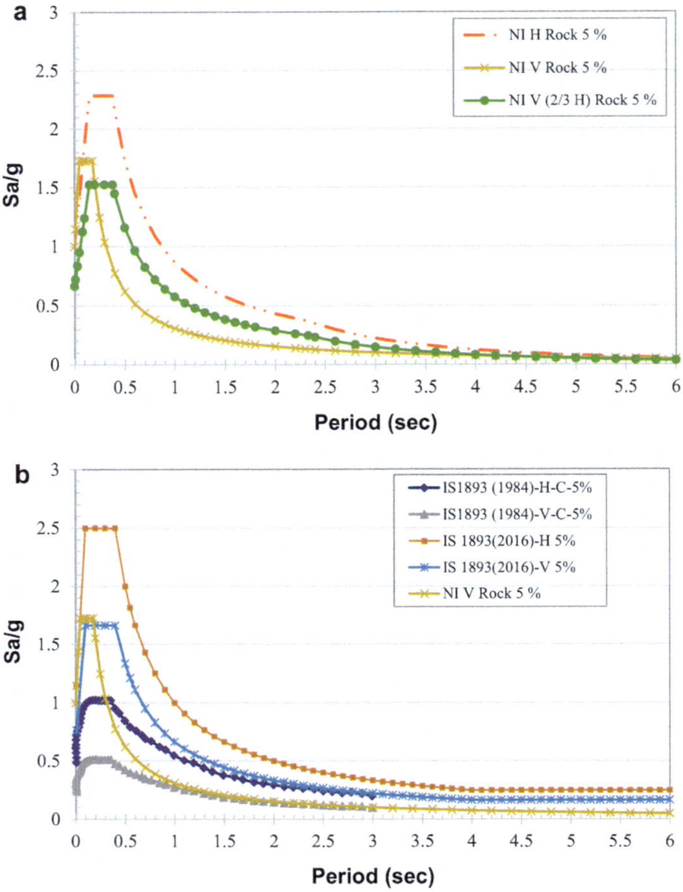

Fig. 1.15 Comparison of **a** proposed vertical (PV) design response spectrum with horizontal design spectrum and 2/3 factor of horizontal spectrum values for bedrock condition with 5% damping for in NI. **b** Proposed vertical design response spectrum (V) compared with horizontal (H) and vertical (V) suggested in IS:1893 [11] and (1985)

south (intraplate) India separately for the first time. We found that spectral amplification factors above unit PGA, PGV, and PGD and cut-off time for acceleration, velocity and displacement-sensitive section of the spectrum are different from the spectrum given in IS 1893. These parameters were also different from South and North India due to variations in seismotectonic characters. Based on regional data analysis first-time design spectrum for bedrock sites with 5% damping was developed and presented in the paper. Normalized elastic design response spectra based on one parameter (effective ground acceleration at rock). For that S, β, T_B, T_C, and T_D have been derived for the proposed seismic site classification to derive the normalized design response spectrum. T_B and T_C which define the constant spectral acceleration region are derived as 0.15 and 0.38 s in the case of bedrock in NI and 0.08 and 0.28 s in the case of bedrock in SI. Similarly, β has been increased by 1.05 times in the case of NI when compared to SI at the bedrock site. It has been observed that the value of the control period and spectral amplification factors derived in the present study based on region-specific seismic data is different from the other region seismic codes. This study can be further refined by adding more data and also for different damping levels. The availability of more and more recorded earthquake data and systematic site response analysis can help to drive the design spectrum for various soil sites in the future.

Acknowledgements The authors thank the Dam Safety (Rehabilitation) Directorate, Central Water Commission for funding the project entitled "Capacity Buildings in Dam Safety" under Dam Rehabilitation and Improvement Project". Authors would thank "Board of Research in Nuclear Sciences (BRNS)", Department of Atomic Energy (DAE), Government of India for funding the project titled "Probabilistic seismic hazard analysis of Vizag and Tarapur considering regional uncertainties" (Ref No. Sanction No. 36(2)/14/ 16/2016-BRNS). The authors also thank the Department of Earthquake Engineering of the Indian Institute of Technology, Roorkee, and the Ministry of Earth Science, India, for making available the recorded earthquake ground motions.

References

1. Koks EE, Rozenberg J, Zorn C et al (2019) A global multi-hazard risk analysis of road and railway infrastructure assets. Nat Commun 10:2677. https://doi.org/10.1038/s41467-019-10442-3
2. Anbazhagan P, Gajawada P, Moustafa Sayed SR, Al-Arifi NSN, Aditya P (2014) Provisions for geotechnical aspects and soil classification in Indian seismic design code IS-1893. Disaster Adv 7(3):72–89
3. Anbazhagan P (2013) Method for seismic microzonation with geotechnical aspects. Disaster Adv 6(4):66–86
4. Anbazhagan P (2007) Site characterization and seismic hazard analysis with local site effects for microzonation of Bangalore. PhD Thesis. Indian Institute of Science https://etd.iisc.ac.in/handle/2005/689
5. BIS, IS 1893 (1962) Indian standard recommendations for earthquake resistant design of structures, Bureau of Indian Standards, New Delhi
6. BIS, IS 1893 (1966) Indian standard criteria for earthquake resistant design of structures, Bureau of Indian Standards, New Delhi
7. BIS, IS 1893 (1970) Indian standard criteria for earthquake resistant design of structures, Bureau of Indian Standards, New Delhi
8. BIS, IS 1893 (1975) Indian standard criteria for earthquake resistant design of structures, Bureau of Indian Standards, New Delhi
9. BIS, IS 1893 (1984) Indian standard criteria for earthquake resistant design of structures, Bureau of Indian Standards, New Delhi
10. BIS, IS 1893 (2002) Indian standard criteria for earthquake resistant design of structures, Part 1—General provisions and buildings, Bureau of Indian Standards, New Delhi
11. BIS, IS 1893 (2016) (Part 1): Indian standard criteria for earthquake resistant design of structures, Part 1—General provisions and buildings. Bureau of Indian Standards, New Delhi
12. https://en.wikipedia.org/wiki/Indian_Plate. Accessed Sept 15, 2021
13. Shen H (2012) Unusual Indian Ocean earthquakes hint at tectonic breakup. Nature. https://doi.org/10.1038/nature.2012.11487

14. Jade S, Shrungeshwara TS, Kumar K et al (2017) India plate angular velocity and contemporary deformation rates from continuous GPS measurements from 1996 to 2015. Sci Rep 7:11439. https://doi.org/10.1038/s41598-017-11697-w
15. Kumar P, Yuan X, Kumar MR, Kind R, Li X, Chadha RK (2007) The rapid drift of the Indian tectonic plate. Nature 449(7164):894–897. https://doi.org/10.1038/nature06214. PMID: 17943128
16. Mohanty S (2011) Crustal stress and strain patterns in the Indian plate interior: implications for the deformation behavior of a stable continent and its seismicity. Terra Nova 23:407–415
17. Mandal HS (2018) Earthquake dynamics in the Central India Tectonic Zone. Nat Hazards 92:885–905. https://doi.org/10.1007/s11069-018-3230-2
18. Stein S, Sella G, Okal EA (2002) The January 26, 2001, Bhuj earthquake and the diffuse western boundary of the Indian plate. In: Stein S, Freymueller JT (eds) Plate boundary zones, Geodyn. Ser., vol 30, pp 243–254, AGU, Washington, DC
19. Fournier M, Chamot-Rooke N, Petit C et al (2008) In situ evidence for dextral active motion at the Arabia-India plate boundary. Nature Geosci 1:54–58. https://doi.org/10.1038/ngeo.2007.24
20. Anbazhagan P, Joo MR, Rashid MM, Al-Arifi NSN (2021) Prediction of different depth amplifications of deep soil sites for potential scenario earthquakes. Nat Hazards 107:1935–1963. https://doi.org/10.1007/s11069-021-04670-4
21. Mallick R, Lindsey EO, Feng L, Hubbard J, Banerjee P, Hill EM (2019) Active convergence of the India-Burma-Sunda plates revealed by a new continuous GPS network. J Geophys Res Solid Earth 124(3):3155–3171. https://doi.org/10.1029/2018JB016480
22. Anbazhagan P, Mog K, Nanjunda Rao KS, Siddharth Prabhu N, Agarwal A, Reddy GR, Ghosh S, Deb MK, Baruah S, Das SK (2019) Reconnaissance report on geotechnical effects and structural damage caused by the 3rd January 2017, Tripura earthquake, India. Nat Hazards 98: 425–450. https://doi.org/10.1007/s11069-019-03699-w
23. Anbazhagan P, Uday A, Moustafa SSR, Al-Arifi NSN (2016) Pseudo-spectral damping reduction factors for the himalayan region considering recorded ground-motion data. PLoS ONE 11(9):e0161137. https://doi.org/10.1371/journal.pone.0161137
24. BSSC (2001) NEHRP recommended provisions for seismic regulations for new buildings and other structures 2000 edition, part 1: Provisions, Report no. FEMA 368. Building seismic safety council for the federal emergency management agency, Washington, DC, USA
25. Anbazhagan P, Arun Kumar K, Reddy GR, Moustafa SSR, Al-Arifi NSN (2018) Seismic site classification and amplification of shallow bedrock sites. Plos One 13(12): e0208226. https://doi.org/10.1371/journal.pone.0208226
26. Anbazhagan P, Kumar A, Sitharam TG (2013) Seismic site classification and correlation between standard penetration test N value and shear wave velocity for Lucknow City in Indo-Gangetic Basin. Pure Appl Geophys 170:299–318
27. Anbazhagan P, Bajaj K (2020) Region-specific correlations between V_S30 and time-averaged V_S and SPT-N values at different depths for the Indo Gangetic Basin. Indian Geotech J 50(3): 454–472. https://doi.org/10.1007/s40098-019-00379-1
28. Bajaj K, Anbazhagan P (2021) Detailed seismic hazard, disaggregation and sensitivity analysis for Indo Gangetic basin. Pure Appl Geophys 178(2021):1977–1999. https://doi.org/10.1007/s00024-021-02762-7
29. Anbazhagan P, Bajaj K, Moustafa SSR, Al-Arifi NSN (2015) Maximum magnitude estimation considering the regional rupture character. J Seismol 19:695–719. https://doi.org/10.1007/s10950-015-9488-x
30. Anbazhagan P, Prabhu G, Aditya P (2012) Seismic hazard map of Coimbatore using subsurface fault rupture. Nat Hazard 60:1325–1345
31. Anbazhagan P, Silas Abraham G (2020) Region-specific seismic hazard analysis of Krishna Raja Sagara Dam, India. Eng Geol 268, AR: 104087. https://doi.org/10.1016/j.enggeo.2020.105512
32. International Conference of Building Officials (ICBO) (1998) 1997 Uniform building code, International Conference of Building Officials, Whittier, California, International Code Council

(ICC), 2003, International Building Code (IBC) 2003, Building Officials and Code Administrators International, Inc., Country Club Hills, Illinois; International Conference of Building Officials, Whittier, California; and Southern Building Code Congress International, Inc., Birmingham, Alabama
33. Seed HB, Idriss IM (1983) Ground motions and soil liquefaction during earthquakes. Earthq Eng Res Inst
34. Trifunac MD (1990) How to model amplification of strong earthquake ground motions by local soil and geologic site conditions. Earthq Eng Struct Dyn 19(6):833–846
35. Trifunac MD (1992) Should peak accelerations be used to scale design spectrum amplitudes? In: Proceedings of the 10th world conference earthquake engineering, vol 10, pp 5817–5822. Madrid, Spain
36. Biot MA (1941) A mechanical analyzer for the prediction of earthquake stresses. Bull Seismol Soc Am 31:151–71
37. Housner GW (1959) Behavior of structures during earthquakes. J Eng Mech Div ASCE 85(EM 4):109–129
38. Housner GW (1970) Design spectrum, Chapter 5. In: Earthquake engineering. Prentice-Hall, R.L Wiegel, New Jersey
39. Newmark NM, Hall WJ (1969) Seismic design criteria for nuclear reactor facilities. In: Proceedings world conference on earthquake engineering, 4th Santiago, Chile, vol B-4, pp 37–50
40. Newmark NM, Hall WJ (1982) Earthquake spectra and design. Earthquake Engineering Research Institute, Oakland, California
41. United States Atomic Energy Commission (1973) Design response spectra for seismic design of nuclear power plants. Regulatory guide no. 1.60. U.S. Atomic Energy Commission, Washington, DC
42. Mohraz B (1976) A study of earthquake response spectra for different geological conditions. Bull Seism Soc Am 66(3):915–935
43. Mohraz B, Hall WJ, Newmark NM (1972) A study of vertical and horizontal earthquake spectra, AEC Report WASH-1255, Nathan M. Newmark Consulting Engineering Services, Urbana, Illinois
44. Seed HB, Ugas C, Lysmer J (1976) Site-dependent spectra for earthquake-resistant design. Bull Seism Soc Am 66(1):221–243
45. Malhotra PK (2006) Smooth spectra of horizontal and vertical ground motions. Bull Seism Soc Am 96(2):506–518
46. Malhotra PK (2001) Response spectrum of incompatible acceleration, velocity and displacement histories. Earthq Eng Struct Dyn 30(2):279–286
47. Bansal BK, Pandey AP, Singh AP, Suresh G, Singh RK, Gautam JL (2021) National seismological network in india for real-time earthquake monitoring. Seismol Res Lett 92(4):2255–2269. https://doi.org/10.1785/0220200327
48. Hall WJ, Mohraz B, Newmark NM (1975) Statistical studies of vertical and horizontal earthquake spectra, Nathan M. Newmark Consulting Engineering Services, Urbana, Illinois
49. Building Seismic Safety Council (BSSC) (1995) NEHRP recommended provisions for seismic regulations for new buildings (1994 edition), Federal Emergency Management Agency, FEMA 222A/223A, Building Seismic Safety Council, Washington, DC
50. International Code Council (ICC) (2012) International Building Code (IBC), Falls Church, VA; 328
51. CEN (European Committee for Standardization) (2004) Eurocode 8: design of structures for earthquake resistance, part 1: general rules, seismic actions and rules for buildings. EN 1998-1:2004. Brussels, Belgium
52. Building Seismic Safety Council (BSSC) (2010) NEHRP recommended provisions for seismic regulations for new buildings and other structures (2009 edition), Federal Emergency Management Agency, FEMA P-749, Building Seismic Safety Council, Washington, DC
53. Freeman JR (1932) Earthquake damage and earthquake insurance. McGraw-Hill, New York
54. Code P (2005) Eurocode 8: Design of structures for earthquake resistance-part 1: general rules, seismic actions and rules for buildings. Brussels: Eur Committee Stand

55. American Society of Civil Engineers (ASCE) (2010) Minimum design loads for buildings and other structures, ASCE Standard 7–10, Reston, VA, p 650
56. American Association of State Highway and Transportation Officials (AASHTO) (2011) LRFD bridge design specifications, 2nd edn, Washington, D.C, p 286
57. Dobry R, Ramos R, Power MS (1999) Site factors and site categories in seismic codes, Technical Report MCEER-99-0010, 81 pp
58. Anbazhagan P, Sagar GI (2021) Status quo of standard penetration test in India: a review of field practices and suggestions to incorporate in IS 2131. Indian Geotech J 51(2):421–434. https://doi.org/10.1007/s40098-020-00458-8

Chapter 2
Development of Relationships to Predict Amplification of SH-Wave Across Two-dimensional Deep-Basins at Fundamental Frequency

J. P. Narayan, Asmita Sharma, and Lav Joshi

Abstract The prediction of amplification of SH-wave at fundamental frequency 'F_{02D}' across 2D deep-basins (basin with shape-ratio >0.2 and response dominated with 2D-resonance phenomenon) is very important for earthquake engineers for retrofitting the existing and economical design of new buildings as well as to take extra measures in designs if double resonance could not be avoided. The paper presents numerical simulations and computation of spectral amplifications of the SH-wave across the considered rectangular and elliptical basins for different shape-ratios, impedance contrast, sediment damping and development of relations to predict the amplification of SH-wave at F_{02D} across the deep rectangular and elliptical basins. An increase of F_{02D} of deep-basin and corresponding amplification with an increase of shape-ratio and impedance contrast is obtained. The relations are developed to predict the amplification at F_{02D} across rectangular and elliptical deep-basins in terms of amplification at the centre of respective basin at F_{02D}, offset 'x' from centre of basin and half-width 'w' of basin. A comparison of numerically computed amplification at F_{02D} across rectangular and elliptical deep-basins with those obtained using developed relations reveals that the % age differences are within the permissible limit.

Keywords Deep and shallow rectangular and elliptical basins · Shape-ratio · Fundamental frequency · Spectral amplifications and development of relations

2.1 Introduction

The day by day increase in population and cost of landmass is pushing the builders to construct structures on sediment filled landmass in many of the Indian megacities. Such filled landmasses may act as mini/micro two dimensional (depressions or elongated filled lake) or three dimensional (pond/ circular lake) deep/shallow

J. P. Narayan (✉) · A. Sharma · L. Joshi
Department of Earthquake Engineering, Indian Institute of Technology Roorkee, Roorkee, India
e-mail: jp.narayan@eq.iitr.ac.in

L. Joshi
e-mail: ljoshi@eq.iitr.ac.in

© Indian Society of Earthquake Technology 2023

R. S. Jakka et al. (eds.), *Earthquake Engineering and Disaster Mitigation*, Springer Tracts in Civil Engineering, https://doi.org/10.1007/978-981-99-0081-7_2

basins **(Basin is defined as a low area in the earth's crust, of tectonic origin, in which sediments accumulate. Sedimentary basins range in size from hundreds of meters to as large as Ganga Basin).** A shallow basin has generally a shape-ratio < 0.2 and a deep-basin has shape-ratio > 0.2. The shape-ratio (*SR*) of basin is simply the ratio of maximum depth of basin to its effective half-width. The effective half-width of basin is the width below which sediment thickness is more than half of the maximum depth of basin. The seismic response of a 2D deep-basin is dominated by a resonance phenomenon and the entire basin vibrates with a single fundamental frequency or higher modes of vibrations [1–9]. The fundamental frequency of 2D deep-basin (F_{02D}) very much depends on its shape and shape-ratio [1, 3, 6, 10–14]. On the other hand, seismic response of shallow basin is dominated by the basin produced surface waves and its fundamental frequency varies from one location to another depending on the S-wave velocity (V_S) and sediment thickness (H), which can be computed using simple relation $F_{01D} = V_S/4H$ [15–20]. So, the development of relationships between the amplification at F_{02D} of deep-basins and geometry, dimension and rheological parameters of basin is very useful for the retrofitting of the existing buildings and for the design of new structures in the 2D deep-basins as well as to take extra measures in the design depending on the location in basin if occurrence of double resonance could not be avoided.

During past earthquakes, unexpected and selective damage to the structures are reported because of the occurrence of double resonance phenomena (The double resonance is defined as matching of resonance frequency of sediment/basin with that of structure) [21–23]. The researchers have developed mathematical relations to forecast the fundamental frequency of 2D/3D basins in terms of minimum F_{01D} of equivalent 1D basin and shape-ratio [1, 3, 6, 10, 24, 25]. It is also reported that the amplification at F_{02D} is largest at the centre of deep-basin and reduces towards the edges of basins [3, 13, 26]. Further, the amplification at the centre of 2D deep-basin at F_{02D} is larger than the equivalent 1D basin at the same location at F_{01D}. So, the damage to a structure due to the occurrence of double resonance will depend on the location of structures in the deep-basin and amplification at frequency F_{02D} where the structure is situated. The extensive literature review reveals that empirical relation is not available to predict the amplification at the F_{02D} at any location within the 2D deep-basins. Therefore, there is a need to develop a relation to predict amplification at F_{02D} across 2D rectangular and elliptical deep-basins considering the shape, shape-ratio, sediment damping and impedance contrast. Since basement shape of most of the 2D deep-basins existing in nature may be represented with simplified mathematical elliptical or rectangular shapes.

In order to develop the desired relationships between the amplification at F_{02D} of deep-basins and geometry, dimension and rheological parameters of basin, the SH-wave responses of the various considered rectangular and elliptical deep and shallow basins with different shape-ratio, sediment damping and impedance contrast are simulated and analyzed. The relationships are developed to predict the amplification at frequency F_{02D} across the deep rectangular and elliptical basins in terms of lowest F_{01D} of equivalent 1D basin and the shape-ratio. The SH-wave responses of the various considered basin models are simulated using fourth-ordered accurate viscoelastic staggered grid finite difference (FD) program developed by Narayan and Kumar [27]. The frequency-dependent damping in the time-domain simulation is incorporated based on the well known **Generalized Maxwell Body-Emmerich and Korn (GMB-EK)** rheological model [28–30].

2.2 Basin Model Parameters

In order to fulfill the aim, fifteen rectangular (SR1–SR15) and fifteen elliptical (SE1–SE15) basin models with varying shape-ratio are considered. Further, for each basin, there are five impedance contrast (IC1–IC5) and three sediment damping (QF1–QF3) cases. Table 2.1 depicts the horizontal dimensions of the rectangular and elliptical basins with a fixed maximum depth of sediment in basins as 32.5 m. The widths of basins are varied in such a way that basin behavior changes from 1D basin to 2D-shallow basin to 2D deep-basin. The range of shape-ratio for the rectangular basins is 0.01–0.55 and that for the elliptical basins is 0.015–0.79 (Table 2.1). Table 2.2 gives the considered five impedance contrasts (IC1–IC5) at the base of basins and three quality factors (QF1–QF3) for the sediment of basins and underlying rock. The impedance contrast is varied by varying the impedance of the bedrock so that the resonance frequency of the basin is unaffected. Finally, there are 225 rectangular (15 $\times$ 5 $\times$ 3 = 225) and 225 elliptical basin models (Tables 2.1 and 2.2). Figure 1a, b depicts the 2D-shallow ($SR < 0.2$) and 2D-deep ($SR \geq 0.2$) rectangular basin (left panels) and elliptical basin (right panels), respectively.

A plane wave-front of the SH-wave was generated in to the FD grid using shear stress σ_{ZY} at various points along a horizontal line at a depth of 200 m. Gabor wavelet is used as a source time function with predominant frequency f_P= 2.5 Hz, oscillatory parameter γ = 1.5, duration parameter t_S = 0.4 s and phase shift φ = 0 [3]. The frequency bandwidth in the Gabor wavelet is 0–10 Hz. Stress imaging technique is used as a free surface boundary condition [31]. Sponge absorbing boundary is implemented on the bottom, left and right edges of the model to avoid edge reflections [32]. The used grid spacing is 5 m in the horizontal direction throughout the model. In the vertical direction grid spacing is 2.5 m up to a depth of 62.5 and 5 m thereafter. The time step of 0.0005 s is taken to avoid instability problem. All the horizontal distances are measured with respect to the centre of basins. Twenty-seven equidistant (25 m apart) receiver points extending from 325 m left to 325 m right of centre of basin are used to record the ground motion at the free surface. The SH-wave responses of all the considered basin models for different impedance contrasts and damping are simulated and analyzed. The responses of an equivalent 1D basin, named as SR0 and SE0, are also simulated for different impedance contrasts and damping to infer the role of shape-ratio on the F_{02D} of basin and corresponding amplification. The equivalent 1D basin is simply a basin with a horizontal sediment layer of thickness 32.5 m overlying the rock.

Table 2.1 A comparison of computed fundamental frequency for different shape-ratio of rectangular and elliptical basins with that computed using empirical relations given by Bard and Bouchon [1] (*Note* Bard and Bouchon [1] is written in short as BB85)

Basin dimension		Rectangular basin					Elliptical basin				
Half-width (m)	Depth (m)	Model name	Shape-ratio	F_{02D} (BB85)	F_{02D} (Comp)	% diff	Model name	Shape-ratio	F_{02D} (BB85)	F_{02D} (Comp)	% diff
3000	32.5	SR1	0.01	2.30	2.31	−0.43	SE1	0.015	2.50	2.33	6.80
1500	32.5	SR2	0.02	2.30	2.29	0.43	SE2	0.030	2.50	2.36	5.60
750	32.5	SR3	0.04	2.31	2.18	5.63	SE3	0.062	2.50	2.50	0.00
600	32.5	SR4	0.05	2.31	2.31	0.00	SE4	0.078	2.51	2.32	7.57
300	32.5	SR5	0.10	2.36	2.39	−1.27	SE5	0.184	2.52	2.56	−1.59
200	32.5	SR6	0.15	2.33	2.28	2.15	SE6	0.245	2.56	2.62	−2.34
150	32.5	SR7	0.20	2.36	2.35	0.42	SE7	0.333	2.60	2.68	−3.08
135	32.5	SR8	0.23	2.37	2.36	0.42	SE8	0.376	2.62	2.72	−3.82
120	32.5	SR9	0.25	2.39	2.38	0.42	SE9	0.424	2.65	2.77	−4.53
100	32.5	SR10	0.30	2.42	2.40	0.83	SE10	0.518	2.72	2.80	−2.91
85	32.5	SR11	0.35	2.47	2.49	−0.81	SE11	0.625	2.79	2.89	−3.58
75	32.5	SR12	0.40	2.51	2.56	−1.99	SE12	0.736	2.87	2.99	−4.18
65	32.5	SR13	0.45	2.58	2.62	−1.55	SE13	0.752	2.96	3.08	−4.05
60	32.5	SR14	0.50	2.62	2.66	−1.53	SE14	0.774	3.06	3.11	−1.63
55	32.5	SR15	0.55	2.68	2.72	−1.49	SE15	0.790	3.17	3.21	−1.26

Table 2.2 The considered model parameters for five impedance contrasts (IC1–IC5) and three quality factors for the rectangular and elliptical basin models

Impedance models	Materials	V_s (m/s)	Quality factors			Density (kg/m^3)	Impedance contrast
			QF1	QF2	QF3		
IC1	Basin	300	24	30	36	1800	6.11
	Rock	1570	125	155	185	2102	
IC2	Basin	300	24	30	36	1800	7.24
	Rock	1800	140	180	215	2172	
IC3	Basin	300	24	30	36	1800	8.33
	Rock	2050	160	205	245	2195	
IC4	Basin	300	24	30	36	1800	9.62
	Rock	2250	180	225	270	2311	
IC5	Basin	300	24	30	36	1800	10.50
	Rock	2400	190	240	285	2362	

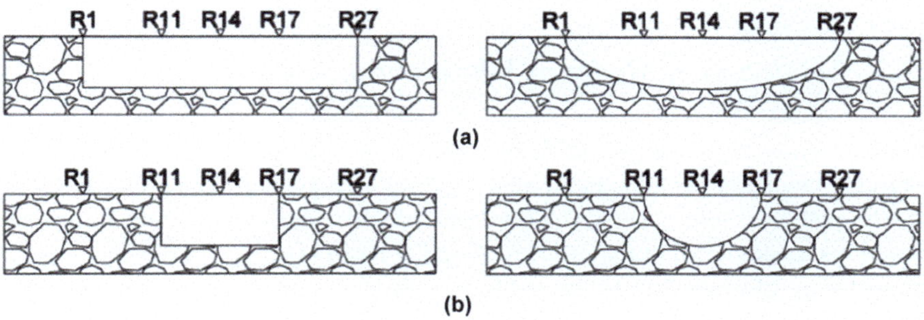

Fig. 2.1 **a** and **b** Shallow and deep rectangular (left panel) and elliptical (right panel) basin models respectively

2.3 SH-Wave Basin Responses and Validation

2.3.1 *SH-Wave Response of Considered Basins*

The left panels of Fig. 2a–c shows the SH-wave responses of SR1, SR10 and SR15 rectangular basins with impedance contrast IC1 and damping QF1 (Table 2.2). Similarly, the right panels of Fig. 2a–c shows the responses of SE1, SE10 and SE15 elliptical basins with impedance contrast IC3 and damping QF2. A drastic change in the SH-wave response with the change of shape-ratio of basin can be inferred. Figure 2.2 illustrates an increase of amplitude of the SH-wave at the centre of both the rectangular and elliptical basins with an increase in shape-ratio. In the case of lesser shape-ratio, the generation of Love waves near the edges can be inferred in both the basins [19]. The basin generated Love waves may be in-phase or out-of-phase at a particular location with the incident SH-wave and its multiples. The incident SH-wave, its multiples and Love waves/horizontally travelling diffracted SH-waves are merging with an increase of shape-ratio, which may be the reason for an increase of amplitude of the SH-wave at the centre of basins with an increase of shape-ratio. The SH-wave response of an equivalent 1D basin with depth 32.5 m was also simulated with and without sediment in the model for computation of spectral amplification and validation. The ratio of spectra of responses with and without sediment is also used to compute the F_{02D} of basin and corresponding amplification. Figure 3a, b shows a comparison of computed spectral amplifications at the centre of SR10 and SR15 rectangular basins and SE10 and SE15 elliptical basins, respectively with that of equivalent 1D basin (Designated as SR0 and SE0 in figure). It is concluded that in both the basins, different modes of vibrations are developed and frequencies corresponding to the different modes of vibrations of basins are varying with shape and shape-ratio. The obtained F_{02D} of basin and corresponding amplification of the rectangular and elliptical basins are larger than those in the case of an equivalent 1D basin.

2.3.2 *Validation of Modes of Vibrations of Deep-Basins*

In order to validate the simulated results, the computed F_{02D} and higher modes of basins are compared with the exact analytical solution given by Bard and Bouchon [1]. The analytical solution given by Bard and Bouchon [1] determines the frequency for various modes of vibration of the rectangular basin ($F_{x(n,m)}$) for the SH-wave in terms of F_{01D} of equivalent 1D basin ($F_{01D} = \frac{V_S}{4h}$).

$$F_{x(n,m)} = F_{01D}\sqrt{\left[(2m+1)^2 + (n+1)^2 \times h^2/w^2\right]} \tag{2.1}$$

where $F_{x(n,m)}$ denotes various vibration modes, 'h' is depth of rectangular basin and 'w' is half-width of the rectangular basin. The vibration modes of horizontal and vertical resonances are characterized by their order 'n' and 'm', respectively. A comparison of numerically obtained frequencies corresponding to fundamental ($F_{0(0,0)}$), first ($F_{1(0,2)}$), second ($F_{2(0,4)}$) and third ($F_{3(0,6)}$) modes of vibration of SR10 rectangular basin as 2.40, 3.15, 4.32, and 5.56 Hz with those obtained using the analytical solution by Bard and Bouchon [1] as 2.41, 3.22, 4.40 and 5.73 Hz reveals an excellent match. Table 2.1 depicts a comparison of analytically and numerically obtained F_{02D} of rectangular basins and the difference is less than 2% for basins with shape-ratio more than 0.2.

Similarly, the fundamental frequency of elliptical basin for SH-wave could be determined as per the analytical solution put forward by Bard and Bouchon [1] (Kumar and Narayan [3]).

$$F_{02D} = F_{01D}\sqrt{(1 + (h/w_e)^2} \tag{2.2}$$

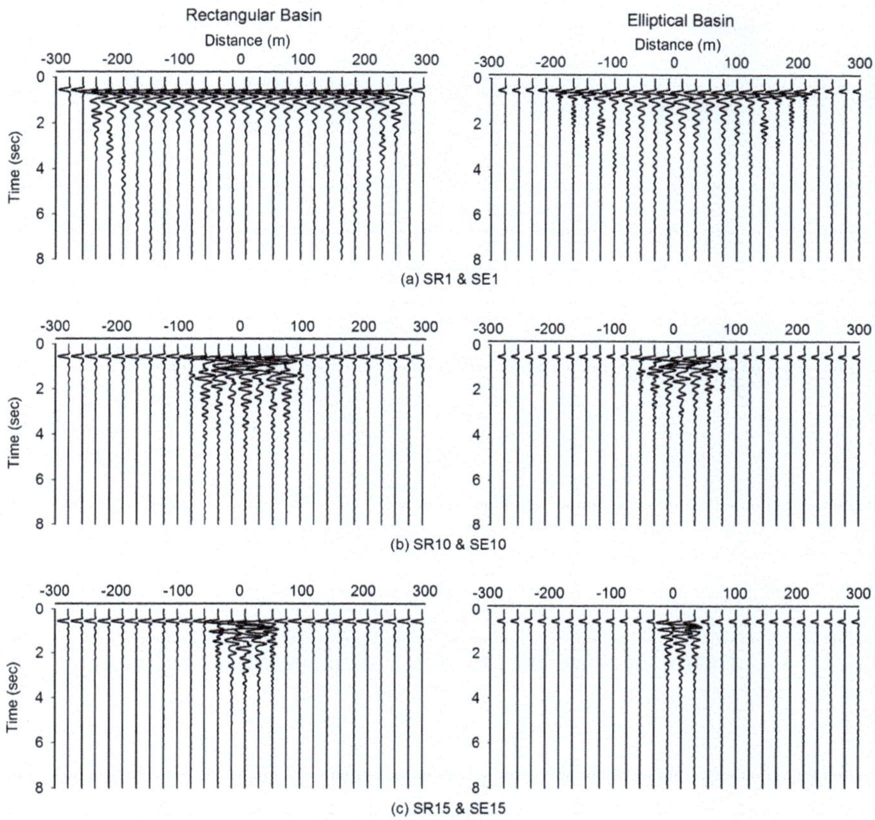

Fig. 2.2 **a–c** The shear wave responses of SR1, SR10 and SR15 rectangular basin with impedance contrast IC1 and damping QF1 (left panel) and SE1, SE10 and SE15 elliptical basins with impedance contrast IC3 and damping QF2 (right panel), respectively

where 'h' is maximum depth of the elliptical basin and 'w_e' is equivalent half-width of the elliptical basin. A comparison of analytically and numerically obtained F_{02D} of elliptical basins is given in Table 2.1. The difference is less than 5% for basins with shape-ratio more than 0.2 (Table 2.1). For example, numerically and analytically obtained F_{02D} of the SE10-elliptical basin are 2.80 Hz and **2.72 Hz**, respectively. The inferred different modes of vibrations of rectangular and elliptical basins are matching but corresponding frequencies are different.

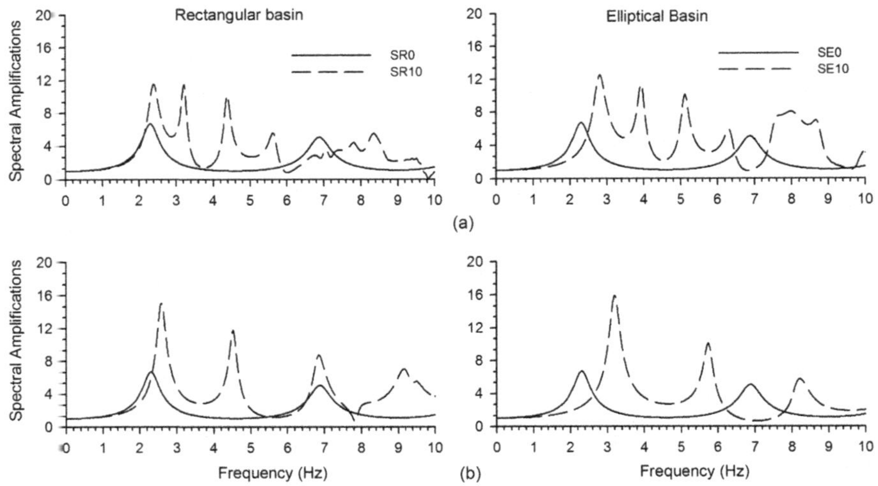

Fig. 2.3 **a** and **b** A comparison of spectral amplifications at the centre of SR10/SE10 and SR15/SE15 basins with that in the case of SR0/SE0 1D basin, respectively (*Note* SR0 and SE0 basins are same)

2.3.3 Validation of Amplification at Fundamental Frequency of 1D Basin

In order to validate the amplification at fundamental frequency taking in to consideration damping (ε) and impedance contrast (IC), the computed amplification (A_{01D}) at frequency F_{01D} of 1D basin are compared with well known analytical relation (Eq. 2.3).

$$A_{01D} = \frac{1}{\frac{1}{IC} + 0.5\pi\varepsilon} \tag{2.3}$$

A comparison of numerically obtained amplification at F_{01D} of equivalent 1D basin for IC3 and QF2 (SR0/SE0 basin) as <u>**6.75 Hz**</u> and that computed using Eq. 2.3 as <u>**6.70 Hz**</u> reveals an excellent match (Table 2.3). The obtained excellent match of the frequencies corresponding to the different modes of vibrations of the deep rectangular and elliptical basins with those analytically computed as well as obtained excellent match of the amplification at frequency F_{01D} of 1D basin with that computed using Eq. 2.3 validates the accuracy of numerically simulated SH-wave responses of the various considered basin models.

Table 2.3 A comparison of variation of amplification at F_{02D} at centre of rectangular and elliptical basins (A_{02D}) for IC3 with that in the case of equivalent 1D basin for different shape-ratio and quality factors

Rectangular basin (A_{02D})				Elliptical basin (A_{02D})			
Shape-ratio	QF1 (A_{01D} = 6.33)	QF2 (A_{01D} = 6.70)	QF3 (A_{01D} = 6.91)	Shape-ratio	QF1 (A_{01D} = 6.33)	QF2 (A_{01D} = 6.70)	QF3 (A_{01D} = 6.91)
0.01	6.42	6.75	7.00	0.015	10.34	11.12	11.62
0.02	6.41	6.89	7.46	0.030	7.77	8.32	8.72
0.04	3.68	3.15	2.74	0.062	3.10	2.98	2.88
0.05	10.33	11.41	12.25	0.078	0.93	0.54	0.72
0.1	0.66	1.55	2.26	0.184	8.14	8.83	9.08
0.15	7.60	8.18	8.61	0.245	9.53	10.10	10.51
0.2	8.60	9.09	9.41	0.333	10.66	11.46	12.06
0.23	9.12	9.63	10.03	0.376	11.19	11.97	12.53
0.25	9.77	10.41	10.88	0.424	11.59	12.50	13.18
0.3	10.61	11.43	12.05	0.518	12.37	13.46	14.29
0.35	11.63	12.57	12.55	0.625	12.97	14.14	15.02
0.4	12.34	13.33	14.07	0.736	13.71	14.86	15.68
0.45	13.08	14.21	15.04	0.752	14.08	15.33	16.25
0.5	13.35	14.40	15.15	0.774	14.06	15.47	16.53
0.55	13.61	14.79	15.83	0.790	14.40	15.73	16.89

2.4 Effects of Shape-ratio on Basin Response

The present sub-section describes the influence of shape-ratio on the F_{02D} of rectangular and elliptical basins and the spatial variation of amplification at F_{02D} across these basins.

2.4.1 *Fundamental Frequency* (F_{02D})

The left panel of Fig. 4a depicts the SH-wave responses at the centre of rectangular basin for different shape-ratio, impedance contrast IC1 and damping QF1. The right panel of Fig. 4a depicts the SH-wave responses at the centre of elliptical basin for different shape-ratio, impedance contrast IC3 and damping QF2. A drastic change in basin response with variation in shape as well as shape-ratio can be inferred. For example, the duration of ground motion is more in the basins with shape-ratio less than 0.15 due to the basin generated Love waves. An increase in amplitude of the SH-wave in both the basins with an increase in shape-ratio can be inferred due to the occurrence of 2D-resonance and less duration due to non-generation of Love waves. The variation of spectral amplifications with the shape-ratio at the centre of rectangular and elliptical basins is shown in the left and right panels of Fig. 4b, respectively. Different modes of vibrations of both the basins can be inferred. There is drastic variation of frequencies corresponding to different modes of vibrations with shape and shape-ratio. The rectangular and elliptical basins are more or less behaving like 1D basin for shape-ratio less than 0.05. In the case of rectangular basin, a transition zone can be inferred when the shape-ratio falls between 0.04 and 0.2, wherein different developed modes are not discernible (Table 2.1). However, different modes are well separated when shape-ratio is more than 0.2 (SR7 model). In the case of elliptical basin, a transition zone can be inferred when shape-ratio falls between 0.05 to 0.15 and different modes are well separated when shape-ratio is more than 0.15 (SE5 model). Table 2.1 depicts that F_{02D} of elliptical basins are more than those of rectangular basins, even for the same shape-ratio.

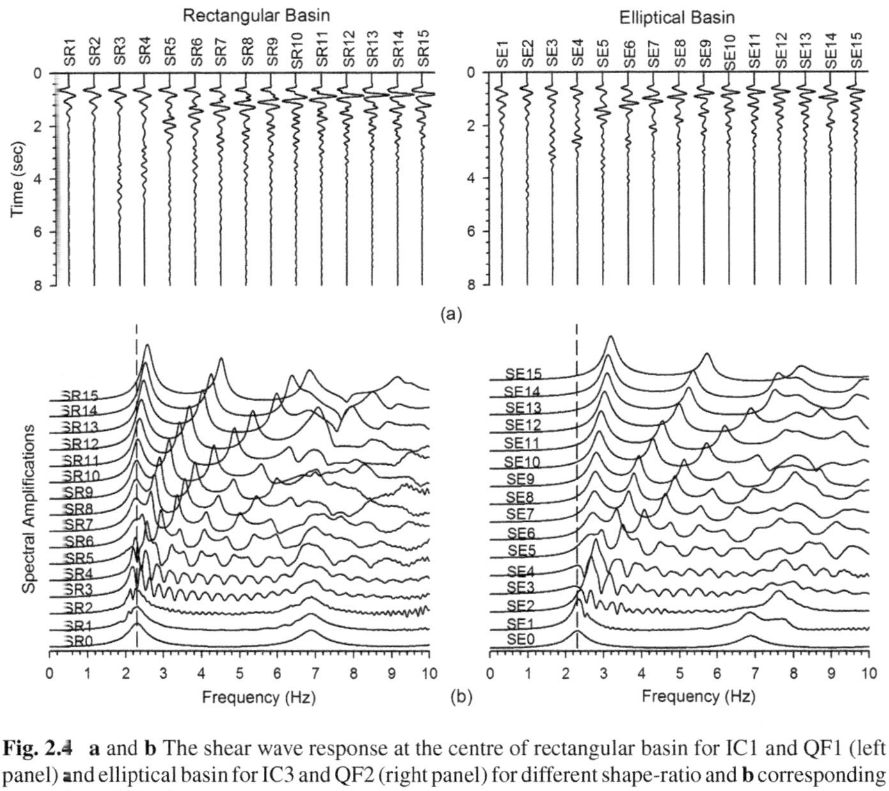

Fig. 2.4 **a** and **b** The shear wave response at the centre of rectangular basin for IC1 and QF1 (left panel) and elliptical basin for IC3 and QF2 (right panel) for different shape-ratio and **b** corresponding spectral amplifications

2.4.2 *Amplification at* F_{02D} *of Basin*

The main objective of this study is to predict the amplification at F_{02D} across the deep-basins. The spectral amplifications are computed at all the receiver points within the basin for all the considered shape-ratios. **It was noticed** that when shape-ratio of basin is more than around 0.2, the F_{02D} of basin is same everywhere in the basin. The amplification at F_{02D} of basin is largest at the centre of basin and reduces towards the edges. The amplification corresponding to F_{02D} of basin were noted at all the receiver points across the basins. Figure 5b represents the variation of amplification at F_{02D} at centre of rectangular basin for IC1 and QF1 and at centre of elliptical basin for IC3 and QF2 with shape-ratio. The obtained larger or smaller amplification at F_{02D} of basin in the cases with shape-ratio less than 0.15 may be due to the Love/diffracted waves falling in-phase or out-of-phase with the incident SH-wave and its multiples (Fig. 5b). In the case of rectangular basin with shape-ratio 0.55, the amplification at F_{02D} at the centre of basin (13.61) is more than twice to that of the equivalent 1D basin (6.33). Similarly, in the case of elliptical basin with shape-ratio 0.79, the amplification at F_{02D} at the centre (14.40) is more than twice that in the case of equivalent 1D basin (6.33). Table 2.3 gives the detail of variation of amplifications at F_{02D} at centre of basins with shape-ratio and damping as compared to that in the case of an equivalent 1D basin.

Figure 6a shows the spatial variation of amplification at F_{02D} across the SR10 and SR15 rectangular basins for IC1 and QF1 as well as SE10 and SE15 elliptical basins for IC3 and QF2. The obtained amplification at F_{02D} of basin is highest

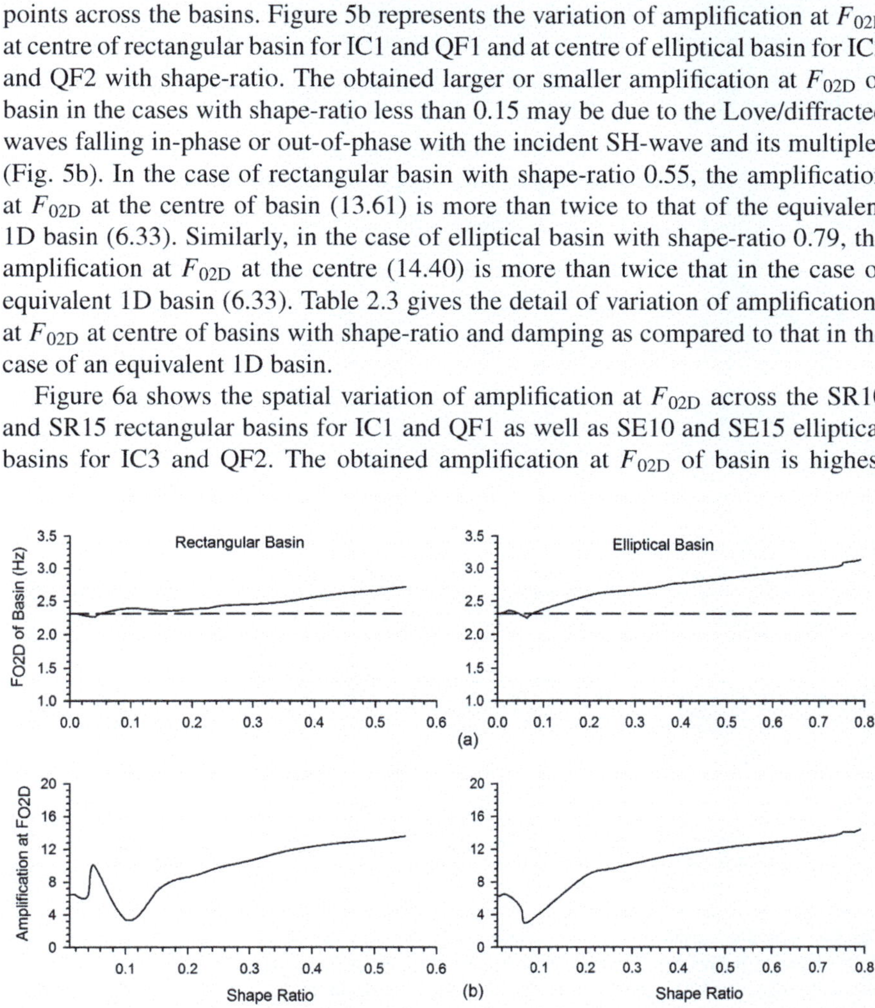

Fig. 2.5 **a** and **b** The variation of fundamental frequency and corresponding amplifications, respectively with shape-ratio in the case of rectangular basin (left panel) and elliptical basin (right panel), respectively for IC3 and QF2 (*Note* Dashed line shows the lowest fundamental frequency of the equivalent 1D basin)

at the centre of basin and gradually decreases towards the edges of basin [3]. An analysis of Fig. 6a and Table 2.3 reveals an increase of amplification at F_{02D} across the deep-basin with an increase of shape-ratio when shape-ratio is more than 0.2. The fundamental modes of vibrations in horizontal as well as vertical directions are accountable for the sizeable amplification at the F_{02D} of deep-basin at the centre and least near the edges (Fig. 6a). For example, the obtained amplification as 14.79 at the F_{02D} at centre of rectangular basin with shape-ratio 0.55 is 325% higher than that as 3.25 at a distance of 40 m. In contrast to this, in the case of shallow basin (shape-ratio < 0.2), there is a lot of spatial variation of amplification at F_{02D} across the basin (result not shown here). The main cause for this variation may be 1D nature of basin and the amplification/de-amplification of a particular frequency due to the interference with the basin generated Love waves and diffracted waves [18, 19].

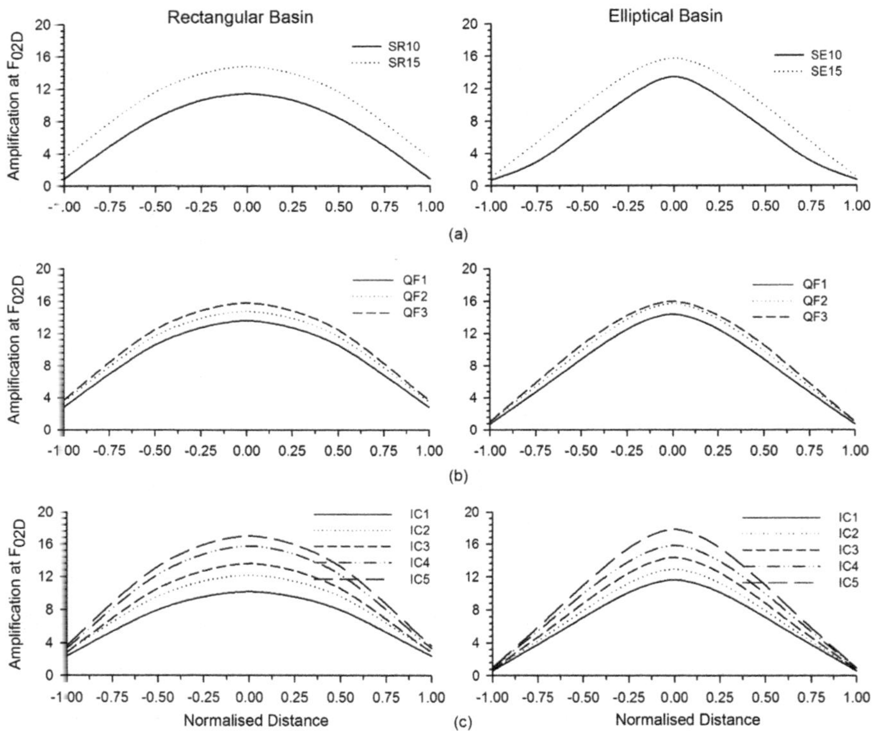

Fig. 2.6 **a** Spatial variation of amplification at F_{02D} of SR10 and SR15 rectangular and SE10 and SE15 elliptical basins for IC3 and QF2; **b** amplification at F_{02D} of SR15 rectangular and SE15 elliptical basins for IC3 and different damping and **c**. Amplification at F_{02D} of SR15 rectangular and SE15 elliptical basins for different impedance contrast and damping QF2 (*Note* Normalised distance is simply the ratio of distance from the centre of basin to the half-width of basin)

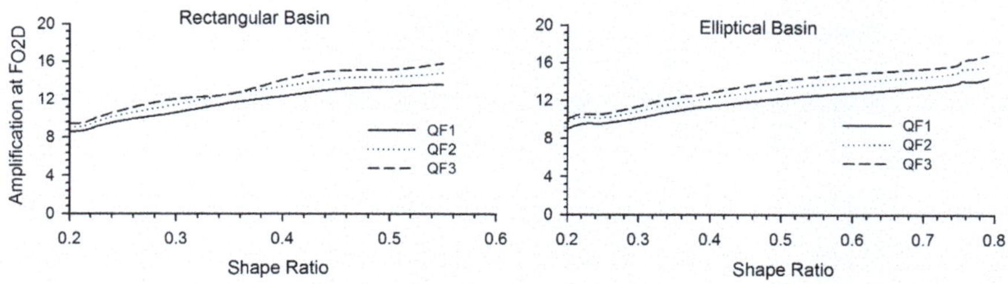

Fig. 2.7 The variation of amplifications at fundamental frequency with shape-ratio in the case of rectangular basin (left panel) and elliptical basin (right) for different quality factors and impedance contrast as IC3

2.5 Effect of Quality Factor

In order to infer the role of sediment damping on the amplification at F_{02D} of 2D deep-basins, the quality factor of the sediment of basin and bedrock are considered as QF1 = 10%, QF2 = 8% and QF3 = 12% of velocity of shear wave in the respective material (Table 2.2). The SH-wave responses of all the considered 45 rectangular and elliptical basins were computed using sediment damping corresponding to QF1, QF2 and QF3. But, in this sub-section, only the role of sediment damping in the spectral amplifications for the impedance contrast IC3 with different shape-ratios of rectangular and elliptical basins are discussed. Figure 2.7 depicts the variation of amplification at F_{02D} of basin with shape-ratio of rectangular and elliptical basins. In general, it has been observed that with the increase in quality factor of sediment the amplification caused by the basin increases. A comparison of the obtained spatial variation of amplification at the F_{02D} of SR15-rectangular and SE15-elliptical basins is shown in Fig. 6b for considered damping and impedance contrast IC3. Table 2.3 reveals an increase of amplification at F_{02D} with increase in quality factor of the sediment at all the locations in the basins.

2.6 Effect of Impedance Contrast

In this sub-section, the effects of impedance contrast on the spectral amplifications across the basin corresponding to the different shape-ratios and quality factors are computed (Table 2.2). The SH-wave responses of SR15-rectangular and SE15-elliptical basins are shown in Fig. 8a–c for impedance contrasts IC1, IC3 and IC5 and damping QF2. The decrease of amplitude of SH-wave in rock as well as in basin due to an increase of impedance contrast may be due to the increase of impedance in the rock since in all the IC models same stress level is used to generate point sources along a horizontal line.

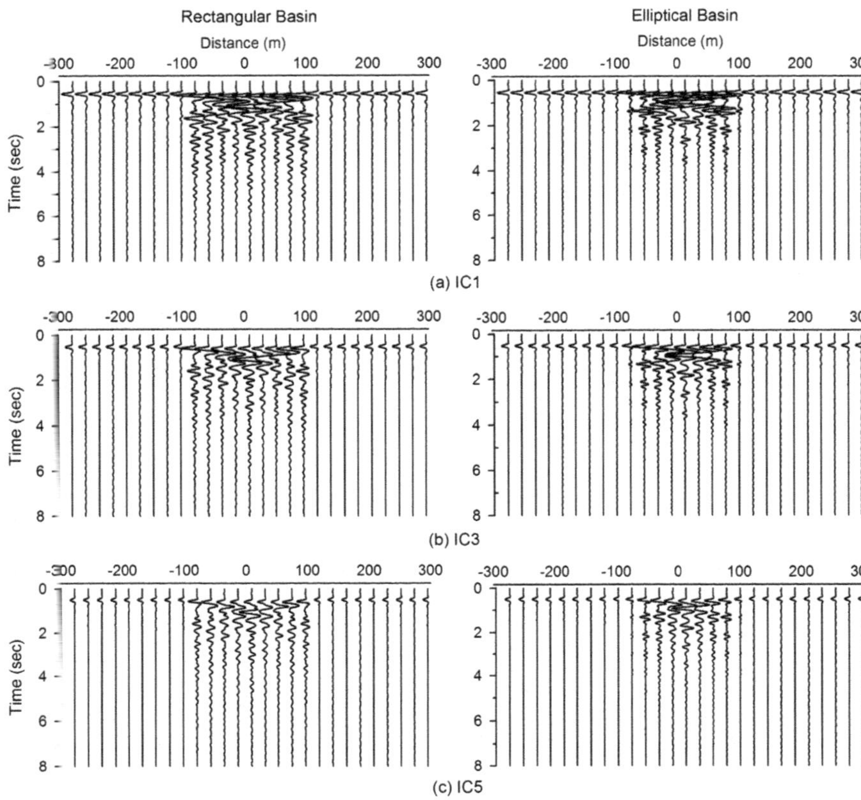

Fig. 2.8 a–c The shear wave responses of the SR10 rectangular (left panel) and SE10-elliptical basins (right panel) with impedance contrast as IC1, IC3 and IC5 and damping QF2

2.6.1 *Variation of* F_{02D} *with IC*

A comparison of variation of F_{02D} of basin with impedance contrast for shape-ratio more than 0.2 is represented in right and left panels of Fig. 9a for the elliptical and rectangular basins, respectively. Analysis of this figure reveals that there is an increase of F_{02D} in basin with increase of impedance contrast in both the basins [13].

2.6.2 *Variation of Amplification at* F_{02D} *of Basin*

A comparison of variation of amplification at F_{02D} of deep-basin with the change of impedance contrast and shape-ratio more than 0.2 is shown in Fig. 9b for the rectangular and elliptical basins. As the impedance contrast increases there is an increase of amplification at F_{02D} across basin (Tables 2.4 and 2.5). The percentage

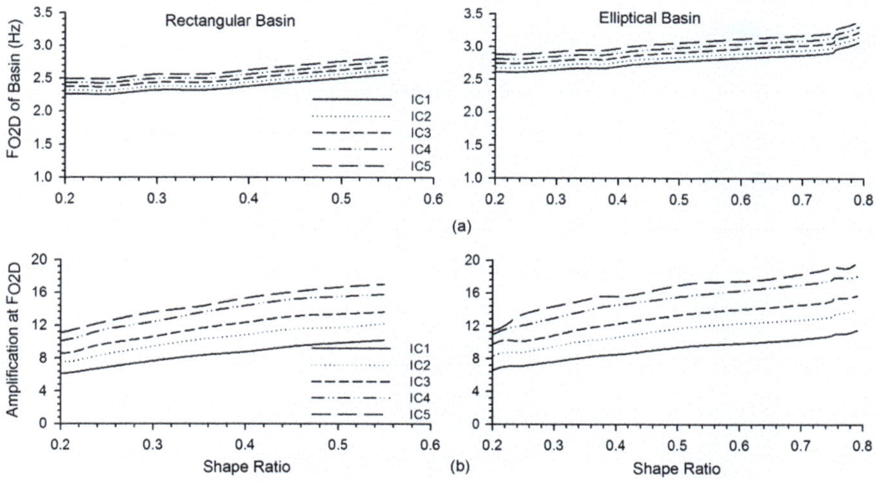

Fig. 2.9 **a** Variation of fundamental frequency of rectangular (left panel) and elliptical basins (right panel) with shape-ratio for different impedance contrast and damping QF2, **b** Variation of amplification at fundamental frequency at Centre of rectangular (left panel) and elliptical basins (right panel) with shape-ratio for different impedance contrast and damping QF2

increase of amplification at F_{02D} at centre of SR10 rectangular basin as compared to that in the case of equivalent 1D basin is 42.5%, 41.1%, 40.3%, 37.8% and 33.6% for impedance contrast IC1–IC5, respectively (Table 2.4). Similarly, the percentage increase of amplification at F_{02D} of SE10-elliptical basin as compared to that in the case of equivalent 1D basin is 52.1%, 51.1%, 50.2%, 48.6% and 44.8% for impedance contrast IC1–IC5, respectively (Table 2.5). The left and right panels of Fig. 6c depict the spatial variation of amplification at F_{02D} of SR15 rectangular and SE15 elliptical basins, respectively. Again generalized amplification trend at the F_{02D} of basin can be inferred i.e., largest amplification at the centre of deep-basin and least near the edge. Further, an increase of amplification at F_{02D} of deep-basin with an increase of impedance contrast is obtained.

2.7 Development of Relationships

In order to develop relationships to predict the amplification (A_{02D}) at F_{02D} at the centre of deep-basin, the obtained amplification at F_{02D} of rectangular and elliptical basins with shape-ratio more than 0.2, impedance contrast IC1–IC5 and sediment damping QF1–QF3 were used. The current version of IBM SPSS (Statistical Packaging for the Social Sciences) software [33] was used to develop the relations. For the purpose of non-linear regression analysis, all the data in form of independent variables, i.e., sediment damping (QF), impedance contrast (IC), shape-ratio (*SR*), the amplification A_{01D} at lowest F_{01D} of equivalent 1D basin were used as input

Table 2.4 The amplification (A_{02D}) at F_{02D} and at centre of the rectangular basins with different shape-ratios, damping QF1 and impedance contrast IC1–IC5

Rectangular basin models	A_{02D} at centre of basin at F_{02D}					Rectangular basin models	A_{02D} at centre of basin at F_{02D}				
	IC5 = 10.5	IC4 = 9.62	C3 = 8.33	IC2 = 7.24	IC1 = 6.11		IC5 = 10.5	IC4 = 9.62	IC3 = 8.33	IC2 = 7.24	IC1 = 6.11
SR0	7.81	7.31	6.33	5.85	5.09	SR8	11.78	10.76	9.12	7.90	6.48
SR1	7.64	6.45	6.42	5.69	5.07	SR9	12.47	11.54	9.77	8.50	6.89
SR2	8.38	6.66	6.41	6.26	4.21	SR10	13.59	12.42	10.61	9.42	7.67
SR3	11.03	10.01	3.68	3.68	2.59	SR11	14.29	13.53	11.63	10.27	8.33
SR4	12.48	11.02	10.33	8.94	9.36	SR12	15.30	14.38	12.34	10.83	8.77
SR5	6.95	6.50	0.66	1.14	1.23	SR13	16.05	15.16	13.08	11.55	9.43
SR6	9.69	8.82	7.60	6.45	5.17	SR14	16.65	15.50	13.35	11.73	9.84
SR7	11.11	10.10	8.60	7.52	6.12	SR15	17.03	15.77	13.61	12.20	10.19

Table 2.5 The amplification (A_{02D}) at F_{02D} and at centre of the elliptical basins with different shape-ratios, damping QF2 and impedance contrast IC1–IC5

Elliptical basin models	A_{02D} at centre of basin at F_{02D}					Elliptical basin models	A_{02D} at centre of basin at F_{02D}				
	IC5 = 10.5	IC4 = 9.62	C3 = 8.33	IC2 = 7.24	IC1 = 6.11		IC5 = 10.5	IC4 = 9.62	IC3 = 8.33	IC2 = 7.24	IC1 = 6.11
SR0	8.23	7.68	6.70	6.08	5.26	SR8	15.58	14.28	11.97	10.36	8.39
SR1	11.53	8.99	11.12	9.13	7.19	SR9	15.76	14.78	12.50	10.87	8.69
SR2	12.52	11.43	8.32	6.64	4.77	SR10	17.18	15.71	13.46	11.85	9.54
SR3	2.44	0.98	2.98	3.00	10.42	SR11	17.59	16.50	14.14	12.41	9.98
SR4	11.25	3.53	0.54	1.29	2.36	SR12	18.88	17.43	14.86	12.97	10.70
SR5	12.23	10.15	8.83	7.60	5.97	SR13	19.26	17.92	15.33	13.38	11.01
SR6	13.28	12.00	10.10	8.75	7.07	SR14	19.07	17.89	15.47	13.68	11.12
SR7	14.87	13.55	11.46	10.00	8.00	SR15	19.79	18.08	15.73	14.04	11.51

Table 2.6 Estimated values of the constants '*a*', '*b*', '*c*', '*d*' and '*e*' for Eq. 2.4 as well as constants '*f*', '*g*' and '*h*' for Eq. 2.5

Values of constants in the case of rectangular basin		Values of constants in the case of elliptical basin	
a	1.81	*f*	1.92
b	−163.81	*g*	10.11
c	134.20	*h*	−8.51
d	−11.15		
e	−7.66		

parameters. The performed regression analysis revealed that A_{02D} mainly depends on shape-ratio (83.6%) and A_{01D} (16.1%). The dependency on the impedance contrast and sediment damping were negligible since A_{01D} already accommodates the effect of impedance contrast and sediment damping. The variation of A_{02D} with shape-ratio and A_{01D} was estimated using curve fitting technique. The non-linear regression analysis was then employed to establish a relationship between A_{02D}, shape-ratio and A_{01D}. The model with highest R-squared value (0.841) and least mean percentage error (3.85%) was selected for the purpose of study. The relationships given in Eqs. (2.4) and (2.5) are developed to predict the amplification at F_{02D} and at the centre of deep rectangular (A^{RB}_{02D}) and elliptical (A^{EB}_{02D}) basins, respectively.

$$A^{RB}_{02D} = a.A_{01D} + b.SR^3 + c.SR^2 + d.SR + e \tag{2.4}$$

$$A^{EB}_{02D} = f.A_{01D} + g.SR + h \tag{2.5}$$

The values of constants *a*, *b*, *c*, *d*, *e*, *f*, *g* and *h* are given in Table 2.6.

In order to predict the amplification at F_{02D} of basin and at an offset 'X' from the centre of basin (A_{0X}) in terms of A_{02D}, shape-ratio and X/w. Similar regression analysis was done using the data set with shape-ratio of more than 0.2. The developed regression relation reveals that the amplification at F_{02D} of basin and at an offset 'X' mainly depends on A_{02D} and the ratio "X/w". The following relationships (Eqs. 2.6 and 2.7) are developed to predict amplification at F_{02D} of basin across the rectangular (A^{RB}_{0x}) and elliptical (A^{EB}_{0x}) basins, respectively.

$$A^{RB}_{0x} = A^{RB}_{02D}\left(1 - 0.553\left(\frac{x}{w}\right)^2\right)^2 \tag{2.6}$$

$$A^{EB}_{0x} = A^{EB}_{02D}\left(1 - 0.771\left(\frac{x}{w}\right)^2\right)^2 \tag{2.7}$$

The %age difference between the predicted amplification at F_{02D} within basin using the developed relationships and that obtained numerically are within the permissible limit.

2.8 Discussion and Conclusions

The obtained excellent match between the numerically computed fundament frequency (F_{02D}) of the deep rectangular and elliptical basins with those computed using analytical solutions given by Bard and Bouchon [1] as well as an excellent match of the calculated amplification at F_{01D} of equivalent 1D basin with that computed using relation $A_{01D} = 1/(\frac{1}{IC} + 0.5\pi\varepsilon)$ validate the accuracy of simulated results. An increase of amplification at F_{02D} of deep—basin was obtained with increase in impedance contrast and shape-ratio. The obtained F_{02D} and corresponding amplification are more in the case of elliptical basin as compared to the rectangular for the same dimension. The obtained increase of F_{02D} of deep-basin with increase of impedance contrast corroborates with the finding of Zhu and Thambiratnam [13]. Most of the field earthquake engineers predict the fundamental frequency of sediment deposit using simple relation $F_{01D} = {}^{Vs}/_{4H}$ to avoid the double resonance, which will be highly variable in the case of elliptical basin. However, for a particular sediment parameters the amplification at F_{01D} will be the same everywhere in the homogeneous elliptical basin. In the 2D deep-basins for retrofitting of the existing buildings, design of new structures and to take extra measures in the design if occurrence of double resonance could not be avoided, it is recommended to predict the fundamental frequency using Eqs. 2.1 or 2.2 and then predict the amplification at F_{02D} at centre of basin using Eqs. 2.3–2.5 and then finally at any location in basin using Eqs. 2.5 and 2.7, depending on the shape of the basin.

References

1. Bard PY, Bouchon M (1985) The two-dimensional resonance of sediment-filled valleys. BSSA 75:519–541
2. Hasal ME, Iyisan R (2014) A numerical study on comparison of 1D and 2D seismic responses of a basin in Turkey. Am J Civil Eng 2:123–133
3. Kumar N, Narayan JP (2018) Quantification of site-city-interaction effects on the response of structure under double resonance condition. Geophys J Int 212(1):422–441
4. Lenti L, Martino S, Paciello A, Scarascia Mugnozza G (2009) Evidence of two-dimensional amplification effects in an alluvial valley (Valnerina, Italy) from velocimetric records and numerical models. BSSA 99:1612–1635
5. Makra K, Chavez-Garcia FJ, Raptakis D, Pitilakis K (2005) Parametric analysis of the seismic response of a 2D sedimentary valley: Implications for code implementations of complex site effects. Soil Dyn Earthq Eng 19:1–22
6. Paolucci R (1999) Shear resonance frequencies of alluvial valleys by Rayleigh's method. Earthq Spectra 15:503–521
7. Raptakis D, Chávez-García FJ, Makra K, Pitilakis K (2000) Site effect at Euroseistest-I. Determination of the valley structure and confrontation of observations with 1D analysis. Soil Dyn Earthq Eng 19:1–22
8. Riga ED (2015) New elastic spectra, site amplification factors and aggravation factors for complex subsurface geology, towards the improvement of EC8. Dissertation, PhD Thesis, Aristotle University of Thessaloniki, Greece

9. Chávez-García FJ, Faccioli E (2000) Complex site effects and building codes: making the leap. J Seismol 4:23–40
10. Meza-Fajardo KC, Semblat JF, Chaillat S, Lenti L (2016) Seismic-wave amplification in 3D Alluvial Basins: 3D/1D amplification ratios from fast multipole BEM simulations. BSSA 106:1267–1281
11. Sánchez-Sesma FJ, Chávez-García FJ, Bravo MA (1988) Seismic response of a class of alluvial valley for incident SH waves. BSSA 78:83–95
12. Semblat JF, Dangla P, Kham M (2002) Seismic site effects for shallow and deep alluvial basins: in-depth motion and focusing effect. Soil Dyn Earthq Eng 22:849–854
13. Zhu C, Thambiratnam D (2016) Interaction of geometry and mechanical property of trapezoidal sedimentary basins with incident SH-waves. Bull Earthq Eng. https://doi.org/10.1007/s10518-016-9938-z
14. Zhu C, Thambiratnam D, Gallage C (2017) Inherent characteristics of 2D alluvial formations subjected to in-plane motion. J Earthq Eng. https://doi.org/10.1080/13632469.2017.1387199
15. Chávez-García FJ, Rodriguez M, Stephenson WR (1998) 1D vs. 2D site effects. The case of Parkway basin, New Zealand. In: 11th European conference on earthquake engineering, Balkema, Rotterdam
16. Bard PY, Bouchon M (1980) The seismic response of sediment-filled valleys. Part 1. The case of incident SH waves, BSSA, 70: 1263–1286
17. Kawase H, Sato T (1992) Simulation analysis of strong motions in the Ashigara Valley considering one-and two-dimensional geological structures. J Phys Earth 40:27–56
18. Kawase H (1996) The cause of the damage belt in Kobe: "The basin-edge effect", constructive interference of the direct S-wave with the basin induced diffracted/Rayleigh waves. Seisomol Res Lett 67:25–34
19. Narayan JP (2005) Study of basin-edge effects on the ground motion characteristics using 2.5-D Modeling. Pure Appl Geophys 162:273–289
20. Raptakis D, Manakou M, Chávez-García FJ, Makra K, Pitilakis K (2005) 3D configuration of Mygdonian basin and preliminary estimate of its site response. Soil Dyn Earthq Eng 25:871–887
21. Kawase H, Aki K (1989) A study on the response of a soft basin for incident S, P, and Rayleigh waves with special reference to the long duration observed in Mexico City. BSSA 79:361–1382
22. Narayan JP, Sharma ML, Kumar A (2002) A seismological report on the 26 January 2001 Bhuj, India earthquake. Seismol Res Lett 73:343–355
23. Romo MP, Seed HB (1986) Analytical modelling of dynamic soil response in Mexico earthquake of September 19, 1985. In: Proceedings, ASCE international conference on Mexico Earthquakes-1985, Mexico City, pp 148–162
24. Ermert L, Poggi V, Burjanek J, Fäh D (2014) Fundamental and higher two-dimensional resonance modes of an Alpine valley. Geophys J Int 198:795–811
25. Le Roux O, Cornou C, Jongmans D, Schwartz S (2012) 1-D and 2-D resonances in an Alpine valley identified from ambient noise measurements and 3-D modelling. Geophys J Int 191:579–590
26. Khanbabazadeh H, Iyisan R (2014) A numerical study on the 2D behaviour of the single and layered clay basins. Bull Earthq Eng 12:1515–1537
27. Narayan JP, Kumar V (2013) A fourth-order accurate finite-difference program for the simulation of SH-wave propagation in heterogeneous viscoelastic medium. Geofizika 30:173–189
28. Emmerich H, Korn M (1987) Incorporation of attenuation into time-domain computations of seismic wave fields. Geophysics 52:1252–1264
29. Kristeck J, Moczo P (2003) Seismic wave propagation in viscoelastic media with material discontinuities—a 3 D 4th order staggered grid finite difference modeling. BSSA 93:2273–2280
30. Narayan JP, Sahar D (2014) 3D viscoelastic finite-difference code and modelling of basement focusing effects on ground motion characteristics. Comput Geosci 18:1023–1047

31. Narayan JP, Kumar S (2008) A fourth order accurate SH-wave staggered grid finite-difference program with variable grid size and VGR-stress imaging technique. Pure Appl Geophys 165:271–294
32. Israeli M, Orszag SA (1981) Approximation of radiation boundary conditions. J Comp Phys 41:115–135
33. IBM SPSS Statistics (2015) www.ibm.com/SPSS/STATISTICS

Chapter 3
Earthquake Induced Landslide Hazard Evaluation for Seismic Microzonation: A Case Study of the Garhwal Himalayas

Ritu Raj Nath, Mukat Lal Sharma, Naveen Pareek, Shilpa Pal, Shweta Bajaj, and Neha Kumari

Abstract Seismic microzonation of the hilly areas requires evaluation and mapping of coseismic landslide hazard. However, most of the landslide susceptibility studies in India lack in incorporating seismic factors. In this study, seismically induced landslide hazard zonation is carried out for a part of the Garhwal Himalayas considering probabilistically generated peak ground acceleration for 10% exceedance probability in 50 years as a landslide triggering factor. For the test bed, 464 landslides are identified, and eight landslide preparatory causative factors are recognized. The results of the study show that around 40% of the study area falls under the zones of high and very high seismically induced landslide hazards. The study adopts a method of mapping seismically induced landslide hazard which is simple in its both conception and application. Thus, the present method can be easily adopted for seismic microzonation of upcoming urban centres in the Himalayan terrain.

Keywords Earthquake · Microzonation · Landslides · Dynamic landslide potential index · Analytical hierarchy process · Garhwal Himalayas

R. R. Nath (✉)
Department of Civil and Infrastructure Engineering, Adani Institute of Infrastructure Engineering, Ahmedabad, Gujarat 382421, India
e-mail: riturajnath14@gmail.com

M. L. Sharma · S. Bajaj · N. Kumari
Department of Earthquake Engineering, Indian Institute of Technology Roorkee, Roorkee, Uttarakhand 247667, India

N. Pareek
Government of India, New Delhi 110042, India

S. Pal
Department of Civil Engineering, Delhi Technological University, New Delhi 110042, India

© Indian Society of Earthquake Technology 2023
R. S. Jakka et al. (eds.), *Earthquake Engineering and Disaster Mitigation*,
Springer Tracts in Civil Engineering, https://doi.org/10.1007/978-981-99-0081-7_3

3.1 Introduction

The population growth and the expansion of settlements and life-lines over hazardous areas are increasing the impact of natural disasters both in the developed and developing world [1, 2]. In many countries, the economic losses and casualties due to natural disasters are of colossal proportion, and generate a yearly loss of property on an unprecedented scale [3, 4]. Casualties and economic losses are increasing because of the greater number of people potentially involved, and the higher value of endangered structures constructed in hazard prone areas. Third world countries have always had difficulties affording the high costs involved in controlling natural hazards through major engineering works and rational land-use planning. Owing to the economic recession, many industrialized societies are now reluctant to invest in structural measures to reduce natural risks. Economic and social considerations suggest that, even if the recurrence of natural disasters remains constant, and it may not be the case, damage caused by catastrophic events is too costly even for industrialized societies [5]. In other words, natural catastrophes occur with higher frequency than our ability to recover from previous events [4]. Such realizations have driven the national disaster management policies of India, where now emphasis has been placed on providing institutional efforts to establish the frameworks for a disaster-resilient society within the wider sphere of sustainable developments. Post the 2001 Bhuj earthquake, the Government of India had identified the requirement to adopt a holistic and integrated approach to disaster management, and thus initiated a paradigm shift in national disaster management policy from the erstwhile relief-centric response to a proactive prevention, mitigation, and preparedness-driven strategy to preserve developmental benefits while minimising losses of lives, livelihoods, and property [6]. Under the new policy, the National Disaster Management Act was introduced in 2005, which defines a disaster as, "*...catastrophe, mishap, calamity or grave occurrence in any area, arising from nature or man-made causes, or by accident or negligence which result in substantial loss of life, of human suffering or damage to, and destruction of property, or damage to, or degradation of environment, and is of such nature or magnitude as to be beyond the coping capacity of the community of affected areas.*" India is vulnerable to a wide range of natural, as well as man-made disasters to varied degrees. As per the National Disaster Management Authority, India [7], 58.6% of the landmass is prone to earthquakes of moderate to very high intensity; 12% of land is prone to floods and river erosion; close to 5700 km long coastline is prone to cyclones and tsunamis; 68% of the cultivable area is vulnerable to drought and 15% areas are at risk from landslides and avalanches. Similarly, vulnerability to disasters/emergencies of Chemical, Biological, Radiological and Nuclear (CBRN) origin also exists. With the ever-increasing population of India, the exposure to both the natural and anthropogenic hazards have increased manifold, and so does the convoluted risk, and propensity to disasters. In this context, it would be pedagogically compelling to examine the current scenario of seismic risk through the prism of microzonation studies carried out for the hilly areas, with emphasis being put on the scope for coseismic landslide hazard microzonation.

One of the major secondary hazards associated with earthquakes in hilly regions is landslides, which accounts for almost 70% of damages which are not caused directly by ground shaking [8, 9]. Coseismic landslides often impact the long-term socio-economic aspects of developing countries [9–11]. Intensification of various developmental activities in earthquake prone hilly areas like the Himalayas (rapid urbanization, installation of capital-intensive hydel power projects, road widening projects etc.) has significantly elevated the seismic risk in these regions. More than 23,000 lives were lost as a result of six large earthquakes in India between 1990 and 2006, which also caused extensive damage to property and public infrastructure [7]. The occurrence of several devastating earthquakes in initially considered low-earthquake-risk places reveals that the country's built-environment is remarkably fragile, with inadequate infrastructure for disaster response, recovery and preparedness. Planners, engineers, and technical administrators now require more exact and micro-level seismic data than a few decades ago, when an earthquake data list for a specific area or a regional seismic zoning map was deemed acceptable. From these past experiences, it is learnt that for effective earthquake disaster mitigation, the pre-earthquake phase needs to be utilized for planning and implementing preparedness measures on the one hand, and working our preparedness activities on the other, with the first step being the assessment of the seismic risk with a holistic approach that includes consideration of secondary coseismic hazards.

An endeavour has been made in the present study to examine the scope of integrating landslide and seismic hazards in seismic microzonation studies for hilly regions. A test bed in the Garhwal Himalayan region was selected considering its active seismicity and high number of landslide incidences. The study used Probabilistic Seismic Hazard Assessment (PSHA) to define the seismic factor. The multi-hazard integration was carried out using the approach proposed by Nath et al. [9].

3.2 Seismic Microzonation Approach

The seismic zonation map of India covers a broader scale perspective, which constrains understanding of the local site behaviour during an earthquake [12–14]. The impact of a catastrophic disaster can be reduced if the zones that are more susceptible to undergo maximum ground motion are identified in pre-disaster preparatory stages [12]. Mitigation and management of earthquake disasters need a systematic and consistent approach in assessing the impact of future earthquakes on people and structures. Such an approach is guided by the principles of four phases [12]: (a) the Probabilistic Seismic Hazard Assessment (PSHA) where probabilistic distribution functions of the source geometry and size are convolved with the rate of seismicity to predict future strong ground motion [15–19]; (b) the estimation of earthquake damage to artificial and natural structures; (c) the translation of seismic hazards into seismic risks by using the selected damage or loss functions; and (d) the formal or informal analysis of earthquake mitigation, where the alternatives, uncertainties, costs, decision criteria, and policymakers' risk aversion are included into the decision logic area [12]. The objectives of these approaches are: (i) evaluation of the pertinent hazards (both primary and secondary); (ii) standardization of a globally recognized policy framework that allows a consistent action plan for implementing urbanization regulation and norms for design and construction practices; and (iii) assessment of seismic vulnerability and risk to facilitate implementation of various preventative measures to reduce the impact. In general, any seismic microzonation procedure comprises of two distinct components [12]: (I) a technical report, which incorporates the various aspects of the perceived hazard, vulnerability and risk; and (II) a regulatory part that enforces the third parties to follow the technical report. A successful implementation of the procedure requires institutional frameworks and specific criteria for both the parts. Finally, the ultimate goal of both the seismic hazard and risk analyses are to provide frameworks for formulating rational decisions on seismic safety, and thereby should invariably incorporate the inherent uncertainties associated with earthquakes' source, path and site characteristics. Modelling of uncertainties would also involve estimation of the same in context of the effects of earthquakes on people and structures, costs of seismic safety and potential losses, and how successfully the risk can be averted. In this scope, evaluation of seismic hazard, and the consequent microzonation of cities would allow for the characterization of the conceivable seismic vulnerability/risks that must be considered when constructing new structures or retrofitting the existing ones. This can be viewed as an effective course for preparing disaster-resilient Indian urban centres.

3.2.1 Definition of Microzonation Problem

Seismic microzonation can be defined as "*the subdivision of a region that has relatively similar exposure to various earthquake related activities or the identification of individual areas having different potential for earthquake effects*" [12]. The urban centres or developing urban centres with high degree of exposure and which falls under high seismic hazard zones require immediate attention in context of microzonation. This may be attributed to the damage pattern due to an earthquake which is largely controlled by the local site conditions (for example: Ahmedabad city during the 2001 Bhuj earthquake, which despite being at a distance of more than 300 kms from the earthquake epicentre, witnessed a very high volume of structural damage). Even with small changes in the local geology, the site responses are observed to be contrasting within short proximities. The local and secondary effects of an earthquake are not incorporated in the existing regional level hazard zonation maps, which renders them ineffective in land-use development and planning, hazard mitigation and management, and structural engineering applications at site-specific terms. It is important to resolve these constraints, particularly in densely populated metropolitan areas where unplanned urbanisation practices have been prevalent. Seismic microzonation is therefore, envisaged to divide an area into homogenous subregions where different preclusive measures must be implemented to reduce and/or avoid damages, loss of life, and social disruptions, in the event of a destructive earthquake. Twenty-seven cities in India have a population of one million or more, accounting for 25.6% of the total urban population of the country. The three metropolises of India: Mumbai, Kolkata, and New Delhi are three of the world's top 20 urban agglomerations. More ominously, 57.1% of the Indian landmass falls under the seismic zones III, IV, and V, within which 63% of the total housing in India are located. The corresponding figure in terms of population is 66%, which is exposed to high to very high degrees of seismic hazard. Thus, it is imperative to suggest that institutional efforts should be put in place to carry out seismic microzonation of the Indian urban centres.

3.2.2 Global Trends of Microzonation Works

The current practice of seismic microzonation works across the globe generally includes assessment of the site-predominant frequency, effective shear wave velocity (V_s^{30}) and deterministic seismic scenarios, while site-specific probabilistic hazard analysis is slowly emerging for engineering decisions. The major aspects of various microzonation works are largely controlled by the availability of data, geological setup of the terrain investigated, and the choice of a particular technique; based on which they can summarily be reviewed as:

a. *Single attribute characterization based on ambient noise/microtremor derived predominant frequency.*
b. *Geotechnical modelling of response spectra and site response based on borehole data.*
c. *Shear wave velocity measurement to achieve* V_s^{30} *zonation.*
d. *Estimation of site response and predominant frequency from earthquake recordings.*
e. *Strong motion simulations for seismic scenario and deterministic hazard assessments*
f. *Probabilistic hazard analysis based on local specific relationships.*

A few global approaches are enumerated in Table 3.1.

3.2.2.1 Microzonation Works Carried Out in India

Seismic microzonation for a few important Indian cities vulnerable to earthquake hazards has been carried out by different workers. The cities include Delhi [20, 21], Sikkim [22, 23], Jabalpur [24], Guwahati [25], Bangalore [13, 26], Dehradun [27], and Mumbai [28]. Table 3.2 summarizes the work carried out for the aforesaid works.

Table 3.1 Seismic microzonation work: global scenario approaches [12]

Approach	Example
Estimation of predominant frequency from ambient noise survey	Bam city Napoli Greater Bangkok
Site response estimation from earthquake data	Alaska
1-D geotechnical modelling of predominant frequencies, and peak amplifications	Bucharest city Yenisehir-Bursa
Probabilistic seismic microzonation maps for various return periods	Tashkent city
Predominant frequency and site amplifications from earthquake records and ambient noise survey	City of Thessaloniki
Site response and response spectra from synthetic seismograms	Santiago de Cuba Beijing Tehran
V_s^{30} based site classifications by means of shear wave velocity measurements	Las Vegas Valley Linares

Table 3.2 Seismic microzonation work summary for Indian cities [12]

Study region	Approaches
Guwahati	Estimation of site response and predominant frequency from strong motion data and geotechnical borehole modelling [25] Seismic scenario for deterministic hazard [25] Landslide hazard and soil liquefaction zonation, site classification, and thematic integration on GIS [25]
Sikkim	Estimation of site response and predominant frequency from strong motion array [22] Deterministic seismic scenario and thematic integration based on analytical hierarchical process on GIS [23]
Delhi	Thematic integration on GIS [20] Synthetic seismogram [21]
Bangalore	Geotechnical data analysis and deterministic seismic scenario, soil liquefaction potential mapping, probabilistic assessment, GIS integration based on AHP [13, 26]
Dehradun	Seismic response from geotechnical modelling on the shear wave velocity derived from MASW and geotechnical data [27]
Jabalpur	Deterministic seismic hazard analysis based on scenario earthquake; site response studies and evaluation of liquefaction potential of the area [24]

3.2.3 Objectives and Outputs of a Typical Seismic Microzonation Project

A typical seismic microzonation study project attempts to achieve and integrate the following components [12], in evaluating the pertinent seismic hazard and configure the levels of seismic microzonation for an urban centre:

i. *Establish the geological and geomorphologic units of the region and its surrounding areas, including the lithological characteristics and Seismic and Engineering Bedrock configurations at different locations.*
ii. *Identify and characterize major/minor faults, lineaments, and seismotectonic units that are seismically active.*
iii. *Characterize seismic activities based on historical seismicity and recorded ground motion data. The attributes pertinent include location of potential sources, magnitude, intensity, focal mechanism and epicentral distances, etc. Selection of suitable attenuation laws.*
iv. *Evaluate spatial variation of shear wave velocity (i.e., average* V_S^{30} *profiles) through geophysical surveys, geotechnical borehole logging, and HVSR modelling to develop a database of shallow subsurface stratigraphy information.*
v. *Evaluate the spatial distribution of predominant frequency of the soil through Nakamura's technique.*
vi. *Establish seismic (ground) response through theoretical and numerical modeling of wave propagation to identify amplification effects associated with near-surface ground motions from alluvial deposits during an earthquake, during which critical facilities and infrastructure are expected to remain operational.*
vii. *Determine the ground motion parameters such as peak ground acceleration/velocity distribution at different locations. Determine the ground response spectrum, duration and the time history of earthquake inputs.*
viii. *Assess seismic stability and estimation of permanent ground deformation within typical structures.*
ix. *Site classification on the basis of the shear wave velocity model, and the geotechnical assessments.*
x. *Assess the liquefaction potential from the detailed borehole geotechnical data and shear wave velocity profiles using suitable model studies. Identification of threshold value, maximum epicentral distance and depth based on field and laboratory studies.*
xi. *Local specific deterministic and probabilistic seismic hazard analyses.*
xii. *Generate thematic hazard maps in terms of site classification, site response, predominant frequency, spectral accelerations/ response spectra, and peak ground accelerations on GIS platform.*
xiii. *Prepare inventory for the demographic, building typology, and land-use patterns.*
xiv. *Identify building hazards from resonance phenomenon on the basis of predominant frequency of the basin to demarcate zones of varying risk.*
xv. *Risk assessment of the urban centre through scenario generations to identify damage levels and achieve a probabilistic account of the same.*

In terms of output, the Ministry of Earth Sciences, India [12] has specified the following maps of the study region:

- *Site characterization/classification maps*
- *Spatial variation of Peak Horizontal Acceleration (PHA) and Spectral Acceleration (SA) values at bed rock level*
- *Spatial variation of PGA and SA values at surface level*
- *Liquefaction Hazard maps*
- *Landslide Hazard maps (if and where required)*
- *Comprehensive seismic hazard map prepared by combining all the above given maps (after giving proper weightage and ranks for each factor and feature in each theme).*

The present study puts forth a case study of mapping the coseismic landslide hazard for a part of the Garhwal Himalayas considering a scenario earthquake of 10% exceedance probability in 50 years.

3.3 Methodology

For an accurate assessment of the earthquake induced landslide hazard in a study area, a comprehensive earthquake induced landslide inventory is required. However, it is possible only if there is a database of coseismic landslides in the study area, and often it is not the case. Many researchers indicated a paucity of earthquake induced landslide inventories worldwide, and for the Himalayan region, only limited works have been carried out till now in this aspect. Most of the coseismic landslide hazard studies for the Himalayan region are restricted to Chamoli Earthquake (1999), Sikkim earthquake (2011) and Nepal Earthquake (2015) only. In an attempt to address this issue, this study considered all the landslides observed in a particular area irrespective of their triggering mechanism under an earthquake scenario of 10% exceedance probability in 50 years. As the triggering earthquake had been considered for a probable earthquake scenario, it allowed to consider all the observed landslides to evaluate the coseismic landslide potential.

The study considered probabilistically generated Peak Ground Acceleration (PGA) as landslide triggering seismic factor, which was integrated with other landslide preparatory static factors. The seismic parameter (PGA) was estimated for a scenario earthquake of 10% exceedance probability in 50 years (corresponding to a return period of 475 years) using PSHA. The other landslide preparatory factors considered in the study were: slope morphometry, elevation profile, distance from existing faults, drains, and roads, and land-use-land-cover pattern. A comprehensive database of 464 landslides was prepared using remote sensing images, Google Earth imagery, and extensive field surveys for proximity analyses.

Integration of seismic factor with landslide preparatory factors produced a numerical value known as the dynamic Landslide Potential Index (LPI_D) [9] for each pixel of the study area. The concept of the dynamic potential index was developed to indicate the likelihood of seismically induced landslide occurrence based on the prevailing causative factors in an area [9]. A higher value of LPI_D indicates a greater susceptibility to seismically induced landslide hazard and vice-versa. The LPI_D was computed as follows:

$$LPI_D = \sum_{i=1}^{n}\sum_{j=1}^{m} R_i \times W_{i,j} + \sum_{k=1}^{l} R_E \times W_{E,k} \tag{3.1}$$

where, R_i and $W_{i,j}$ denote the rank of the ith thematic layer of landslide causative factor and the weightage of the jth class of the ith layer respectively. Similarly, R_E and $W_{E,k}$ denote the rank of the PGA thematic layer and the weightage of the kth class of the PGA layer respectively. The dynamic LPI_D in Eq. (3.1) has two parts: the static part, and the dynamic part. Introducing a dummy variable, a combined equation was derived as

$$LPI_D = \sum_{i=1}^{n}\sum_{j=1}^{m} R_i \times W_{i,j} + \left(\sum_{k=1}^{l} R_E \times W_{E,K}\right)\delta(e) \tag{3.2}$$

where, $\delta(e) = \begin{cases} 1, \text{ if seismicity is considered} \\ 0, \text{ otherwise} \end{cases}$

If seismicity is not considered, LPI_D will become static and could be computed for landslide preparatory factors only. In that case, the calculated *LPI* values would indicate the likelihood of landslide occurrence excluding the seismic factor. Readers may further refer to Nath et al. [9] for an elaborate discussion on *LPI*.

The ranks and weights of various landslide thematic layers (both static and dynamic) and their subclasses could be decided based on the previous knowledge of the terrain, scope, and scale of the study and professional judgment. Thus, the process involved a certain degree of subjectivity [9, 29, 30]. Many researchers have used semi-quantitative methods like Analytical Hierarchy Process (AHP) to reduce the subjectivity of the weight assignment process [9, 31–34]. In AHP decisions are taken using matrix-based pair-wise relative comparisons of the contribution of different causative factors [35]. The consistency of the decision matrix is examined using Consistency Ratio (*CR*), which is defined as the ratio of Consistency Index (*CI*) and Random Consistency Index (*RI*). Mathematically, *CI* is defined as

$$CI = \frac{\lambda_{\max} - N}{N - 1} \tag{3.3}$$

$$CR = CI/RI \tag{3.4}$$

where, $\lambda_{\max}$ is the largest Normalized Principal Eigen Vector and N is the order of the decision matrix. The value of *RI* is predefined for different values of N. If $CR \leq$

10%, the decisions made (i.e., the ranks assigned) are consistent. If not, the process should be repeated.

3.3.1 Generation of Seismic Factor (PGA) Using PSHA

Incorporation of probabilistic PGA values as landslide triggering factors becomes common [9, 34, 36–38]. PSHA is a platform, where the probabilistic distribution functions of the geometry of the source and size distributions are convolved with the rate of seismicity to predict future strong ground motion. Therefore, the predicted PGA values are not generated from single events but are representatives of the expected deformations and stress releases in the future in a finite time-period; and hence can be utilized as one of the landslide causative factors along with conventionally used other static landslide causative factors. PSHA combines probability distribution functions for Strong Ground Motion (SGM) parameters due to the total expected seismicity at a particular site [9, 15, 16, 19]. If $\upsilon(M_j,R_i)$ defines the annual rate of occurrence of earthquake(s) for a set of magnitude (M_j) and distance (R_i), and λ $(Z > z)$ defines the rate of occurrence of SGM parameter Z exceeding a value z (a value fixed for Z based on engineering judgment), then λ $(Z > z)$ can be mathematically expressed as the linear combination of υ (M_j, R_i) for n number seismogenic sources considered for that particular site.

$$\lambda(Z > z) = \sum_{n=1}^{n}\sum_{i=1}^{i}\sum_{j=1}^{j} q(Z > z|M_j, R_i) \times \upsilon_n(M_j, R_i) \tag{3.5}$$

where, q defines the probability of SGM parameter Z exceeding the value z for an earthquake of magnitude M_j at a distance R_i. So, the probability of SGM parameter Z exceeding the value z, due to all the events in all the source zones during an exposure period of T years can be defined by Poissonian distribution, and is expressed as:

$$P(Z > z|T) = 1 - \exp(-\lambda(Z > z) \times T) \tag{3.6}$$

where T is the finite time-period.

The catalogue for the area was compiled from published literature and various databases like National Centre for Seismology, National Earthquake Information Centre, National Geophysical Data Centre, Indian Meteorological Department, United States Geological Survey, and Japan Meteorological Agency. The treatment required in terms of homogenization, declustering, and completeness was carried out as per the requirements of PSHA methodology [39–42]. Seven seismogenic source zones were identified to estimate the probabilistic seismic hazard using the doubly truncated Gutenberg Richter relationship. The attenuation relationship of NGA-WEST 1 [43] was used. The results were presented in the form of the PGA map for an exceedance probability of 10% in 50 years (corresponding to 475 years' earthquake return period). The result of the PSHA in terms of the PGA map is shown in Fig. 3.2i, which is categorized into 5 classes for further analysis.

3.4 Seismically Induced Landslide Hazard Zonation: Results and Discussion

The study considered a part of the Garhwal Himalayas as a test bed (Fig. 3.1) which falls in the central seismic gap of the 1905 Kangra earthquake and the 1934 Bihar Nepal earthquake. The correlation between the seismicity and landslide hazard in the Garhwal Himalayas was indicated in the literature [9, 29, 37, 44–47]. Pachauri and Pant proposed a Landslide Susceptibility Zonation (LSZ) map for the Aglar river catchment area which indicated a positive relationship between the Aglar fault and landslide activities of the area [44]. Champati ray et al. used a fuzzy set-based approach for LSZ mapping in parts of the Garhwal Himalayas and concluded that the long-term slope stability is largely controlled by the seismic activities in this area [45]. Pareek et al. assessed the impact of seismic factors in LSZ mapping for the Chamoli region quantitatively and concluded that moderate to major earthquakes rendered a pattern shift in the demarcated zones towards higher landslides susceptibility [37]. Bhattacharya et al. concluded that the present day Mansadevi landslide was a cumulative effect of slow surface movements of the landslide relative to the neighboring area from 1992 to 1998 [46]. Nath et al. remarked that the tectonic environment as a landslide causative factor should be included in all the landslide hazard zonation studies carried out for the Himalayan region due to its active seismicity [29]. Nayak and Geda investigated the seismic landslide hazard for the central seismic gap region of the Himalayas and observed that for a scenario earthquake of magnitude (M_w) 8.5 there exists high probability of coseismic slope failure in this region [47]. Nath et al. quantified the effect of various scenario earthquakes on the landslide hazard zonation for the lower Himalayan region and proposed a novel approach to integrate seismic parameter in landslide hazard zonation [9]. A critical review of the available literature reveals that there is a requirement of carrying out seismically induced landslide hazard studies for the lower Himalayan region. Especially, with more developmental activities now planned for these mountain chains, it is necessary to evaluate the coseismic landslide hazard for the lower Indian Himalayas in the context of seismic microzonation. This acts as a major motivation in selecting the test bed as well.

The test bed encompasses approximately 5000 km^2 area in the lower Himalayan region. It was considered for the study due to its high seismicity and high number of landslide events. Both the Main Boundary Thrust (MBT) and the Main Frontal Thrust (MFT), along with numerous transverse lineaments traverse the study area [18, 19, 48]. Dehradun, Mussoorie, Haridwar, Rishikesh, Devprayg, and Chakrata are some of the major cities of the study area, with an approximate population of more than 5 lakhs [49]. NH7 and NH72(A) are the two major national highways passing through

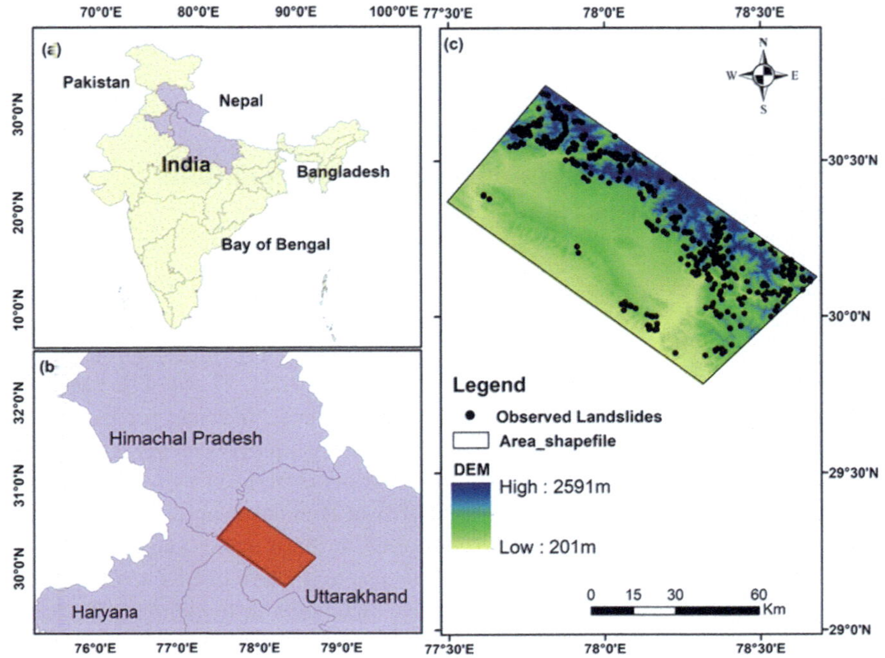

Fig. 3 1 The study area map showing the map of **a** India and its boundaries, **b** location of the study area within the Indian state of Uttarakhand, and **c** the elevation profile, observed landslides and the study area box

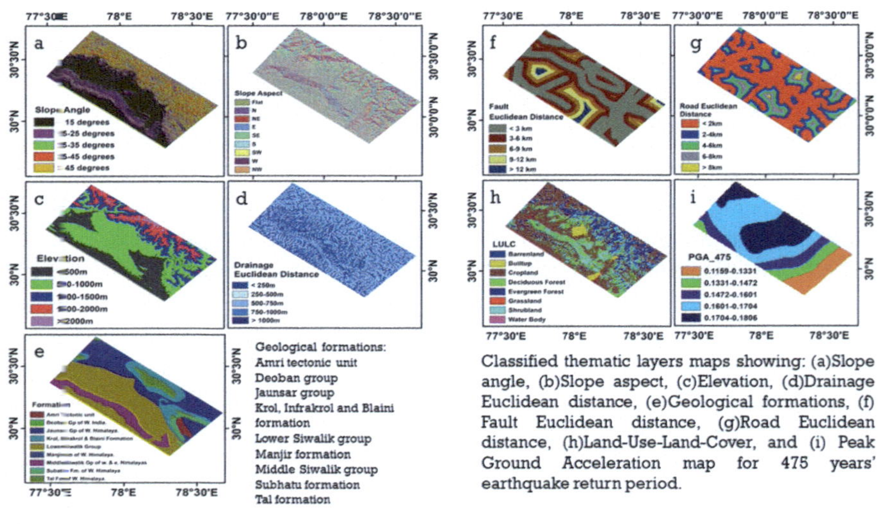

Fig. 3.2 The thematic layers of various landslide causative factors recognized for the study area

the study area. The study area exhibits an assorted matrix of geological formations (Fig. 3.2e) viz. Lower Siwalik group, Middle Siwalik group, Manjir formation of Western Himalayas, Subhatu formation of Western Himalayas, Krol, Infrakrol and Blaini formation, Deoban group of Western Himalayas, Jaunsar group of Western Himalayas, Amri tectonic unit, and Tal formation of Western Himalayas. A total of 464 landslides were identified, delineated and mapped in the study area (at a scale of 1:12,000) using geo-referenced LISS IV (Linear Imaging Self-Scanning System-IV) image from Resoucesat 2 (2011), Google Earth platform, and extensive field surveys. The methodology of Nath et al. [9] was adopted in this study to identify the existing landslides in the area. The study recognized eight (8) landslide preparatory factors viz. slope angle, slope aspect, elevation profile, geology, distance from drains, distance from faults, distance from roads, and land-use-land-cover pattern in this area. All the factors were mapped in a Geographical Information System (GIS) environment to generate the database for multi-hazard integration.

3.4.1 Preparation of the Database for Multi-hazard Integration

3.4.1.1 Digital Elevation Model (DEM) and its Derivatives

Data for elevation, slope angle, slope aspect, and drainage Euclidean distance, derived for the study area were extracted from the Advanced Spaceborne Thermal Emission and Reflection Radiometer (ASTER) Global Digital Elevation Model (GDEM) v2.0 with 30 m resolution (Fig. 3.1c). It is categorized into 5 classes at an interval of 500 m for the ease of classification (Fig. 3.2c). The slope angle map (Fig. 3.2a) of the study area was categorized into five classes 0° to 72°. The aspect map of the study area was categorized into 9 classes (Fig. 3.2b). The drainage network for the study area was extracted from the DEM. The obtained drainage network was further verified with digitized streamlines from Survey of India Toposheets at 1:50,000 scale. The drainage network map was used to generate the drainage Euclidean distance (Fig. 3.2d), which was categorized into 5 subclasses at an interval of 250 m.

3.4.1.2 Fault Euclidean Distance

In the present study, fault Euclidean distance was used as a thematic layer. The fault Euclidean distance map of the study area was categorized into 5 subclasses at an interval of 3 km (Fig. 3.2f).

3.4.1.3 Road Euclidean Distance

There exists a sporadic distribution of road networks throughout the study area. While the urban centres of the area have a high concentration of road networks, the overall density of roads is low to moderate due to large forest areas. In the present study, road Euclidean distance was used as a thematic layer. The road Euclidean distance map of the study area was categorized into 5 classes (Fig. 3.2h).

3.4.1.4 Land-Use-Land-Cover (LULC) Pattern

The LULC map of the study area was extracted from Copernicus Global Landcover database (Fig. 3.2h). It was observed that the study area was mostly (~57%) covered with sparsely vegetated landscape (cropland and grassland), followed by Evergreen Forest (~15%), Deciduous Forest (~12%) and Shrub Land (~12%). Barren lands constituted only 1.2% of the total study area, yet around 3% of the total observed landslides were mapped within this LULC pattern. Similarly, 1% of total observed landslides were identified in the vicinity of urban build up areas that constituted ~1% of the total area. No landslide was observed at waterbodies which were ~0.5% of the total study area.

3.4.2 Multi-hazard Integration

The rank (R_i) of each thematic layer was decided based on extensive field surveys and experience of the terrain to ascertain the perceived control of the causative factor on the overall landslide hazard in the study area. It was observed that owing to the fragile Himalayan geology, chances of slope failure were higher in steep slopes, especially slopes with inclination greater than 45°. Therefore, it was perceived that slope angle and geological units had the highest control on landslide occurrence in the study area. Similarly, the earthquake parameter (PGA) was considered a major landslide triggering factor due to the hyper-active seismicity of the study area. Further, due to its active tectonism, the transitory earth-stresses are very high in the study area, and therefore, distance from faults was considered a major landslide preparatory factor [50]. Effects of drainage, elevation and slope aspect were perceived to be moderate to high in the study area. Although, road widening projects in the Himalayan region have induced additional instability [51], due to a sporadic distribution of road networks in the study area, its effect was observed to be low. Similarly, the effect of LULC patterns was deemed to be low to very low as forest and grassland covered most of the study area, with only few barren lands and built-up areas. Based on these observations, the ranks of thematic classes were decided, and the AHP decision matrix was formed to check the consistency of the decisions made (Table 3.3). For the decision matrix formed, 36 pair-wise comparisons were made, and the consistency ratio (CR) was calculated as 0.0256 (2.6%). As CR < 10%, the decisions made (i.e., ranks assigned) were consistent.

The weights of the various subclasses of the thematic layers were decided based on the proximity analyses of landslide causative factors and the prepared landslide database. The maximum and minimum PGA values observed for the study area were 0.18 g and 0.13 g, respectively. Such ground acceleration would invariably induce landslides of multitude proportion, and therefore, the landslide susceptibility of the PGA thematic class was perceived to be moderate to very high. On the contrary, the effects of slope aspect were observed to be low to moderate. The perceived susceptibility of the other thematic subclasses ranged from very low to very high. Tables 3.4 and 3.5 shows the weights ($W_{i,j}$) assigned to the various thematic subclasses, their perceived landslide susceptibility, formation of the AHP decision matrix and the consistency of the weights assigned. For all the assigned weights, the decisions made were found to be consistent.

Numerical integration of the assigned Ranks and weights of the landslide thematic layers and their subclasses in GIS environment produced the dynamic Landslide Potential Index (LPID) for each pixel of the study area. The LPI_D values were segregated and categorized using the natural breaks (Jenks) classification method to produce the Landslide Hazard Zonation map (Fig. 3.3) of the study area under seismic conditions.

For the study area, the maximum and the minimum dynamic landslide potential indices (LPI_D) were 99 and 338, respectively. The prepared landslide hazard zonation map was further analyzed spatially (Success rate curve) and statistically (Frequency Ratio analysis) to measure the prediction accuracy of the model. The results are shown in Table 3.6.

Table 3.5 shows that although the very high landslide hazard zone constitutes only 4.75% of the total study area, it contains 26% of the observed landslide pixels. 81% of the total landslide pixels are observed in the high and the very high landslide hazard zones, indicating a frequency of 6.65 observed landslide pixels per square km in these two zones. On the other hand, very low and low landslide hazard zones contain only 6% of the observed landslide pixels in the study area, with a combined frequency of just 0.65 observed landslide pixels per square km. In between lies the moderate landslide hazard zone that constitutes about 32% of the study area, and contains ~12.6% of the observed landslide pixels. The higher FR values of the high and very high landslide hazard zones confirm that the distribution patterns of the observed landslides are in concurrence with the obtained results, indicating that the areas with higher landslide density were demarcated as these two zones. Similarly, the success rate curve was used to measure the prediction accuracy [9, 25], which was computed as 0.7853. This donates that the overall success rate of the Landslide Hazard Zonation map is an acceptable 78.53%.

Table 3.3 Assignment of Ranks (R_i) to various landslide causative factors based on their perceived control on landslide hazard and formation of the AHP decision matrix to check the consistency of the process

Landslide causative factors	Perceived control level on landslide hazard	Ranks (R_i)	Formation of AHP decision matrix									Normalized principal Eigen vector
Slope angle	Very high	9	1									0.274
Geological units	High to very high	8	0.50	1								0.181
Peak ground acceleration	High to very high	8	0.50	1	1							0.181
Fault Euclidean distance	High	7	0.33	0.50	0.50	1						0.117
Drain Euclidean distance	Moderate to high	6	0.25	0.33	0.33	0.50	1					0.075
Elevation	Moderate to high	6	0.25	0.33	0.33	0.50	1	1				0.075
Slope aspects	Moderate	5	0.20	0.25	0.25	0.33	0.50	0.50	1			0.050
Road Euclidean distance	Low	3	0.14	0.17	0.17	0.20	0.25	0.25	0.33	1		0.026
Land-Use-Land-Cover	Very low to low	2	0.13	0.14	0.14	0.17	0.20	0.20	0.25	0.50	1	0.019

Table 3.4 Assignment of weights ($W_{i,j}$) to thematic layers' (Slope angle, PGA, fault Euclidean distance, drainage Euclidean distance, elevation, and road Euclidean distance) subclasses, their perceived landslide susceptibility, and formation of the AHP decision matrix to check the consistency of the process

Thematic classes	Perceived susceptibility	Weight ($W_{i,j}$)	Formation of AHP decision matrix					Consistency ratio (CR)	Normalized principal Eigen vector
Slope angle									
>45°	Very high	9	1					0.0527	0.513
35°–45°	High	7	0.33	1					0.261
25°–35°	Moderate	5	0.20	0.33	1				0.129
15°–25°	Low	3	0.15	0.20	0.33	1			0.063
<15°	Very low	1	0.11	0.15	0.20	0.33	1		0.034
Peak ground acceleration									
0.1704–0.1806	Very high	9	1					0.0163	0.419
0.1601–0.1704	High to Very high	8	0.50	1					0.263
0.1472–0.1601	High	7	0.33	0.50	1				0.160
0.1331–0.1472	Moderate to High	6	0.25	0.33	0.50	1			0.097
0.1159–0.1331	Moderate	5	0.2	0.25	0.33	0.50	1		0.062
Fault Euclidean distance									
<3 km	Very high	9	1					0.0527	0.513
3–6 km	High	7	0.33	1					0.261
6–9 km	Moderate	5	0.20	0.33	1				0.129
9–12 km	Low	3	0.15	0.20	0.33	1			0.063
>12 km	Very low	1	0.11	0.15	0.20	0.33	1		0.034

(continued)

Table 3.4 (continued)

Thematic classes	Perceived susceptibility	Weight ($W_{i,j}$)	Formation of AHP decision matrix					Consistency ratio (CR)	Normalized principal Eigen vector
Drainage Euclidean distance									
<250 m	Very high	9	1					0.0527	0.513
250–500 m	High	7	0.33	1					0.261
500–750 m	Moderate	5	0.20	0.33	1				0.129
750–1000 m	Low	3	0.15	0.20	0.33	1			0.063
>1000 m	Very low	1	0.11	0.15	0.20	0.33	1		0.034
Elevation									
500–1000 m	Very high	9	1					0.0527	0.513
1000–1500 m	High	7	0.33	1					0.261
1500–2000 m	Moderate	5	0.20	0.33	1				0.129
<500 m	Low	3	0.15	0.20	0.33	1			0.063
>2000 m	Very low	1	0.11	0.15	0.20	0.33	1		0.034
Road Euclidean distance									
<2 km	High	7	1					0.0452	0.406
2–4 km	Moderate to high	6	0.50	1					0.263
4–6 km	Moderate	5	0.33	0.33	1				0.188
6–8 km	Low	3	0.25	0.25	0.25	1			0.094
>8 km	Very low	1	0.20	0.20	0.20	0.20	1		0.048

Table 3.5 Assignment of weights ($W_{i,j}$) to thematic layers' (Geological units, slope aspect, and land-use-land-cover) subclasses, their perceived landslide susceptibility, and formation of the AHP decision matrix to check the consistency of the process

Thematic classes	Perceived susceptibility	Weight ($W_{i,j}$)	Formation of AHP decision matrix							Consistency ratio (CR)	Normalized principal Eigen vector
Geological units											
Krol, Infrakrol and Blaini formation	Very high	9	1							0.0156	0.290
Subathu formation	High to very high	8	0.50	1							0.183
Manjir formation	High to very high	8	0.50	1	1						0.183
Jaunsar group	High	7	0.33	0.50	0.50	1					0.106
Tal formation	High	7	0.33	0.50	0.50	1	1				0.106
Middle Siwalik group	Low	3	0.14	0.20	0.20	0.33	0.33	1			0.044
Lower Siwalik group	Low	3	0.14	0.20	0.20	0.33	0.33	1	1		0.044
Amri Tectonic unit	Very low	1	0.11	0.14	0.14	0.20	0.20	0.33	0.33		0.022
Deoban group	Very low	1	0.11	0.14	0.14	0.20	0.20	0.33	0.33		0.022
Slope aspects											
Flat	–	–	–							0.0001	–
West	Moderate	3	1								0.214
North West	Moderate	3	1	1							0.214

(continued)

Table 3.5 (continued)

Thematic classes	Perceived susceptibility	Weight ($W_{i,i}$)	Formation of AHP decision matrix							Consistency ratio (CR)	Normalized principal Eigen vector
East	Moderate	5	1	1	1						0.214
South East	Low	3	0.33	0.33	1	1					0.071
South	Low	3	0.33	0.33	0.33	1	1				0.071
South West	Low	3	0.33	0.33	0.33	0.33	1	1			0. 071
North	Moderate	5	0.33	0.33	0.33	0.33	0.33	1	1		0. 071
North East	Moderate	5	0.33	0.33	0.33	0.33	0.33	0.33	1		0. 071
Land-Use-Land Cover											
Barren land	Very high	9	1							0.0293	0.398
Built-up	High	7	0.33	1							0.220
Crop land	Moderate	5	0.20	0.33	1						0.104
Deciduous forest	Moderate	5	0.20	0.33	1	1					0.104
Evergreen forest	Low-moderate	4	0.17	0.25	0.50	0.50	1				0.064
Grassland	Low	3	0.14	0.20	0.33	0.33	0.50	1			0. 043
Shurb land	Low	3	0.14	0.20	0.33	0.33	0.50	1	1		0. 043
Waterbody	Very low	1	0.11	0.14	0.20	0.20	0.33	0.33	0.33		0. 023

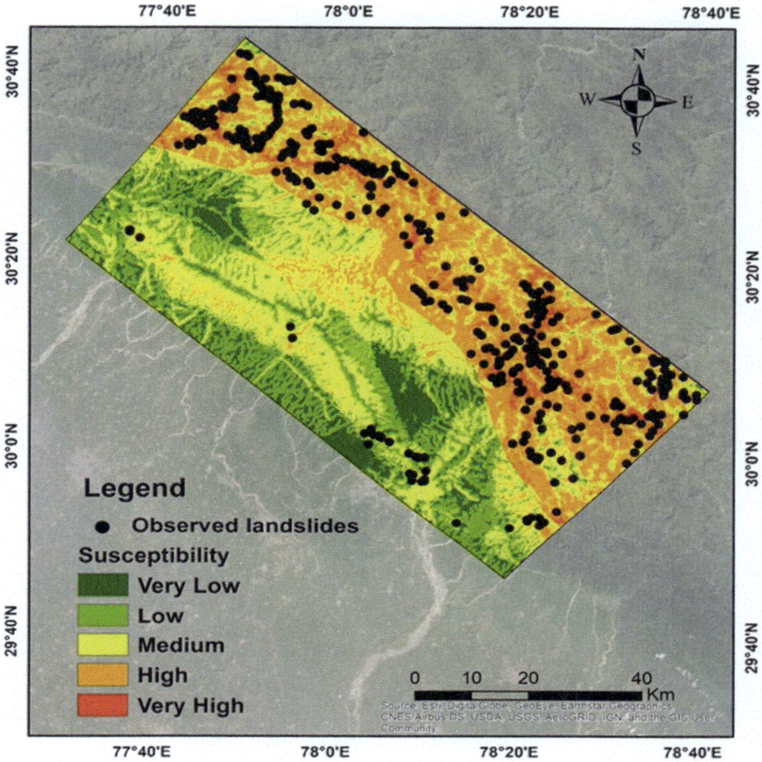

Fig. 3.3 The prepared seismic landslide hazard zonation map showing five distinct classes of landslide susceptibility and spatial distribution of the observed landslides within the classes

3.5 Conclusion

The Ministry of Earth Sciences, India has categorically specified Landslide Hazard map(s) of an area as one of the six outputs of a seismic microzonation exercise. With new urban centres now being established in the hilly regions of the Himalayas, seismic microzonation studies must evaluate the potential coseismic landslide hazard pertaining to the study area. However, this is a challenging issue due to the absence of a globally recognized method for carrying out seismically induced landslide hazard zonation. The problem becomes more complicated for regions like the lower Himalayan belt, where there is a paucity of a comprehensive seismically induced landslide database due to the non-occurrence of major earthquakes in the last 100 years. Nevertheless, considering the active seismicity of the Himalayas, and owing to the presence of major thrusting systems, one must not exclude the potential risk associated with the coseismic landslides in the Garhwal Himalayas. This can be further validated from the studies which demonstrate that zones demarcated as moderately susceptible to landslides under preparatory conditions become

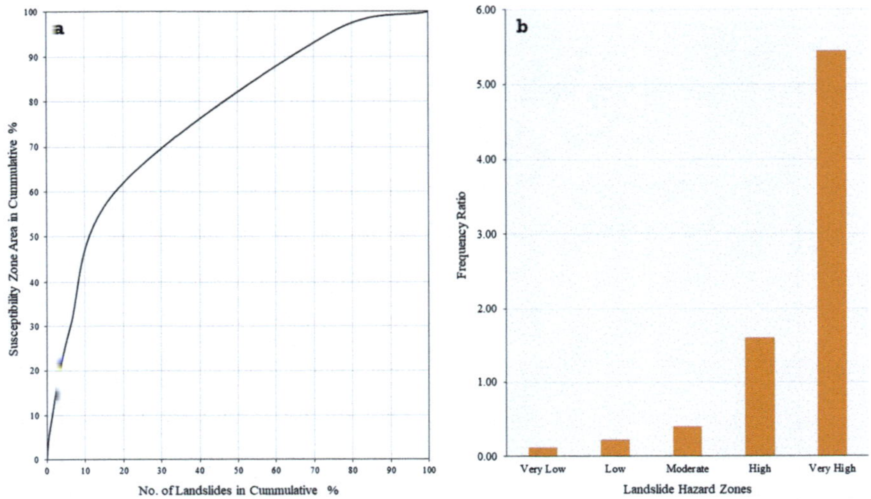

Fig. 3.4 Spatial and statistical analyses of the landslide hazard zonation map showing: **a** the success rate curve, and **b** frequency ratios of the hazard zones

Table 3.6 Analyses of the seismically induced landslide hazard zonation map

Landslide hazard zone	Zone area in km^2	Percentage of the zone area in terms of the total study area	Number of landslide pixels in the zone area	Percentage of landslide pixel in terms of the total landslide pixels	Frequency ratio (FR)
Very low	371	7.28	150	0.91	0.13
Low	1134	22.20	840	5.11	0.23
Moderate	1591	31.16	2066	12.57	0.40
High	1768	34.62	9132	55.57	1.61
very high	242	4.74	4244	25.83	5.45

highly/very highly susceptible to landslide under seismic conditions. This necessitates a comprehensive assessment of coseismic landslide hazard as an integral part of seismic microzonation studies carried out for the hilly areas. The present study examines the various challenges of evaluating the co-seismic landslide hazard for a part of the Garhwal Himalayas in the prism of seismic microzonation.

One of the major drawbacks of landslide susceptibility studies in the Himalayan region is the lack of seismic factors, mainly due to the paucity of earthquake induced landslide inventories. This necessitates consideration of all the observed landslides in the study area irrespective of their triggering mechanism, which in turn reduces the accuracy of the predictive model. With an ever-increasing database of seismically induced landslides worldwide, a more robust method can be proposed in the future. The study observes that in absence of a comprehensive earthquake induced landslide database, probabilistically generated peak ground acceleration can be considered as an effective seismic parameter to evaluate coseismic landslide hazard. In this context, the study proposes a novel method, which is simple in both conception and application, and can be successfully implemented for multi-hazard integration. However, assignments of ranks and weights to various landslide causative factors and their subclasses involve certain degree of uncertainty, which can be reduced by adopting structured decision-making viz. Analytical Hierarchy Process. Based on the results, the study recommends that seismic microzonation of hilly areas must evaluate the co-seismic landslide hazard, and the proposed method will be of practical importance for that purpose.

Acknowledgements The authors would like to commemorate the body of work carried out by Prof. D. K. Paul in the field of earthquake engineering and his immense contribution towards the Indian Society of Earthquake Technology. All authors were indeed fortunate to have met and learnt from Prof. Paul at various stages of their careers. The authors would also like to acknowledge the support extended by the Department of Earthquake Engineering, Indian Institute of Technology, Roorkee and the Ministry of Earth Sciences, India.

References

1. Rosenfeld CL (1994) The geomorphological dimensions of natural disasters. In: Proceedings of the 25th Binghamton symposium in geomorphology in SUNY, Binghamton, USA
2. Alexander ED (1995) Vulnerability to landslides in landslide risk assessment. Wiley, New York
3. Schuster RL, Fleming RW (1986) Economic losses and fatalities due to landslides. Bull Assoc Eng Geol 23(1):11–28
4. Guzzetti F, Carrara A, Cardinali M, Reichenbach P (1999) Landslide hazard evaluation: a review of current techniques and their application in a multi-scale study, Central Italy. Geomorphology 31(1–4):181–216
5. Nath RR (2019) Seismically induced landslide hazard analyses for a part of lower Indian Himalaya. PhD Thesis, Department of Earthquake Engineering, Indian Institute of Technology Roorkee
6. National Disaster Management Authority of India (2005) The Disaster Management Act, 2005 (No. 53 of 2005). The Gazette of India, Part II, Section 1, Published by Authority, New Delhi, December 26, 2005
7. National Disaster Management Authority (2009) National policy on disaster management. A Publication of the National Disaster Management Authority, Government of India, New Delhi
8. Marano KD, Wald DJ, Allen TI (2010) Global earthquake casualties due to secondary effects: a quantitative analysis for improving rapid loss analyses. Nat Hazards 52(2):319–328
9. Nath RR, Pareek N, Sharma ML (2022) Implication and inclusion of size-dependent scenario earthquakes on landslide hazard zonation: a case study of the Indian Himalayas. CATENA 212:1–18
10. Korup O (2006) Effects of large deep-seated landslides on hillslope morphology, western Southern Alps, New Zealand. J Geophys Res Earth Surf 111:1–18
11. Tang C, van Westen CJ, Tanyaş H, Jetten VG (2016) Analysing post-earthquake landslide activity using multi-temporal landslide inventories near the epicentral area of the 2008 Wenchuan earthquake. Nat Hazard 16(12):2641–2655
12. Ministry of Earth Sciences (2011) Seismic microzonation manual. A Publication of the Geoscience Division, Ministry of Earth Sciences, Government of India, New Delhi

13. Sitharam TG, Anbazhagan P (2008) Seismic microzonation: principles, practices and experiments. Electr J Geotech Eng Special Volume Bouquet 08:1–58
14. Nath SK, Thingbaijam KKS (2009) Seismic hazard assessment–a holistic microzonation approach. Nat Hazard 9(4):1445–1459
15. Bajaj S, Sharma ML (2019) Modeling earthquake recurrence in the Himalayan seismic belt using time-dependent stochastic models: implications for future seismic hazards. Pure Appl Geophys 176(12):5261–5278
16. Choudhury C, Sharma ML (2017) probabilistic models for earthquakes with large return periods in Himalaya region. Pure Appl Geophys 174:4313–4327
17. Choudhury C, Sharma ML (2018) Global strain rates in western to central Himalayas and their implications in seismic hazard assessment. Nat Hazards 94(3):1211–1224
18. Herbindoo A, Kumar S, Sharma ML (2014) Earthquake ground motion predictive equations for Garhwal Himalaya, India. Soil Dyn Earthq Eng 66:135–148
19. Sharma ML, Lindolhm C (2012) Earthquake hazard assessment for Dehradun, Uttarakhand, India, including a characteristic earthquake recurrence model for the Himalaya Frontal Fault (HFF). Pure Appl Geophys 169:1601–1617
20. Rao KS, Mohanty WK (2001) Microzonation of Delhi region: an approach. J Indian Build Congr 8:102–114
21. Parvez IA, Vaccari F, Panza GF (2004) Site-specific microzonation study in Delhi metropolitan city by 2-D modeling of SH and P-SV waves. Pure Appl Geophys 161:1165–1184
22. Nath SK (2004) Seismic hazard mapping and microzonation in the Sikkim Himalaya through GIS integration of site effects and strong ground motion attributes. Nat Hazards 31:319–342
23. Pal I, Nath SK, Shukla K, Pal DK, Raj A, Thingbaijam KKS, Bansal BK (2008) Earthquake hazard zonation of Sikkim Himalaya using a GIS platform. Nat Hazards 45:333–377
24. Mishra PS (2004) Seismic hazard and risk microzonation of Jabalpur. In: Workshop on seismic hazard and risk microzonation of Jabalpur in National Geophysical Research Institute, Hyderabad, India
25. Nath SK et al (2008) Site amplification, Qs, and source parameterization in Guwahati region from seismic and geotechnical analysis. Seismol Res Lett 79(4):526–539
26. Sitharam TG, Anbazhagan P, Ganesha Raj K (2006) Use of remote sensing and seismotectonic parameters for seismic hazard analysis of Bangalore. Nat Hazards 6:927–939
27. Mahajan AK, Slob S, Ranjan R, Sporry R, Ray PKC, van Westen CJ (2007) Seismic microzonation of Dehradun City using geophysical and geotechnical characteristics in the upper 30m of soil column. J Seismol 11:355–370
28. Raghukanth STG, Iyengar RN (2006) Seismic hazard estimation for Mumbai city. Curr Sci 91:1486–1494
29. Nath RR, Sharma ML, Tyagi A (2020) A review of the current practice on inclusion of seismicity in landslide susceptibility zonation: a case study for Garhwal Himalaya. Himalayan Geol 41(2):222–233
30. Kanungo DP et al (2009) Landslide susceptibility zonation (LSZ) mapping–a review. J South Asia Disaster Stud 2(1):81–105
31. Long NT, De Smedt F (2012) Application of an analytical hierarchical process approach for landslide susceptibility mapping in A Luoi district, Thua Thien Hue Province, Vietnam. Environ Earth Sci 66(7):1739–1752
32. Kayastha P, Dhital MR, De Smedt F (2013) Application of the analytical hierarchy process (AHP) for landslide susceptibility mapping: a case study from the Tinau watershed, west Nepal. Comput Geosci 52:398–408
33. Papadakis M, Karimalis A (2017) Producing a landslide susceptibility map through the use of analytic hierarchical process in finikas watershed, North Peloponnese, Greece. Am J Geogr Inf Syst 6(1(A)): 14–22
34. Nath RR, Pal S, Sharma ML (2022) Use of probabilistically generated scenario earthquakes in landslide hazard zonation: a semi-qualitative approach in impact of climate change, land use and land cover, and socio-economic dynamics on landslides. disaster risk reduction (Methods, approaches and practices). Springer Publication, Singapore

35. Saaty TL (1980) The analytic hierarchy process: planning, priority setting, resource allocation. McGraw-Hill Book Co, New York
36. Nadim F et al (2006) Global landslide and avalanche hotspots. Landslides 3(2):159–173
37. Pareek N, Sharma ML et al (2013) Inclusion of earthquake strong ground motion in a geographic information system-based landslide susceptibility zonation in Garhwal Himalayas. Nat Hazards 65(1):739–765
38. Massey CI et al (2020) Landslides triggered by the M_W 7.8 14 November 2016 Kaikōura earthquake: an update. Landslides 17:2401–2408
39. Stepp JC (1973) Analysis of completeness of the earthquake sample in the Puget Sound area. In: Harding ST (ed) Contributions to seismic zoning, NOAA Technical Report ERL 267-ESL 30, pp 16–28
40. Scordilis EM (2006) Empirical global relations converting M_S and M_b to moment magnitude. J Seismol 10:225–236
41. Uhrhammer R (1986) Characterization of northern and central California Seismicity. Earthq Notes 57(1):21–27
42. Wooessner J, Wiemer S (2005) Assessing the quality of earthquake catalogues: estimating the magnitude of completeness and its uncertainty. Bull Seismol Soc Am 95(2):685–698
43. Boore DM, Atkinson GM (2008) Ground-motion prediction equations for the average horizontal component of PGA, PGV, and 5%-damped PSA at spectral periods between 0.01 s and 10.0s. Earthq Spectra 24(1):99–138
44. Pachauri AK, Pant M (1992) Landslide hazard mapping based on geological attributes. Eng Geol 32(1–2):81–100
45. Champati-ray PK, Dimri S et al (2007) Fuzzy-based method for landslide hazard assessment in active seismic zone of Himalaya. Landslides 4(2):101–109
46. Bhattacharya A, Vöge M et al (2013) Surface displacement estimation using multi-temporal SAR Interferometry in a seismically active region of the Himalaya. Georisk Assess Manage Risk Eng Syst Geohazards 7(3): 184–197
47. Nayek PS, Gade M (2021) Seismic landslide hazard assessment of central seismic gap region of Himalaya for a Mw 8.5 scenario event. Acta Geophys 69(3):747–759
48. Kumar S, Sharma ML, Das J (2018) Consistent scaling laws for thrusting environment: a case study for Himalayan region. Int J Geotech Earthq Eng 9(2):46–62
49. Cencus India (2011) Office of the Registrar General & Census Commissioner, India. www.census2011.co.in/census/city; https://censusindia.gov.in/census_and_you/area_and_population.aspx, accessed on 10/09/2020
50. Nath RR, Sharma ML et al (2021) Landslide susceptibility zonation with special emphasis on tectonic features for occurrence of landslides in lower Indian Himalaya. J Indian Soc Remote Sens 49(5):1221–1238
51. Nath RR, Das N, Satyam ND (2021) Impact of main boundary thrust (MBT) on landslide susceptibility in Garhwal Himalaya: a case study. Indian Geotech J 51:746–756

Chapter 4
Role of Uncertainties in Site Response Analysis and Their Incorporation in Seismic Hazard Workflow

C Shreyasvi and K Venkataramana

Abstract The inherent randomness in the underlying geotechnical and geological formations makes its characterisation highly site-specific. While mapping the hazard at a larger scale, the site characteristics sampled at individual locations are collectively used to represent an entire region. Maps representing local geology, soil amplification, and hazard of a region are created through spatial interpolation of the site-specific estimates. Since the half-space is highly unpredictable, the estimates are probabilistic with a certain amount of uncertainty. This uncertainty can be partly modelled during the computation of site response while a fraction of it will remain as a standard error or deviation of the estimated values. The latter is known as aleatory uncertainty while the former is known as epistemic uncertainty. When the uncertainties are systematically partitioned and accounted for in the hazard integral, an ergodic approach (conventional PSHA with no site-specific information) transforms to a non-ergodic (completely site-specific) and partially non-ergodic approach (partially site-specific). The present article outlines the scope of a non-ergodic PSHA in the Indian scenario and the existing practices in capturing the uncertainties introduced by the site component. The authors believe that the article can provide insight towards improving the existing site-specific PSHA practices in the country.

Keywords Local site effects · Site component logic tree · Variability in site parameters

4.1 Introduction

A Probabilistic Seismic Hazard Analysis (PSHA) aids in determining the intensity of ground shaking that can be expected in a region by considering all the seismically contributing factors. The outcome of PSHA is mainly hazard curves, uniform hazard

C. Shreyasvi (✉)
Seismic Hazard Scientist, GEM Foundation, 1 c/o Eucentre, 27100 Pavia, PV, Italy
e-mail: shreyasvic@globalquakemodel.org

K. Venkataramana
Department of Civil Engineering, NITK Surathkal, Mangalore, Karnataka 575025, India

© Indian Society of Earthquake Technology 2023

R. S. Jakka et al. (eds.), *Earthquake Engineering and Disaster Mitigation*,
Springer Tracts in Civil Engineering, https://doi.org/10.1007/978-981-99-0081-7_4

spectrum and seismic hazard maps. These outputs are found to be exceptionally useful to the engineering community and hence, more efforts are applied in improving the methodology and thereby the accuracy of the results. The contributing factors for a PSHA are broadly categorised into three components such as source, path and site. The source component of the PSHA focusses on all the faults/seismogenic sources that are capable of causing an earthquake at the site of interest. These sources undergo rupture from time to time resulting in an earthquake. Hence, a rupture is defined by the magnitude (M_W) and hypocentral depth (Z_{hyp}), focal mechanism, depth to the top of the rupture (Z_{TOR}) and other seismological parameters related to the source. The path component focuses on the seismic wave propagation, spreading of the ground vibration and attenuation of the ground motion intensity. The path component is defined by the distance from the source to the site of interest and can be expressed in multiple distances metrics such as R_{epi} (epicentral distance), R_{rup} (closest distance to rupture), R_{JB} (Joyner-Boore distance) and R_{hyp} (hypocentral distance). These different distance metrics are explained with an illustration in Fig. 4.1.

The site component defines the characteristics of the local geology/soil that may have an impact on the intensity of ground shaking and its associated secondary effects. The surface geology is responsible for three phenomena such as amplification, liquefaction and landslide. Soil amplification also termed as local site effect can enhance or attenuate the seismic waves being transmitted from the underlying source to the surface. On the other hand, liquefaction and landslides are categorised as secondary geotechnical damages. The previous earthquake experiences have proven that the local site amplification and its secondary effects are more devastating than the earthquake itself. The impact of the local site effect during the Loma Prieta earthquake has been demonstrated in Fig. 4.2. The picture represents the difference in the amplitude recorded at three different stations with different soil site conditions.

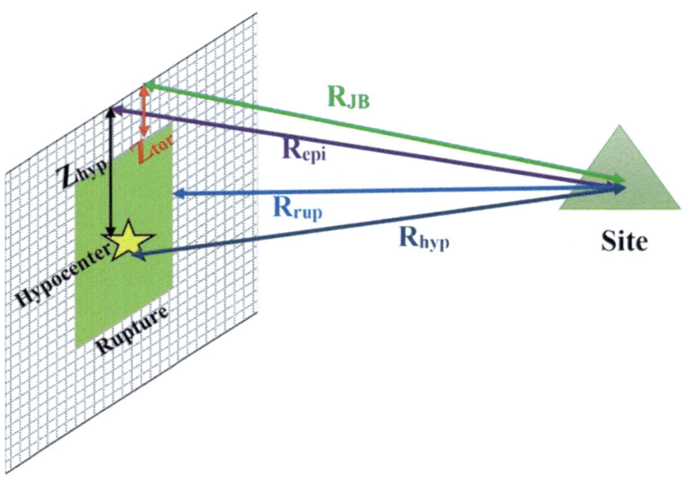

Fig. 4.1 Distance metrics used in defining the source to site distance in PSHA. (*Note* The rectangular grid—fault surface, the green rectangle—rupture and the yellow star—hypocentre/focus)

The amplitude in the bedrock site (MAS) is relatively negligible as compared to the other two sites. Hence, it is very important to understand the concept of local site effects and design strategies to effectively undermine its ill effects. It is possible to quantify these local site effects and include it in the framework of PSHA. However, the secondary effects need to be assessed individually and they are not integrated into the PSHA workflow. Hence, the present article focusses only on the site component of PSHA, discussing about the different methods that are available for its quantification and inclusion in the hazard.

The site component of PSHA is characterised through amplification which is defined as the ratio of the ground motion Intensity Measure (IM) at the surface to that at the bedrock. This amplification can be measured through installation of seismometers at the surface as well as at a depth (down hole) below the surface. The ratio of the IM observed by the seismometers at the surface and the downhole is calculated. This kind of set up has been installed in Japan and seismically active parts of America. However, the downhole installations are not always practical and may not be feasible financially everywhere. Hence, the ground motion recordings from multiple stations situated on different site conditions are used to calculate the amplification. Shreyasvi and Venkataramana [1] outlined the different techniques that are available to characterise the local site effects. The numerical simulation is a commonly used method to estimate the local site amplification and generate surface level response spectra. The output from the simulation is extremely important to the engineering community and is one of the expected outputs from microzonation studies. There have been numerous studies on improving the site response simulation [2–6, 7] and providing guidelines [8] to perform site specific hazard analysis.

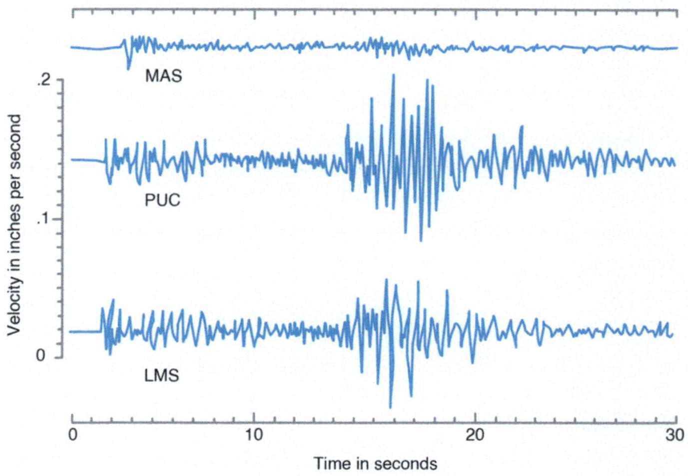

Fig. 4.2 Seismic waveform recorded during the M_W 4.6 aftershock of the Loma Prieta earthquake at two stations situated on soil sites (PUC, LMS) and one station situated on bedrock site (MAS). (https://www.usgs.gov/)

The present article focuses on the site component of PSHA, discussing the methods to quantify the local site effects and the associated uncertainties. The objective is to provide information about the uncertainties involved in modelling the soil, simulating the site response and incorporating it into the PSHA workflow. The article describes different methods with varying levels of complexity to include the local site response into PSHA. The article also presents an example of one of the partially non-ergodic approaches while the rest of the methodologies are discussed in detail. The present article serves as a source of information in making an informed decision about the methods that can be used for PSHA and microzonation studies.

4.2 Site Response Analysis (SRA)

The simulation of wave propagation through the layered soil media to determine its amplification characteristics is known as site response analysis. The phenomena of wave propagation can be visualised in 1, 2 and 3 dimensions of which 1-dimension has been widely used. The reason for such widespread usage is the limited availability of data, ease of use, accuracy and robustness. In 1-dimensional SRA, the soil layers are assumed to be horizontally supported by the bedrock which can be modelled as an elastic half-space or assigned a rigid boundary condition. The input for a typical SRA calculation can be broadly categorised as soil modelling and input motion. The site response can be determined by considering the linear, nonlinear (NL) or equivalent linear behaviour of the soil material. In terms of input motion, SRA can be conducted using Random Vibration Theory (RVT) based input spectra or time series motion. Additionally, the analysis can be carried out in the time or frequency domain. Depending on the availability of the data as well as the platform used for the SRA, a suitable technique can be used. Irrespective of the choice of method for SRA, the required input parameters remain fixed. The different methods available for SRA have been represented in Fig. 4.3.

In soil modelling, the soil profile information describing individual layer thickness, unit weight, Shear wave velocity (V_S) or SPT 'N' value, Atterberg's limits, and overburden pressure are used. The soil profile information and the simulation of the wave propagation have been illustrated in Fig. 4.4. The V_S values are used to calculate the maximum shear modulus ($G_{\max}$) using its correlation with density ($G_{\max} = \rho * V_s^2$). In the absence of information on V_S, the SPT 'N' values can be used to estimate the V_S values for each layer. There have been studies that have estimated the V_s profile for a borehole using the correlation between N and V_s values for different soil categories (Table 1 of [10]).

The dynamic behaviour of the soil is modelled in the form of modulus reduction (M-R) and damping (D) curves. These curves vary as a function of strain induced in the soil. These curves can be generated with the aid of the soil samples from the site and relevant lab tests. In case of unavailability of the soil samples or laboratory set up, there are standard M-R and D curves generated for standard soil types such as sand, clay (with varying plasticity index), weathered rock and rock. The availability of

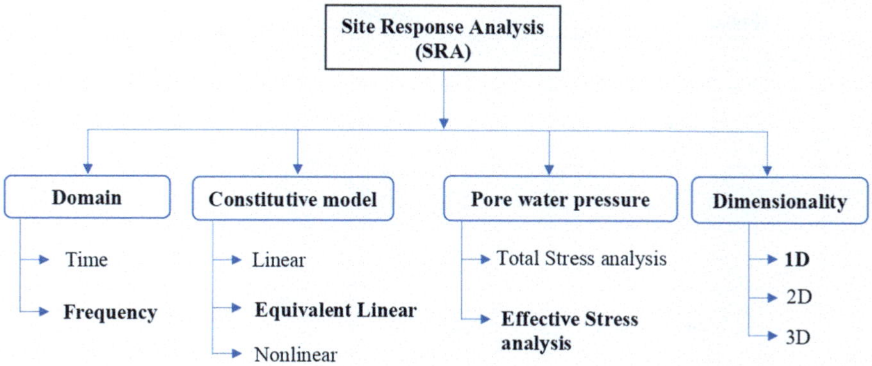

Fig. 4.3 Classification of site response methods [9]. The terms highlighted in **BOLD** are the main focus of the article

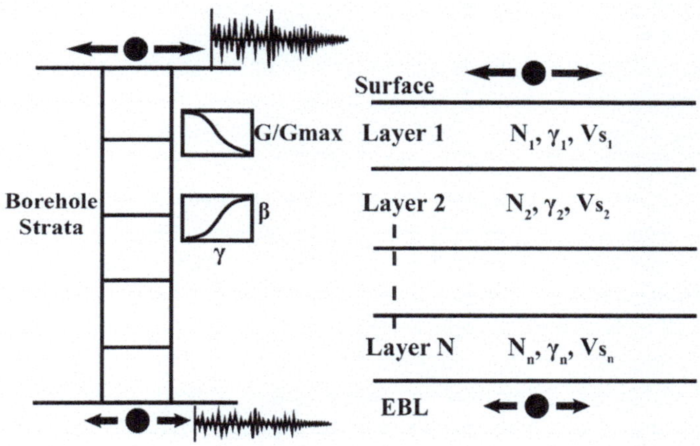

Fig. 4.4 Illustration of the layered soil medium and the profile information used in SRA [1]

detailed information about the geotechnical parameters aids the selection of standard M-R and D curves. Some of the standard curves used for modelling the dynamic characteristics of the soil in SRA have been presented in Fig. 4.5.

The selection of input motion plays a major role in SRA. Depending on the availability of strong-motion records in the study region, time series or Random Vibration Theory (RVT) analysis can be chosen. RVT based response spectra are favoured as input motions when there is sufficient knowledge about the regional hazard but insufficient ground motion data available. On the other hand, time series analysis is highly efficient when ground motion records are available. There have been studies wherein the merits and demerits of the two methods are investigated. Kottke and Rathje [11] proved that the site amplification predicted by the RVT method offers 20–50% higher values as compared to the time series method at the

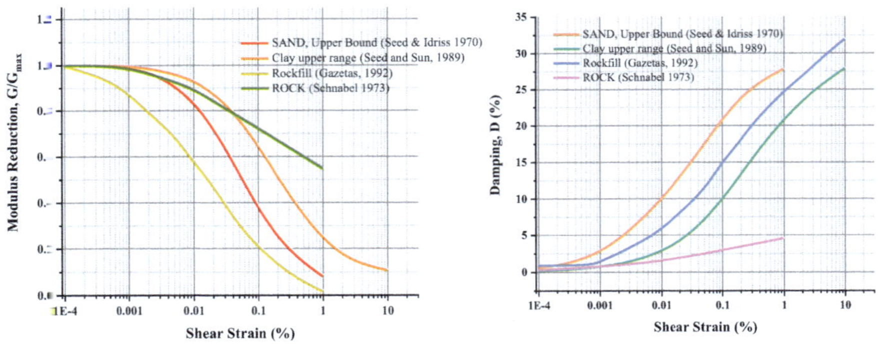

Fig. 4.5 Sample standard Modulus Reduction (M-R) and Damping (D) curves used in soil modelling for SRA [10]

natural frequency of the site. They further revealed that the overprediction observed is significant for sites with a smaller natural frequency.

The soil modelling and the input motions complete the input requirements for an SRA. The choice of method, as well as the domain, is up to the discretion of the user. The equivalent linear (EQL) method is usually preferred as it combines the advantages of both linear and nonlinear methods. The EQL method estimates the backbone curve for each layer by using the estimated G_{max} and the assigned M-R curves. This is an iterative scheme wherein the strain compatible secant moduli and damping are computed in each cycle and the iteration is said to be converged when the difference in the computed parameters between two successive iterations is less than 5%. The difference between EQL and NL analysis is that the parameters estimated during each iteration are time-invariant in the former. The input motions are applied at the base of the modelled soil profile. The response simulation can be conducted on numerous platforms both commercial (SHAKE2000 [12], FLAC3D [13], PLAXIS2D [14]) and open source (DEEP SOIL [25], STRATA [16] or PySRA). The output of an SRA is usually amplification (AF) spectrum defined by the ratio of the IM at the surface to that at the bedrock. Other additional output parameters are response spectrum at the surface, strain and peak ground acceleration induced at each layer.

4.3 Uncertainty in Site Response

The recorded ground motion time history offers first-hand information about the local site effects. The simulation techniques that have been developed and extensively used are based on the observations from these recorded ground motion time histories. The overall objective of the site response simulation technique is to capture the local site amplification to the maximum possible accuracy. Hence, site response estimation methods are constantly improved with the availability of new information and advancement in modelling techniques. The presence of large uncertainty in the modelling phase usually impacts the PSHA estimates at longer return periods.

The unpredictable nature of the soil medium adds more complexity to the estimates of site response and amplification cannot be defined in a deterministic manner. In other words, the amplification estimated through site response will always be in the form of mean ± standard deviation (SD). The SD account for the randomness inherent in the soil medium. There are two types of uncertainty and they are aleatory and epistemic. Aleatory variability represents the natural randomness in the site response and can be reduced to a certain extent with the availability of additional information. On the other hand, epistemic uncertainty refers to the lack of scientific knowledge in modelling and ideally, can be reduced to zero with additional information. A major challenge in understanding uncertainty in site response is differentiating the factors contributing towards aleatory and epistemic components. For instance, the spatial variability in V_s is treated as aleatory and inaccuracy in measuring a V_s profile due to instrumental and human error or shortcoming of the method used for characterising V_s can be treated as epistemic. Generally, it is not feasible to have the complete spatial coverage of measured V_s over a large area and hence, the aleatory component can be slightly reduced with more measurements at closer intervals. On the other hand, the V_s profiles at a given site can be visualised with several realisations. The present article focuses on modelling the epistemic uncertainty in site response.

The local site amplification is sensitive to its input parameters and hence, a small variation in the input may vary the mean as well as the standard deviation of AF. The variability in site response can be captured by analytically combining the variability in the independent input parameters. In soil modelling, the parameters such as depth to bedrock, V_s profile, unit weight and thickness of each soil layer cannot have an exact estimate. In practice, these parameters are defined with an expected value and variability. Using the statistical parameters (expected and variability), the Monte Carlo simulation technique [17] can be used to create multiple realisations of a soil profile. Further, the site response is computed for each realisation and then the final output combines the outcome for all the realisations thereby reducing the variability component. A randomised V_s profile has been shown in Fig. 4.6 and each realisation is represented by a unique colour. A similar attempt was made by Shreyasvi et al. [18] wherein the sensitivity of site amplification to the variability in the soil profile was studied by considering multiple cases with a varying number of realisations. The study found that a minimum of 22 realisations is necessary for obtaining a stable estimate. Similar studies by Barani et al. [19], Dejphumee and Sasanakul [20] investigated about the uncertainties in modelling the soil and its impact on the hazard estimates.

However, the Monte Carlo simulation requires the soil properties to be randomised. In this regard, there have been two significant models developed by Toro [21] and Darendeli [22]. These models attempt to capture the statistical distribution and properties of the soil. The model proposed by Toro [21] is extensively used in randomising unit weight, shear velocity, the thickness of each layer and depth to bedrock. On the other hand, Darendeli [22] explains the statistical distribution of the soil nonlinear properties such as G/G_{max} and D.

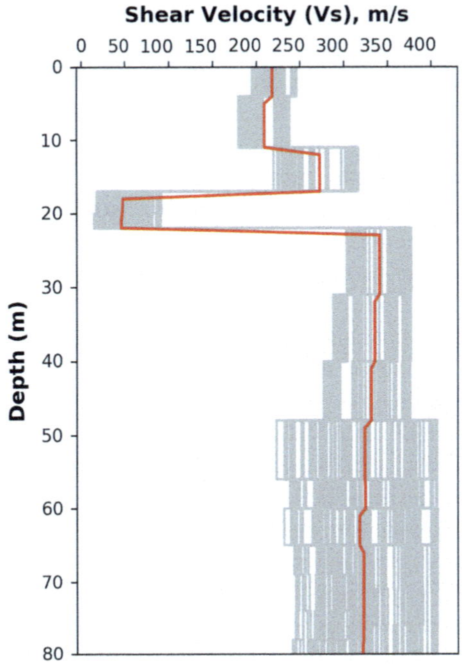

Fig. 4.6 Randomised V_s profile consisting of 50 realisations (grey) and the median estimate (red)

The epistemic uncertainty in site response is captured by considering the variability in V_s profile, depths to bedrock, the thickness of individual layers and many other contributing factors. A logic tree is formed specifically for the epistemic component in site response as shown in Fig. 4.7. The weights are assigned for each of the considered parameters for variability based on the relative impact of each branch on the overall results. The logic tree of site response forms a branch in the original PSHA logic tree.

Similarly, a single input motion cannot effectively represent the ground motion from all the possible earthquake scenarios at a given site. Hence, multiple-input motions from a wide range of scenarios are selected. However, there are no specific guidelines in regard to the selection and scaling of ground motion records for SRA. The common practice is to choose ground motions that are recorded locally or at a regional level. If the study area has no records of previous earthquakes, then ground motions from a similar tectonic regime can be selected. The selected ground motions must be from the scenarios that are significant for a given region. The influential earthquake scenarios for a region can be found from the de-aggregation of its seismic hazard. A sample de-aggregation result has been depicted in Fig. 4.8 and Shreyasvi [24] suggested that earthquakes in the M_W range 5–6 at distances <30 km can cause significant impact in South West India. In general, earthquake records are chosen from such identified scenarios and are used for SRA. The main concern is the number of records needed to completely capture the variability induced due to input motions. Shreyasvi et al. [18] suggested that a minimum of 7 ground motion records are

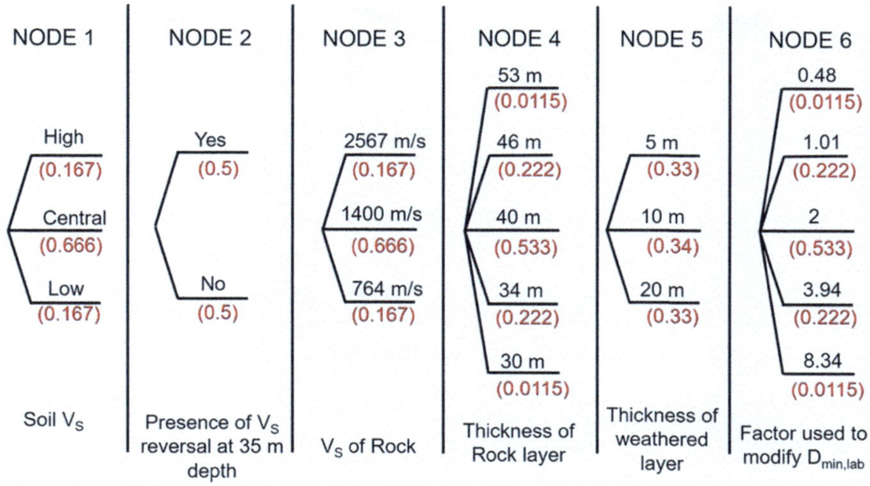

Fig. 4.7 A sample logic tree specifically designed to account for epistemic uncertainty in site response [23]

necessary and the variability associated with input motion in time series analysis is higher than the variability in the soil profile. Hence, there have been studies that prefer to use RVT analysis with the knowledge of the seismic hazard. Similar to the time series analysis, in RVT, several input motion spectra are chosen and SRA is performed. The final site amplification will be a combination of multiple ground motions originating from different scenarios.

4.4 Incorporation of Site Response in PSHA

Ground motion prediction equations (GMPEs) are one of the necessary input parameters in PSHA. They are the empirically or stochastically derived models representing the source, path and site characteristics of a region. Empirically derived GMPEs are developed for regions with frequent earthquakes and sufficient records of ground motions. On the other hand, stochastic or a hybrid GMPE is developed for regions with limited ground motion data. In the Indian scenario, there are few GMPEs that are developed for the Himalayan and North East part of India. However, the GMPEs applicable for the Peninsular India are limited and hence, most of the PSHA studies in India make use of the GMPEs developed for other data rich regions.

While estimating the hazard at the surface level, the site specific information such as $V_{S(30)}$ is applied to the GMPE. This is known as ergodic PSHA and this is a straight forward approach. In the absence of PSHA, the hazard can be estimated for standard site conditions as defined by NEHRP [25] or IS1893 [26]. In this method, the site amplification is quantified and included in the PSHA through GMPE and may or

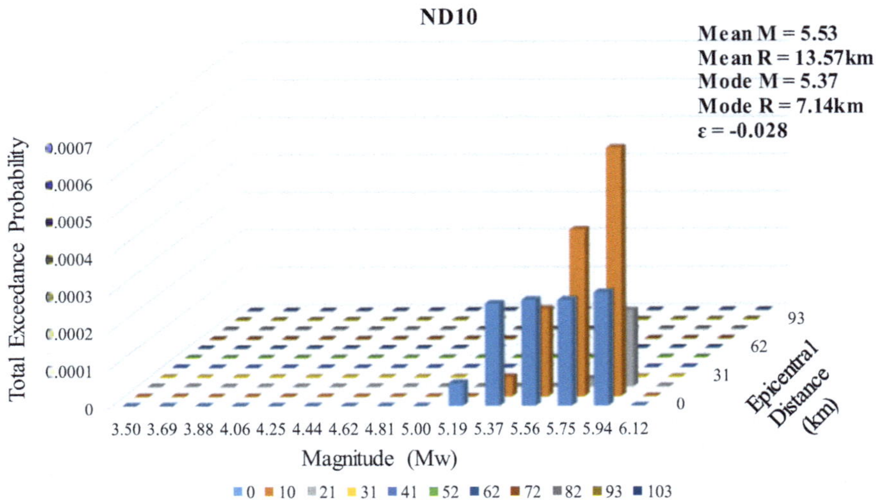

Fig. 4.8 De-aggregation plot for South West India with intensity 0.17 g for 2% probability of exceedance [24]

may not represent the actual site behaviour. With the availability of the region or site-specific details, there is an opportunity to transform a PSHA from an ergodic to partially non-ergodic or a complete non-ergodic approach. Also, non-ergodic PSHA is the future as more site-specific information becomes available. When the site response is incorporated into PSHA, it can be termed as partially non-ergodic as there is still a path component in ground motion prediction that is not region-specific.

The PSHA estimates can be made site specific by numerically modelling the site conditions and simulating the site response. The output from the site response (ln $Y(T)$) is multiplied with the PSHA estimate corresponding to bedrock or a reference site condition (ln $X_{\text{ref}}(T)$). This is referred to as partially non-ergodic. In this case, the hazard estimate at the bedrock is probabilistic but the same at the surface is deterministic. This is due to the fact that the site amplification used for multiplying with the bedrock estimate is a deterministic value and does not account for the uncertainty in site response. Also, this is the simplest method to obtain site specific PSHA estimates.

$$\ln Z(T) = \ln X_{\text{ref}}(T) + \ln Y(T) \tag{4.1}$$

where $Z(T)$ represents the time-dependent IM at the surface, X represents the IM at a reference site condition and Y represents the site amplification quantified through site response. Hence, it can be said that GMPE predictions are used up to bedrock level and from there onwards the site-specific amplification factors are used to modify the ground motion. However, this method has faced criticism due to its questionable accuracy at different levels. Additionally, the Hazard can be overestimated (due to neglecting nonlinear soil behaviour) or underestimated (due to resonance effects not being captured).

A completely probabilistic site specific PSHA can be estimated by considering a probability distribution of the amplification function and integrating it with PSHA. These probabilistic approaches involve either modifying a GMPE or incorporating site response through convolution method. The convolution method was proposed by Bazzurro and Cornell [27] wherein a probability density function of the site amplification convolves with the hazard at the bedrock level to generate hazard at the surface for a given probability of exceedance. Shreyasvi and Venkataramana [10] demonstrated a completely probabilistic approach by modifying the GMPE and estimating region specific hazard estimates for South West India.

The different site response techniques lead to generating site specific design spectra and in principle, this is similar to the design spectra provided by IS 1893 [26] code. Hence, most of the engineers adopt the code spectra as opposed to site specific spectra generated from PSHA. The drawback of using the code specified spectra is that it is generic and may not capture the spatial variability in geotechnical formations. Also, these spectra were derived from a limited number of ground motion records and may not be as relevant as a site specific spectra.

A completely non-ergodic PSHA is possible through systematic quantification of the repeatable path and site characteristics. There have been numerous studies explaining non-ergodic PSHA [28, 29] and targeting some of the data rich regions around the world to demonstrate a PSHA that is close to non-ergodic [7]. However, it is still a long way to a completely non-ergodic PSHA but with reducing application of ergodic PSHA.

4.5 Conclusion

The non-ergodic approach is the future in the domain of probabilistic seismic hazard analysis. With the availability of data and advancement in the modelling techniques, the classical PSHA is slowly transforming from an ergodic to a non-ergodic approach. Availability of region-specific and site-specific information makes a huge difference and has proven to improve the hazard estimates. Previously the epistemic and the aleatory variability was considered only for source and ground motion modelling. With the advancement in the field of site response analysis, the researchers are making an effort to quantify the uncertainty in local site effects as well. The reason behind uncertainty becoming a major topic of interest is the need for more reliable hazard estimates and improve confidence in engineers to rely on PSHA for the design of infrastructures. Further, increased uncertainty affects the PSHA outputs at a longer return period (2% probability of exceedance in 50 years) which is of main interest for major infrastructures such as dams and nuclear power plants. The current standard practice for quantifying epistemic uncertainty in site response is to identify the input parameters that may have a significant impact on the final hazard outcome. Assign weights to alternative models in a logic tree for each of the considered input parameters and combine all the possible modelling strategies to provide a more reliable estimate. The present article outlines the existing methodologies for incorporating uncertainty in site response analysis and integrating the site response into PSHA.

References

1. Shreyasvi C, Venkataramana K, Chopra S (2019) Local site effect incorporation in probabilistic seismic hazard analysis–A case study from southern peninsular India, an intraplate region. Soil Dyn Earthq Eng 123:381–398
2. Nath RR, Jakka RS (2012) Effect of bedrock depth on site classification. In: 15th world conference on earthquake engineering 15WCEE. Lisbon, Portugal, pp 24–28
3. Régnier J, Bonilla LF, Bard PY, Bertrand E, Hollender F, Kawase H, Verrucci L et al (2018) PRENOLIN: international benchmark on 1D nonlinear site-response analysis—validation phase exercise. Bull Seismol Soc Am 108(2):876–900
4. Rodriguez-Marek A, Kruiver PP, Meijers P, Bommer JJ, Dost B, van Elk J, Doornhof D (2017) A regional site-response model for the Groningen gas field. Bull Seismol Soc Am 107(5):2067–2077
5. Stewart JP, Afshari K, Hashash YM (2014) Guidelines for performing hazard-consistent one-dimensional ground response analysis for ground motion prediction. Peer Rep 16:117
6. Tönük G, Ansal A (2022) Factors affecting site-specific response analysis. J Earthq Eng 1–18
7. Kotha SR, Bindi D, Cotton F (2017) From ergodic to region-and site-specific probabilistic seismic hazard assessment: method development and application at European and Middle Eastern sites. Earthq Spectra 33(4):1433–1453
8. Hanks TC, Abrahamson NA, Boore DM, Coppersmith KJ, Knepprath NE (2009) Implementation of the SSHAC guidelines for level 3 and 4 PSHAs—experience gained from actual applications. US Geol Surv Open-File Rep 1093:66
9. Shreyasvi C, Venkataramana K (2022) Estimation of local site effects in Indian scenario: lessons from past earthquakes, current practices, and future trends. In: Civil engineering for disaster risk reduction. Springer, Singapore, pp 209–226
10. Shreyasvi C, Venkataramana K (2021) Nonlinear soil amplification models for a moderately active seismic zone in India. In: Local site effects and ground failures. Springer, Singapore, pp 39–50
11. Kottke AR, Rathje EM (2013) Comparison of time series and random-vibration theory site-response methods. Bull Seismol Soc Am 103(3):2111–2127
12. Ordonez GA (2000) SHAKE2000: A computer program for the 1D analysis of geotechnical earthquake engineering problems. Geomotions, LLC, USA
13. Itasca Consulting Group, Inc. (2019) FLAC3D—Fast Lagrangian analysis of continua in three-dimensions, Ver. 7.0. Itasca, Minneapolis
14. Brinkgreve RBJ, Swolfs WM, Engin E, Waterman D, Chesaru A, Bonnier PG, Galavi V (2010) PLAXIS 2D 2010. User manual, Plaxis bv
15. Hashash YMA, Musgrove MI, Harmon JA, Ilhan O, Xing G, Numanoglu O, Groholski DR, Phillips CA, Park D (2020) Deep Soil 7, user manual. IL, Board of Trustees of University of Illinois at Urbana-Champaign, Urbana
16. Kottke AR, Rathje EM (2009) Technical manual for Strata. Berkeley, California: Pacific Earthquake Engineering Research Center
17. Kottke AR, Rathje EM (2009) Technical manual for Strata. Report No.: 2008/10. Pacific Earthquake Engineering Research Center, University of California, Berkeley
18. Shreyasvi C, Badira Rahmath N, Venkataramana K (2020) Influence of variabilities of input parameters on seismic site response analysis. In: Advances in computer methods and geomechanics. Springer, Singapore, pp. 233–244

19. Barani S, De Ferrari R, Ferretti G (2013) Influence of soil modeling uncertainties on site response. Earthq Spectra 29(3):705–732
20. Dejphumee S, Sasanakul I (2021) Evaluation of uncertainties in site response analysis of deep soil profiles in South Carolina Coastal Plain. Bull Seismol Soc Am 111(4):1974–1988
21. Toro GR (1995) Probabilistic models of site velocity profiles for generic and site-specific ground-motion amplification studies. Technical Rep No 779574 Brookhaven National Laboratory, Upton NY
22. Darendeli MB (2001) Development of a new family of normalized modulus reduction and material damping curves [Ph. D. dissertation]. Austin: University of Texas
23. Rodriguez-Marek A, Bommer JJ, Youngs RR, Crespo MJ, Stafford PJ, Bahrampouri M (2021) Capturing epistemic uncertainty in site response. Earthq Spectra 37(2):921–936
24. Shreyasvi C (2019) Probabilistic seismic hazard assessment and site characterisation of Southwest India. Doctoral dissertation, National Institute of Technology Karnataka, Surathkal
25. BSSC (2003) NEHRP recommended provisions for seismic regulations for new buildings and other structures. FEMA 450 and Building Seismic Safety Council
26. IS 1893 (2016) Criteria for earthquake resistant design of structures, part 1—general provisions and buildings (Fifth Revision). Bureau of Indian Standards, New Delhi
27. Bazzurro P, Cornell CA (2004) Nonlinear soil-site effects in the probabilistic seismic hazard analysis. Bull Seismol Soc Am 94(6):2110–2123
28. Walling MA, Abrahamson NA (2012) Non-ergodic probabilistic seismic hazard analyses. In: Proceedings of 15th world conference earthquake engineering, Lisbon, Portugal, Paper No 1627
29. Stewart JP, Afshari K, Goulet CA (2017) Non-ergodic site response in seismic hazard analysis. Earthq Spectra 33(4):1385–1414
30. Stewart JP, Afshari K (2021) Epistemic uncertainty in site response as derived from one-dimensional ground response analyses. J Geotech Geoenvironmental Eng 147(1):04020146

Chapter 5
Dynamic Characterisation of Soil Profiles in Southern Calicut Region in Kerala State

K. Rangaswamy, P. G. Abhishek, L. Abhijith, and Gloria Campilongo

Abstract The present study has focused on characterising the dynamic properties of the ground soil profiles existing in the southern Calicut city nearby the NIT Calicut region. Field MASW tests were carried out at 48 representative sites by using a geode seismograph with 24 channels. It has a geophone capacity of 4.5 Hz and a sledgehammer source. MASW testing facility has provided testing and analysis software to obtain the shear wave velocity profiles. During testing, a spacing of 1.5 m between the successive receiver geophones is controlled to receive the signals once after inducing an energetic hammer shot at an offset distance of 5 m from the trigger geophone. Results indicate that the average shear wave velocity (V_{s30}) of the uppermost 30 m depth of soil strata is within the range of 273–615 m/s.

Keywords NEHRP site class · Shear wave velocity · MASW · Seismic site classification · V_{s30}

5.1 Introduction

The dynamic characterisation can be considered a primary step for estimating ground response and the seismic microzonation of a particular region. It helps determine the dynamic behavioural properties of deposits of soils and monitor the surficial level site response. Measurement of dynamic properties of earthen fills, embankments, subsurface soils etc., plays a crucial role in designing the seismic-resistant structures, i.e. buildings, road bridges, dams, slopes and earth retaining structures. Multichannel analysis of surface wave (MASW) is a universally accepted non-destructive geophysical technique to obtain the dynamic properties of the subsurface materials. It works

K. Rangaswamy (✉) · P. G. Abhishek · L. Abhijith
NIT Calicut, Calicut, India
e-mail: ranga@nitc.ac.in

G. Campilongo (✉)
University of Calabria (DiBEST), Rende, Italy
e-mail: gloria.campilongo@unical.it

© Indian Society of Earthquake Technology 2023

R. S. Jakka et al. (eds.), *Earthquake Engineering and Disaster Mitigation*,
Springer Tracts in Civil Engineering, https://doi.org/10.1007/978-981-99-0081-7_5

based on seismic surface wave propagation which gives the shear wave velocity (V_s) profiles with the depth of soil strata.

The most commonly adopted technique for site characterisation is using borehole data, like SPT and seismic cross/downhole testing data, which, for a long time, is considered the standard way of attaining the characteristics of the soil layers in the past. These kinds of methods are the most reliable as they provide the data as close to reality as possible, but are too expensive. For this reason, other methods, such as the one used in this article, are used. Currently, the survey focuses on surface wave dispersion analysis, leading to a more economically feasible alternative. Non-destructive geophysical techniques have been used to interpret the physical behaviour of the soil, such as stiffness, liquefaction potential etc., by estimating the shear wave velocities of the soil layers. The usage of these geophysical methods started in the early 1980s. Surface wave methods, including those generated by the active and passive source, have begun since the 1980s and are still in existence at the present time [1].

Multichannel analysis of surface wave (MASW) is a practical geophysical approach widely accepted due to its non-intrusive nature. The abundance in data gaining involves taking less time without drilling boreholes [2]. MASW is considered an accurate method for predicting the shear wave velocity, which is used to determine parameters of engineering purposes. These can be used for estimating the 1D and 2D seismic shear wave velocities of different sections [3].

Based on the requirement of a surface wave for the analysis, MASW surveys are classified into active and passive methods. In an active MASW, the source, i.e., the surface waves, are produced by an impulsive or a seismic vibratory source. In passive MASW, seismicity is generated by some natural sources or by using any induced magnetic activities [4]. The MASW, indirectly, interprets the lithology, quality of rock, and geotechnical specifications of the area under consideration, which can be used for many civil engineering purposes [5, 6].

The most crucial parameter for dynamic ground response study and the characterisation of the site is the shear wave velocity (V_s). Their values are in proportion with shear modulus, density and damping ratios. As the V_s value is small, lower is the shear modulus, and higher will be the damping ratio [7]. The MASW will measure the shear wave velocity, thereby providing a one-dimensional profile of the subsurface layers of the soil beneath.

The average shear wave velocity for the uppermost 30 m depth of the soil profiles is considered for the primary site classification under seismic conditions as per the seismic design guidelines (Table 5.1) given by the National Earthquake Hazard Reduction Program [8]. It provides the site classifications as A, B, C, D, E and F that related to the average shear wave velocity (V_{s30}) values. E and F site classifications are considered easily liquefiable in these classes; thus, they require proper evaluation and detailed investigations.

The current study focuses on assessing the dynamic characterization of underlain soil profiles for the spatially distributed sample sites across the study area by calculating the average shear wave velocity V_{s30} and classifying the site according to the NEHRP provisions. The shear wave velocity gives a basis for the ground motion

Table 5.1 Site classification under seismic loading conditions [8]

Site classification	General description of soil profile	Average shear wave velocity, V_{s30} (m/s)
A	Hard rock	>1500
B	Rock	760 – 1500
C	Very dense soil and soft rock	360 – 760
D	Stiff soil $15 \leq N \leq 50$ or $50\ kPa \leq S_u \leq 100\ kPa$	180 – 360
E	Soil or any profile with more than 3 m of soft clay defined as soil with; $PI > 20$, $w \geq 40\%$, and $S_u < 25\ kPa$	≤180
F	Soils requiring for site-specific evaluation	–

response, potential effects on liquefaction, and an idea on the natural frequency with soil amplifications during an earthquake event.

5.2 Literature Review

A few published papers below discuss the site characterisation with shear wave velocity profiles, ground response analysis, and micro zonation mapping.

Joyner and Fumal [9] estimates the local site effects on strong ground motion as functions of local shear wave velocity measurements arrived from the downhole testing. It presents the PHA, velocity and response spectra at 5% damping. Luna and Jadi [10] reveal that dynamic soil properties are strain-dependent and that achieving compatibility between the results of several approaches when the strain levels overlap is difficult. As a result, the strain level must be kept consistent to compare dynamic soil properties using different methodologies. The measurement of dynamic soil parameters contributes to estimating shear modulus and shear wave velocity at low strain levels.

Mohammed et al. [11] conducted an active seismic survey to quantify the near-surface seismic properties. It has been found that The V_{s30}, which is obtained from the MASW technique, plays a critical role in evaluating the site response of the upper 30 m depth. The distribution of the obtained values of V_{s30} through the studied area demonstrates site classes of C and D, according to the NEHRP (National Earthquake Hazard Reduction Program) and IBC (International Building Code) standards.

Akin et al. [12] carried out the dynamic analysis. The soil amplification and predominant periods were evaluated by ground response analyses of one-dimensional equivalent linear and empirical approaches that supported shear wave velocity rate profiles. The study results produce amplification and predominant period maps of Turkey's seismically active Erbaa settlement region. These maps of the research area have been compared to evaluate different site response studies. Olafsdottir [13] organized for Multichannel analysis of surface wave records using the MASWamethod and understanding the efficiency of MASWaves software. The field tests illustrate the maximum investigation depth of the survey and its resolution at shallow depth by combining dispersion data gathered by receiver spreads of different lengths. The dispersion analysis was done to extract Rayleigh wave dispersion curves from the acquired data.

Lin et al. [14] compared the dispersion analysis between two stations SASW and multi-station MASW approach. Limited to the two-channel data, conventional dispersion analysis of SASW suffers from possible phase un-wrapping errors, inefficient data filtering and synthesis, and the inability to distinguish multiple modes. The MASW approach, on the other hand, can avoid these problems and provide visualization of dispersion relation through 2D wavefield transformation. As opposed to multiple shots in the SASW testing, the MASW testing typically uses only a single shot, limiting the available wavelengths due to near field and far-field effects.

Sairam et al. [15] carried out site characterization of intraplate regions of western India using the Multichannel analysis of surface wave method and P-S logging method. For granites and Deccan traps, V_{S30} values were in the range of 760–1500 m/s, thus belonging to class B-class. The Tertiary, Cretaceous, Jurassic and Paleoproterozoic sediments had V_{s30} values between 360 and 760 m/s, and hence, were assigned C-class. Moreover, the Quaternary soils were categorized as D-class, as they show V_{s30} in the range of 180–360 m/s. Also, the Holocene tidal flat and Rann sediments were classified E-class since the V_{s30} values are less than 180 m/s. In addition to these, the site response revealed that the D and E type soils significantly have higher amplification than that in the B and C category soils.

Abudeifa et al. [16] evaluated the geotechnical characteristics of New Akhmim city in Egypt using the MASW technique. It found that the degree of competence of the compacted shale layer is higher with more incredible shear wave velocity than that of unconsolidated gravel or sand material at the site of construction.

5.3 Study Area

The Calicut (Kozhikode) district is located in Kerala, the southernmost state in India. It lies between N 11° 08′ and N 11° 50′ latitudes and E 75° 30′ and E 76° 8′ longitudes. Calicut is enclosed on the northern side by district Kannur, on the southern side by district Malappuram, on the eastern side by district Wayanad and the western side by Lakshadweep Sea. Area of study demarcated in Fig. 5.1.

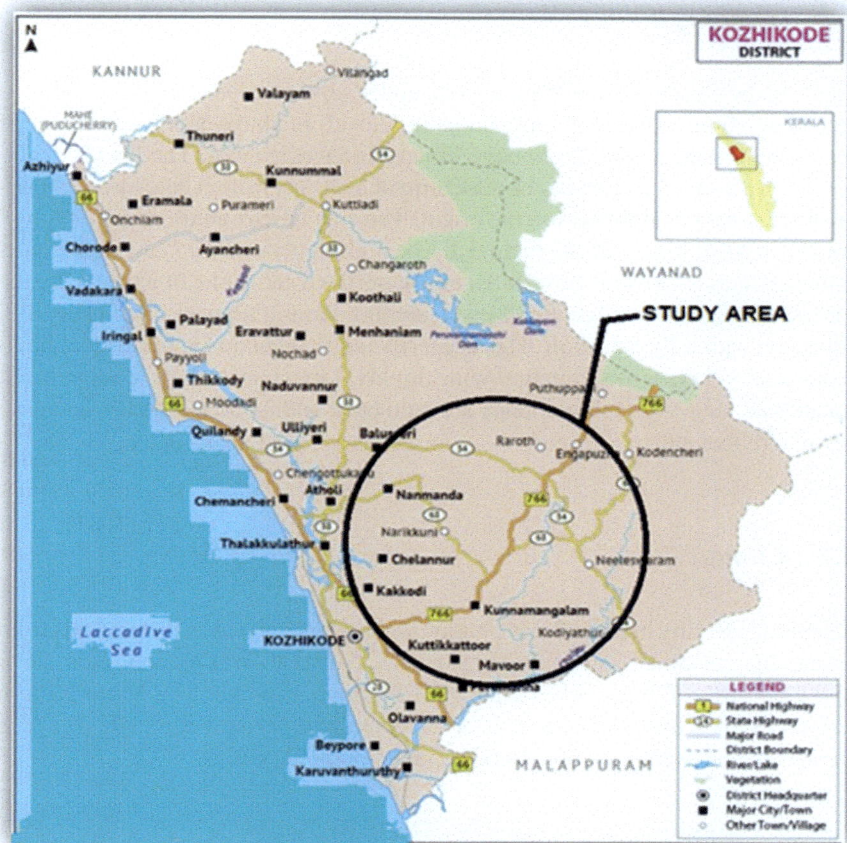

Fig. 5.1 Location of the study area in Kerala state (*Source* Wikimedia Commons)

5.3.1 Geology

Calicut is geologically classified into three belts, including: (i) A linear NW–SE gneissic belt along the middle ranging from north to south, (ii) A narrow coastal belt, and (iii) A charnockite belt that occupies the northeast and southern area presents as pockets within the gneissic terrain. Granite gneiss is one of the oldest units of the area that belongs to the peninsular gneissic complex and occurs to the north of Alampore. Charnockite has an extensive distribution available, especially in the northeast and southern region districts, with varieties like biotite-hornblende- hypersthene gneiss. Magnetite quartzite occurs as narrow linear bodies within the charnockite belt. Migmatite complex extending from north to south having hornblende-biotite gneiss with well-foliated structure.

5.3.2 Geomorphology

Calicut can be classified into physiographical divisions as coastal plain, namely; (i) low land (<7.6 m above MSL), (ii) midland (7.6–76 m above MSL) and (iii) high land—hilly terrain (above 76 m above MSL) from west to east. The land is gently sloped with a most exceptionally slender form of 5–10 km wide. These have deposits of marine, stream, and fluvio-marine origin. The area having an elevation on top of 76 m above MSL is termed the highland. The highest peak of the Calicut district is Nilamala which is 1935 m above MSL on the north-eastern side. In Kozhikode, the landmass is of a linear ridge, flood plain, sloping terrain, alluvial plain valley fill, rocky slope with hilly landform. The valley fill and flood plain are the critical fluviatile terrains. Whereas moderately sloping, highly sloping, and others are significant denudational land. Agricultural lands are cultivated with paddy, coconut, and cash crops like rubber etc.

5.3.3 Climate

Calicut is a generally humid climatic region with a summer season that extends from March to May month. The monsoon season, viz., the South-Western monsoon, starts in the first week of June and extends until September month. The North-Eastern monsoon extends from the latter half of October up to November month. The area shows an annual average rainfall in the range of 300–326 cm.

5.4 MASW Testing

For the present investigation, MASW tests were performed at the spatially selected locations in the Calicut region. The tests were conducted using geometrics 24 channelled geode seismic recorder with seismograph as shown in Fig. 5.2. The vertical geophones with a natural frequency of 4.5 Hz (24 numbers) were used to receive the wave fields produced by an active source of 8 kg sledgehammer on a metal plate. The main quest in this data acquisition is to attain the shear wave velocity of the underlain sub surficial layers to reach at least 30 m below; thus, the frequency record has to be lower so that the phase velocity can be obtained at longer wavelengths. Lower frequency of signal indicates longer will be the wavelength of the surface wave recorded, thus larger the depth of investigation. So low-frequency geophones of 4.5 Hz were used for the study.

The most important part of the MASW test's field configuration is the spacing between the geophones and the offset distance. Although data acquisition beneath the ground is an easy task, the configurations to be considered for the field need to be optimized based on the requirement. The planar characteristics of surfaces

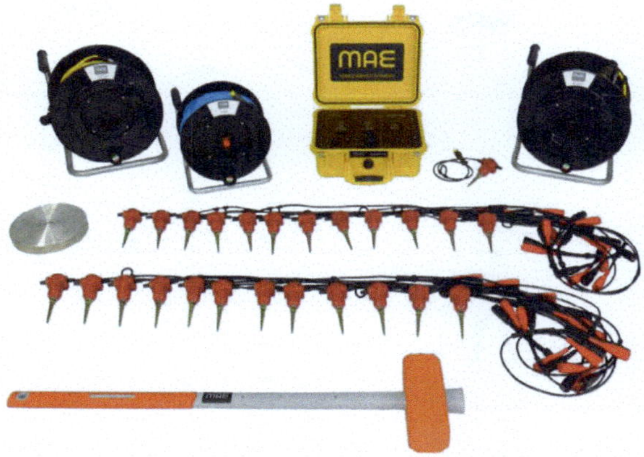

Fig. 5.2 Photograph on MASW testing facility available at NIT Calicut

waves evolve solely when at a particular distance from the source. This distance must be higher than half of the maximum intended wavelength in the vast majority of instances. However, the amplitude of body waves and higher Rayleigh waves at high frequencies may dominate the fundamental frequency mode.

In the current study, 12 receiver geophones are rigidly placed in the field of the ground surface as a series along a straight stretch at an interval of 1.5 m. The offset distance between the trigger geophone and the first receiver geophone is fixed to 5 m. The geophones are interconnected by using a cable that, in turn, is connected with the geode seismograph. The apparatus and acquisition setup used in this investigation are shown in Figs. 5.2 and 5.3. The seismograph is connected with a computer that runs the testing software SYSMATRACK, which records the shot gathered and store the data in SEG.2 format in the user-chosen file. A sledgehammer of 8 kg with a circular steel disc is used to create the impact of an energy impulse wave. A sledgehammer is used as a source, triggered on a circular steel plate of 200 mm diameter connected to the seismograph using a trigger geophone, as shown in Fig. 5.3. The properties for the acquisition are given in Table 5.2.

The test setup in the field has presented in Fig. 5.4. The test is repeated three times to avoid uncertainty, thereby minimising the possible errors. Once the 1st phase of acquisition is finished, the 12 geophones are shifted within the same linear array without changing the trigger point. Then the process will be repeated, and then the results can be combined, which gives a more significant number of channel records, i.e., 24 channelled values. The schematic diagram shown in Fig. 5.5 gives the details of arrangements of the positioning of geophones.

The SYSMATRACK software will provide the raw seismic plots obtained from the 12 channels as shown in Fig. 5.6 by making use of which the data have to be processed.

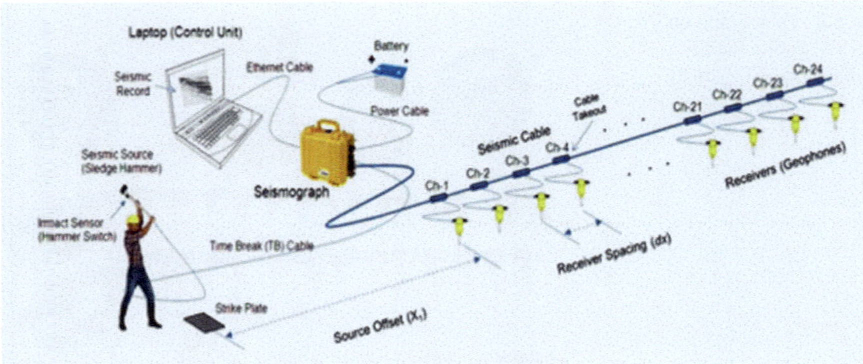

Fig. 5.3 Schematic on MASW data acquisition setup (*Source* ParkSEIS Software Manual; www.masw.com)

Table 5.2 Data acquisition parameters of MASW testing

Parameters	Range
Number of channels	12 in each phase
Geophone spacing	1.5 m
Sampling rate	0.5 s
Array length	36 m
Receiver geophones	4.5 Hz vertical
Impact source	8 kg sledge-hammer
Offset distance from source	5 m

Fig. 5.4 Test setup in the field

5.4.1 Data Processing

MASW raw seismic data obtained from the seismograph is processed by using an analysis software named EASYMASW. The data processing involves importing

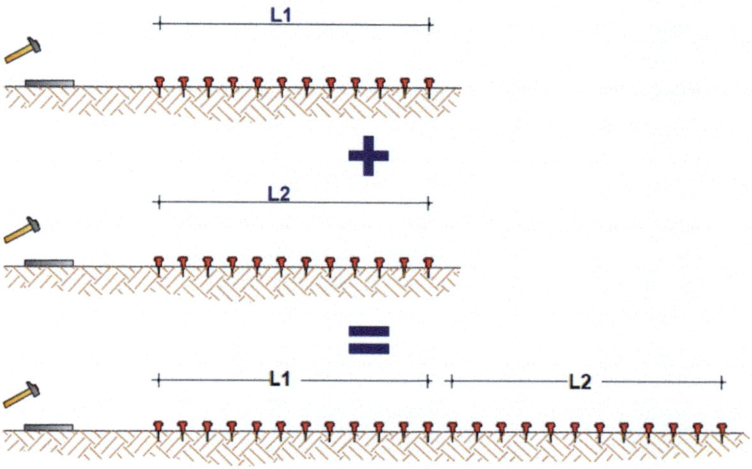

Fig. 5.5 Schematic arrangements of the positioning of geophones

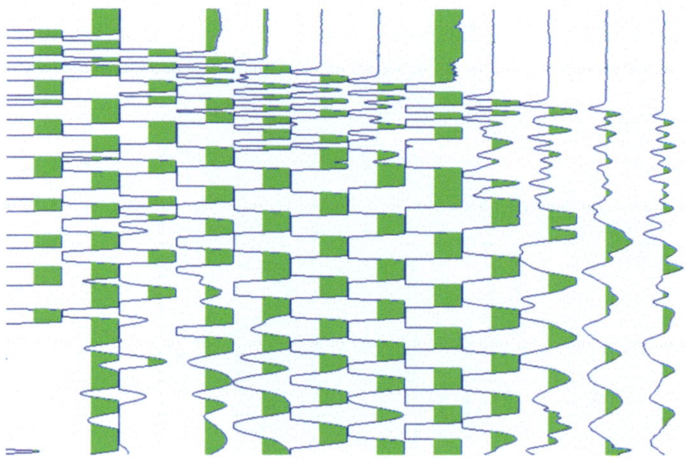

Fig. 5.6 Typical raw seismic plots obtained from seismograph

tracks, spectral analysis, inversion, and determination of velocity profiles. Average shear wave velocities of different soil profiles have been obtained after conducting a series of field MASW testing. Shear wave velocities are obtained after processing the MASW testing data using testing and analysis software's provided along with the MASW facility. Figures 5.7, 5.8, 5.9 and 5.10 shows the stages of arriving at a typical one-dimensional shear wave velocity profile for one of the representative sites near Pottaseri located at 10 km from NIT Calicut. The track record imported from the analysis of raw data is shown in Fig. 5.7. The spectral analysis is the process where the experimental dispersion curve will be obtained by performing a double discrete

Fourier transform within time and space. For this process, a synthetic model based on the previous experience or existing bore log data at a particular location is taken as a theoretical dispersion curve, which can be varied as a trial-and-error procedure to fit the actual experimental curve. Generally, the surface wave dispersion is related to subsurface soil characteristics, as various frequencies represent different soil thicknesses and, consequently, moves at different velocities. Through a wave Field Fourier Transformation method (FFT), the dispersion properties of different waves have been obtained on both the body and surface with the multichannel approach.

Dispersion curves obtained from the geophones are essential steps in generating the necessary shear wave velocity profile. The dispersion curve for the specific sample site of Pottaseri is shown in Fig. 5.8. Once forming the experimental and theoretical dispersion curves, then the curves have to be inverted. It is an iterative process where the starting shear wave profile is compared with the theoretical curve established. The process is continued several times to attain the one with a minor error percentage. The fitted dispersion curve with the experimental curve after inversion is shown in Fig. 5.9.

The software applies a heuristic algorithm during the inversion process to determine the soil stratigraphy model that generates a dispersion curve as obtained from the survey experiment. From the model given by the user, the solver builds a group of compatible models and, for every one of them, compares the dispersion curve

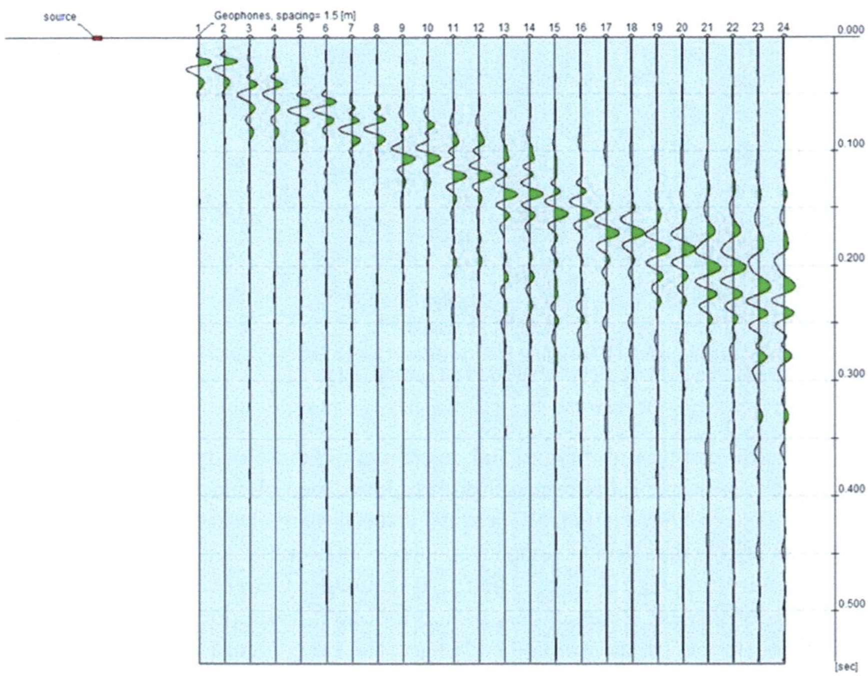

Fig. 5.7 Track records imported from 24 geophone channels

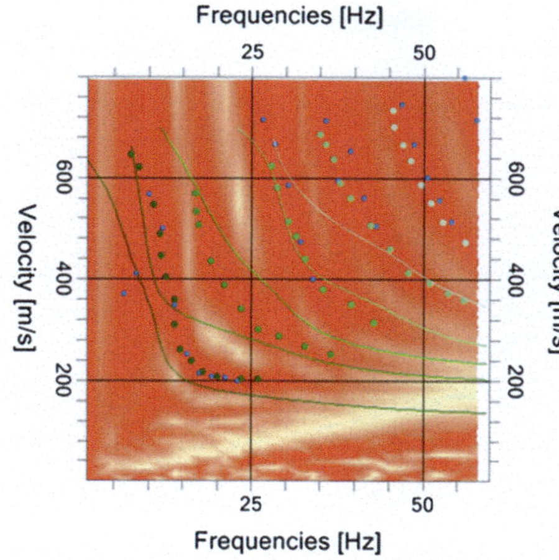

Fig. 5.8 Phase velocity frequency spectrum (dispersion curve)

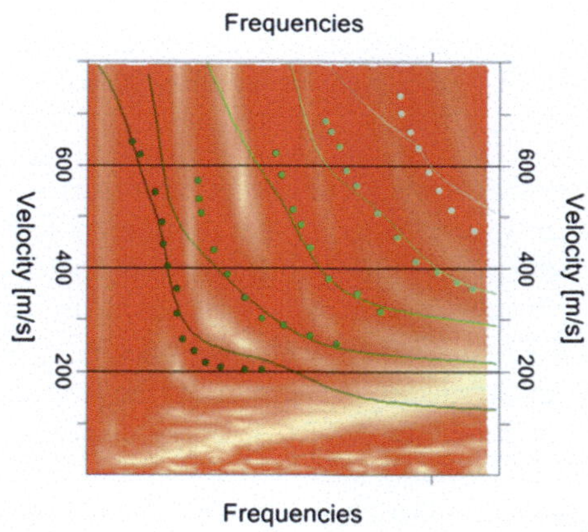

Fig. 5.9 Dispersion curve fit after inversion process

generated thereupon of the experimental test. The generation of models formulated to attenuate the value of an objective function that evaluates the model's results efficiency. It is possible to assess the performance of the target function's worth directly; this parameter may be an honest indicator for selecting the number of models to generate.

5.4.2 Shear Velocity Profiles and Site Classification

For a seismic site study, shear wave velocity is a deciding parameter in evaluating the dynamic behaviour of the soil. Shear wave velocities are obtained after processing the MASW testing data using testing and analysis software provided along with the facility. A typical one-dimensional shear wave velocity profile for one of the

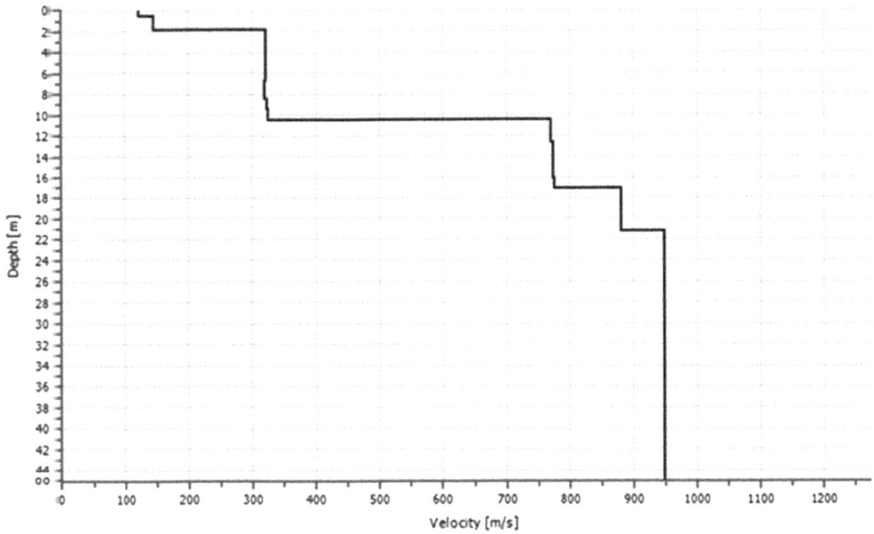

Fig. 5.10 Shear wave velocity profile obtained for representative site at Pottaseri

representative sample sites of Pottaseri is shown in Fig. 5.10. Similarly, average shear wave velocity profiles of different soil profiles for 48 sites were obtained after conducting a series of Field MASW testing are reported in Tables 5.3 and 5.4.

Shear wave velocities were derived from the corresponding dispersion spectra by inversion, and the shear wave velocity profiles for 48 sample sites were obtained. From the dispersion spectrum, the average shear wave velocities for the initial 30 m depth are deducted. Every dispersion curve is inverted to develop the 1 D shear wave velocities. Soil sites were categorized into three regions with locations labels MK (Mukkam), MR (Mavoor Road), PM (Poolakode-Malayamma Road), NC (NIT campus surroundings), CK (Chethukadavu), CM (Chathamanagalam), and KM (Kunnamangalam). Figure 5.11 shows the shear wave velocity distribution for the entire region covered with 48 sites.

From the interpretation of the data, for region I, the MK site had the average shear wave velocity in the range of 427.75–519.76 m/s. MR site the values were coming in between 273.47 and 357.01 m/s. Region II showed the values in the range of 454.98–559.27 m/s. The V_{s30} was in the range of 341.22–407.89 m/s for the site CM. In region III, CH and KM sites, the average shear wave velocity falls within the range of 491.14–615.18 m/s. Figure 5.12 shows the NEHRP based site classification map of the study area in the present study. It revealed that regions I and III belong to site classification D (stiff soil) out of the three distinguished regions within the study area. All the other areas belonged to site class C (very dense soil and soft rock) based on the NEHRP classification.

Table 5.3 Average shear wave velocities for various representative samples (Phase I)

Code	Site	Latitude	Longitude	V_{s30} (m/s)
MK-1	Mukham	N 11°19′16.32″	E 75°59′46.06″	489.94
MK-2	Agasthyamozhi	N 11°17′35.16″	E 75°59′9.86″	518.8
MK-3	Nairkuzhi	N 11°17′33.06″	E 75°58′23.06″	479.86
MK-4	Manasseri	N 11°19′1.973″	E 75°57′32.34″	514.22
MK-5	West Manaseri	N 11°19′8.46″	E 75°57′30.58″	519.76
MK-6	Pottaseri	N 11°18′58.87″	E 75°58′1.877″	475.88
MK-7	Chennamagalore	N 11°18′31.82″	E 75°57′43.41″	427.75
MR-1	Mavoor road-1	N 11°17′4.36″	E 75°55′49.06″	302.4
MR-2	Mavoor road-2	N 11°16′55.17″	E 75°55′29.27″	291.11
MR-3	Mavoor road-3	N 11°16′42.29″	E 75°54′59.29″	273.47
MR-4	Mavoor road-4	N 11°16′57.80″	E 75°55′56.23″	285.83
MR-5	Sivapuram	N 11°20′22.70″	E 75°59′29.38″	361.26
MR-6	Muthalam Road	N 11°20′17.38″	E 75°56′04.35″	357.01
MR-7	Pullavoor	N 11°20′31.81″	E 75°55′43.41″	320.68
MR-8	Thazhecode	N 11°20′5.49″	E 75°57′54.35″	318.59
MR-9	Chuloor	N 11°17′51.97″	E 75°56′25.21″	323.33
PCM-1	Balasadhan	N 11°18′35.77″	E 75°56′17.84″	313.47
PCM-2	Mini-Stadium	N 11°18′42.52″	E 75°56′34.37″	524.36
PCM-3	Malayamma-1	N 11°20′11.22″	E 75°57′20.78″	632.84
PCM-4	Malayamma-2	N 11°20′19.08″	E 75°58′09.88″	505.66
PCM-5	Temple road	N 11°18′35.93″	E 75°54′49.23″	524.49
NC-1	Spring valley	N 11°19′27.76″	E 75°56′14.56″	468.20
NC-2	Guest house	N 11°19′7.16″	E 75°55′54.623	487.30
NC-3	Mega LH	N 11°18′57.86″	E 75°55′50.862	512.80
NC-4	East campus	N 11°19′24.56″	E 75°56′16.79″	539.20
NC-5	Men's hostel	N 11°19′12.71″	E 75°56′15.13″	545.29
NC-6	Survey field	N 11°19′19.93″	E 75°56′2.812″	460.42

5.5 Conclusion

Shallow seismic geophysical testing is a powerful tool for the identification of subsurface soil layers. MASW acts as an effective tool that provides rapid and cost-efficient means for extracting the shear wave velocities from the Rayleigh waves for determining various properties that assist in engineering needs. This paper presents the MASW system of field surveying and its application in the geotechnical means of site classification. The study classified the southern part of the Calicut region based on the field measurement of the shear wave velocity on the topsoil deposits up to 30 m suggested by the NEHRP guidelines. The entire study area with 48 representative

Table 5.4 Average shear wave velocities for various representative samples (Phase II)

Code	Site	Latitude	Longitude	V_{s30} (m/s)
NC-7	Bhaskara hall	N 11°19′15.18″	E 75°56′1.03″	530.58
NC-8	Architecture block	N 11°19′20.34″	E 75°56′12.58″	454.98
NC-9	Water treat plant	N 11°18′54.88″	E 75°55′46.74″	591.21
NC-10	Water resources lab	N 11°19′23.11″	E 75°56′9.594″	583.31
NC-11	Geotechnical lab	N 11°19′20.76″	E 75°56′6.318″	460.69
NC-12	Admin. building	N 11°19′12.59″	E 75°55′48.64″	510.83
NC-13	SBI junction	N 11°19′31.06″	E 75°55′39.37″	559.27
CK-1	Manipuram	N 11°20′45.39″	E 75°53′40.78″	382.42
CK-2	12th mile	N 11°18′25.38″	E 75°54′40.31″	407.89
CK-3	Chethukadavu-1	N 11°18′28.96″	E 75°54′9.323″	334.65
CK-4	Chethukadavu-2	N 11°18′27.54″	E 75°54′20.16″	341.22
CM-1	Chathamangalam	N 11°17′37.03″	E 75°54′15.85″	508.23
CM-2	Poyyali	N 11°19′5.32″	E 75°54′37.35″	528.25
CM-3	Varittyakkal	N 11°18′35.54″	E 75°54′2.042″	530.45
CM-4	Chathankadavu	N 11°17′59.42″	E 75°55′49.73″	491.14
CM-5	Variyyakkal	N 11°17′55.16″	E 75°53′12.67″	524.47
KM-1	PWD guest house	N 11°18′21.62″	E 75°53′19.66″	558.86
KM-2	Peringalam	N 11°17′48.79″	E 75°52′44.37″	615.18
KM-3	Mundakkal	N 11°17′2.89″	E 75°53′1.314″	578.80
KM-4	Kurikathor	N11°17′10.66″	E 75°52′49.52″	553.49
KM-5	Peruvazhikadavu	N 11°17′58.6″	E 75°54′51.16″	608.18

locations came under site classification of C and D based on the NEHRP classification system. The study aid in assessing the ground-level response of these areas and, hence finding the vulnerable points within the region. The current study results can be further used for structural design under cyclic or earthquake loading scenarios.

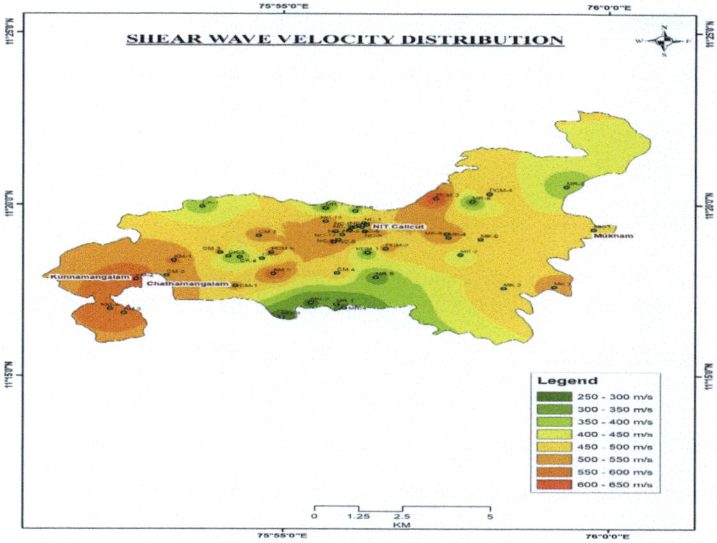

Fig. 5.11 Shear wave velocity distribution map of the study area

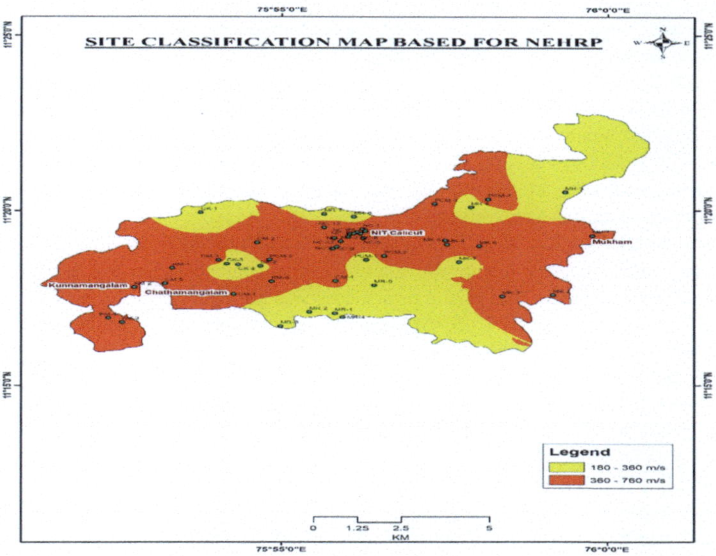

Fig. 5.12 Site classification map of the study area (NEHRP classification)

Acknowledgements G.C. acknowledges support from the program "POR Calabria FSE/FESR 2014/2020" related to her SIACE PhD at the University of Calabria.

All the authors thank the Geostru Company for having licensed the software EASYMASW.

References

1. Everett ME (2013) Near-surface applied geophysics. Cambridge University Press
2. Park CB, Miller RD, Xia J (1999) Multichannel analysis of surface waves. Geophysics 64(3):800–808
3. Anbazhagan P, Sitharam TG (2008) Mapping of average shear wave velocity for Bangalore region: a case study. J Environ Eng Geophys 13(2):69–84
4. Park CB, Miller RD, Xia J, Ivanov J (2007) Multichannel analysis of surface waves (MASW)—active and passive methods. Lead Edge 26:60–64
5. Xia J, Miller RD, Park CB (1999) Estimation of near surface shear-wave velocity by inversion of Rayleigh wave. Geophysics 64(3):691–700
6. Xia J, Miller RD, Xu Y, Luo Y, Chen C, Liu J, Ivanov J, Zeng C (2009) High frequency Rayleigh-wave method. J Earth Sci 20:563–579
7. Karl L, Haegeman W, Degrande G (2006) Determination of the material damping ratio and the shear wave velocity with the seismic cone penetration test. Soil Dyn Earthq Eng 26:1111–1126
8. NEHRP (2001) NEHRP recommended provisions for seismic regulations for new buildings and other structures (FEMA 368 and 369), 2000. Building Seismic Safety Council, National Institute of Building Sciences, Washington, DC
9. Joyner WB, Fumal TE (1984) Use of measured shear wave velocity for predicting geologic site effects on strong ground motion. In: Proceedings of the 8th world conference on earthquake engineering, San Francisco, CA, pp 777–783
10. Luna R, Jadi H (2000) Determination of dynamic soil properties using geophysical methods. In: First international conference on the application of geophysical and NDT methodologies to transportation facilities and infrastructure, pp 1–15
11. Mohamed AM, El Ata AA, Azim FA, Taha M (2013) Site-specific shear wave velocity investigation for geotechnical engineering applications using seismic refraction and 2D multi-channel analysis of surface waves. NRIAG J Astron Geophys 2:88–101
12. Akin MK, Kramer SL, Topal T (2016) Dynamic soil characterization and site response estimation for Erbaa, Tokat (Turkey). Natural Hazards Springer International Publishing, vol 82, pp 1833–1868
13. Olafsdottir EA, Erlingsson S, Bessason B (2018) Tool for analysis of multichannel analysis of surface waves (MASW) field data and evaluation of shear wave velocity profiles of soils. Can Geotech J 55(2): 217–233
14. Lin C-P, Lin C-H, Chien C-J (2017) Dispersion analysis of surface wave testing – SASW vs. MASW. J Appl Geophys
15. Sairam B, Singh AP, Vandana P, Sumer C, Ravi KM (2019) V_S30 mapping and site characterization in the seismically active intraplate region of Western India: implications for risk mitigation. Near Surf Geophys 17: 533–546
16. Abudeifa AM, Helbaryb RE, Mohammed MA, Alkhashaba HM, Masoud MM (2019) Geotechnical engineering evaluation of soil utilizing 2D multichannel analysis of surface waves (MASW) technique in New Akhmim city, Sohag, Upper Egypt. J Afr Earth Sci 157

Chapter 6
Frequency-dependent Dynamic Properties of Saturated Brahmaputra River Sand Based on Cyclic Triaxial Tests

Shiv Shankar Kumar, Arindam Dey, and A. Murali Krishna

Abstract This paper reports a comparative influence of the two different types of loading mechanisms on the frequency-dependent dynamic properties of the cohesionless Brahmaputra sandy soil. The shear modulus and damping ratio of the chosen soil are largely influenced by the loading frequencies of the two different testing types. Based on the strain-controlled and stress-controlled loading conditions, the loading frequency is found to influence the shear modulus and damping ratio with a contradicting response between the parameters. In case of strain-controlled tests, the damping ratio increases with a loading frequency of up to 1 Hz followed by a decrement for higher frequencies; while for stress-controlled testing, the shear modulus increases up to a loading frequency of 1 Hz followed by a decrement for higher frequencies. Further, it can be stated that, during strain-controlled loading, the frequency of loading is found to influence significantly the damping ratio only, whereas for stress-controlled loading, both the damping ratio and shear modulus are found to be affected significantly by the loading frequency.

Keywords Frequency of loading · Saturated sand · Strain-controlled tests · Stress-controlled tests · Cyclic triaxial tests · Shear modulus · Damping ratio

---

S. S. Kumar
Department of Civil Engineering, National Institute of Technology Patna, Patna 800005, India
e-mail: k.shiv.ce@nitp.ac.in

A. Dey (✉)
Department of Civil Engineering and Center for Disaster Management & Research (CDMR), Indian Institute of Technology Guwahati, Guwahati 781039, India
e-mail: arindamdeyiitg16@gmail.com

A. M. Krishna
Department of Civil and Environmental Engineering, Indian Institute of Technology Tirupati, Tirupati 517506, India
e-mail: amk@iittp.ac.in

© Indian Society of Earthquake Technology 2023

R. S. Jakka et al. (eds.), *Earthquake Engineering and Disaster Mitigation*, Springer Tracts in Civil Engineering, https://doi.org/10.1007/978-981-99-0081-7_6

6.1 Introduction

Seismic waves, characterized by time-variant strain energy and frequency content, lead to noticeable alteration in the temporal response of soils while propagating through the substrata. Governed by the dynamic characteristics of soil, namely strain-dependent shear modulus degradation and damping behaviour, the temporal response of soil can even lead to widespread destruction of the supported structures. Such destructions are mostly incurred due to the occurrence of induced resonance when the natural or predominant fundamental frequency of the structure is incited by the forcing frequency within the bandwidth of the seismic motion [1, 2]. A successful seismic resistant design depends on the proper estimation of the strain-dependent as well as frequency-dependent dynamic properties of the underlying soil. As the supported structures have different natural frequencies, exhibiting different modes of vibrations due to the variations in the stiffness and modal mass participations, the effect of earthquake frequency on the dynamic response of soil needs to be imperatively characterized for a better understanding of the behaviour of geotechnical structures under such circumstances [3]. Several researchers have investigated the effect of frequency on the dynamic response of soils and reported that higher stiffness is manifested in high-plasticity soils when subjected to higher frequencies of vibration, whereas the low-plasticity soils exhibited marginal dependency [4–7]. Stoll [8] observed that within a frequency bandwidth of 2–100 Hz, the dynamic properties of the saturated soils exhibited frequency dependency, while for the dry Ottawa sands, the response was found to be largely frequency independent. Although literature is available on the frequency-dependent response of soils, literature pertaining to the dynamic properties and response of soils subjected to strain-controlled and stress-controlled cyclic tests are literally scanty. This paper reports the frequency-dependent dynamic properties and dynamic response of sandy soil obtained from the Brahmaputra riverbed, Assam, when subjected to stress-controlled and strain-controlled loadings at different harmonic frequencies.

6.2 Previous Studies on Frequency Dependent Dynamic Properties of Soil

Considering that the seismic motions have a wide bandwidth that can significantly influence the substrata response as well as the response of the supported structures, several researchers have studied the response of soils at different loading frequencies. Bhattacharya [9] has reported that the predominant frequency range of earthquakes is 0.5–10 Hz, although there are studies that have considered the effects of much higher frequencies in the studies. Lee and Fitton [10] reported that a lower loading frequency would manifest slightly lower strength of the soil in comparison when the same is tested against a higher loading frequency. However, contrary to the abovementioned findings, Wang [11] and Wong et al. [12] reported that the lower loading frequency gives slightly higher strength. Hence, there exists a contradiction regarding the influence of frequency on the strength of the cyclically loaded soils. Based on the both resonant column and cyclic torsional shear tests, Lin et al. [13, 14] observed that the shear modulus was not significantly affected by the excitation frequency, while the damping ratio was significantly influenced. Researchers have also reported that the dynamic properties of cohesionless soil do not exhibit significant frequency dependency. However, the cohesive soils exhibited a moderate increase in shear modulus and a decrease in damping ratio with an increase in frequency up to 1 Hz [15–19]. Based on torsional shear and resonant column tests, Iwasaki et al. [16] reported that there are no significant variations between the shear modulus of dry and saturated sands within a frequency range of 0.1–100 Hz. Based on the resonant column tests, Bolton and Wilson [17] have also found that the dynamic properties of dry sand are almost independent of the frequency of loading within the range of 45–95 Hz. Based on cyclic Triaxial tests conducted on clay soil at a frequency range of 0.1–10 Hz, Teachavorasinskun et al. [20] reported that the shear modulus was not significantly affected by loading frequency, while the damping ratio was observed to slightly decrease with the increase in loading frequency. Similar observations have also been reported by other researchers [21, 22]. Yilmaz et al. [23] have performed cyclic triaxial tests on silt–clay mixtures with the aid of a cyclic axial load over a frequency range of 0.5–1.0 Hz to reflect the influence of viscosity on the dynamic soil properties. Wichtmann et al. [24] have reported that there is an insignificant change in the dynamic properties of sand when excited within a frequency range of 0.1–2 Hz. Govindaraju [25] has reported that the shear modulus of sand is not significantly affected by frequency, whereas the damping ratio is only marginally affected.

There are quite some literature where the stress-controlled cyclic triaxial tests are adopted to identify the dynamic properties of soils when subjected to specific frequency and ranges. Wong et al. [12] have conducted stress-controlled cyclic triaxial tests on the reconstituted specimens of Monterey zero sand at a relative density of 60% to study the effect of frequency ($f = 0.017$ and 0.33 Hz) on its liquefaction potential. It was observed that the liquefaction of saturated sand was negligibly affected by the loading frequency. Similar observations were reported by Wang and Kavazanjian [26] based on stress-controlled cyclic triaxial tests at frequencies of 0.25 and 1.0 Hz. However, many other researchers have reported that the loading frequency has a marginal or negligible effect on the liquefaction behaviour of soil [27–29]. Rascol [30] has performed stress-controlled cyclic triaxial tests on saturated sand at different frequencies ($f = 0.2$, 0.8, 1.6 and 6.4 Hz) and, it was reported that in comparison to the lower frequencies $f = 0.2$ and 0.8 Hz, the shear modulus is higher at $f = 1.6$ and 6.4 Hz due to less PWP development. Choudhary et al. [31] have performed stress-controlled undrained cyclic triaxial tests to evaluate the liquefaction resistance of Solani sand and reported that at higher frequencies of excitation, the chances of liquefaction increases. There exists a debate about the influence of cyclic frequency on the dynamic properties of saturated clay. Some researchers have indicated that there is a significant increment in pore water pressure (PWP) along with a decrement in axial strain with the increase in cyclic frequency [32, 33], while others have reported that the pore water pressure and the axial strain are independent of the frequency [34]. Chen et al. [35] reported that the cyclic frequency had a strong effect on the pore water pressure and axial strain, although the dynamic strength of soil is found negligibly affected by the cyclic frequency.

Strain-controlled Triaxial tests have also been in use to understand the frequency-dependent dynamic parameters and dynamic response of soils. Govindaraju [25] has studied the effect of loading frequency ($f = 0.2$–3.0 Hz) on the liquefaction potential of soils through strain-controlled cyclic tests and reported that the excess PWP generation causing the liquefaction phenomenon is independent of the frequency. Zhang et al. [36] have performed dynamic triaxial tests on saturated cohesive soil at different frequencies ($f = 0.1$, 1.0, 4.0 and 6.0 Hz), and it was found that with increasing frequency from 0.1 to 6.0 Hz, the dynamic intensity of cyclic loading increases. It was also found that the PWP ratio increases in the frequency range 0.1–1.0 Hz, while the same decreases in the frequency range 1.0–6.0 Hz. At the same time, both the shear modulus and damping ratio were found to increase with the increase in frequency in the entire range of 0.1–6.0 Hz. Dash and Sitharam [37, 38] have reported that the rate of excess PWP generation increases with the increase in frequency and magnitude of loading. The effect of cyclic loading frequency on the dynamic properties of saturated soft marine clay has also been studied by Jiang et al. [39]. It was reported that the accumulative axial strain and normalized average PWP with the number of cycles are strongly affected by cyclic frequency, but the effect of frequency on the accumulated axial strain is not significant. Khan et al. [40] have reported that the shear modulus and damping ratio obtained from resonant column tests increases with frequency, whereas the same obtained from cyclic triaxial tests are negligibly affected by the frequency. Lombardi et al. [41] have used a frequency of 0.1 Hz for the loading cycle with the assumption of the minimal effect of viscosity on the cyclic behaviour of soils. Chattaraj and Sengupta [42] have evaluated the strain-dependent dynamic properties of soil along with the liquefaction potential of sandy soil, considering the frequency of 1 Hz, for the assessments of geotechnical problems involving dynamic loading. Dash and Sitharam [43] have reported that the rate of generation of excess PWP increases with an increase in frequency, thereby leading to the reduction in the liquefaction resistance of saturated sand. It was also found that the shear modulus decreases and the damping ratio increases with an increase in frequency.

It is understood that different natural and anthropogenic events are bound by different frequency bandwidths. Table 6.1 lists the frequency ranges of uniform sinusoidal loadings mostly used in a cyclic triaxial test to approximate different dynamic loadings.

Although there are several literature and studies are available on the response of soils under different frequency contents, conclusive observations were not found because of many contradictory observations, remarks and inferences. Therefore, this study is focused on the behaviour of saturated sand under strain-controlled and stress-controlled cyclic loading at different frequency contents.

Table 6.1 Typical frequency range for cyclic loading [44]

Loading type	Typical test frequency (Hz)
Wave action	0.1
Wind action	0.1–1
Earthquake	1
Rail transit	>1
Vibrating machinery	≤20

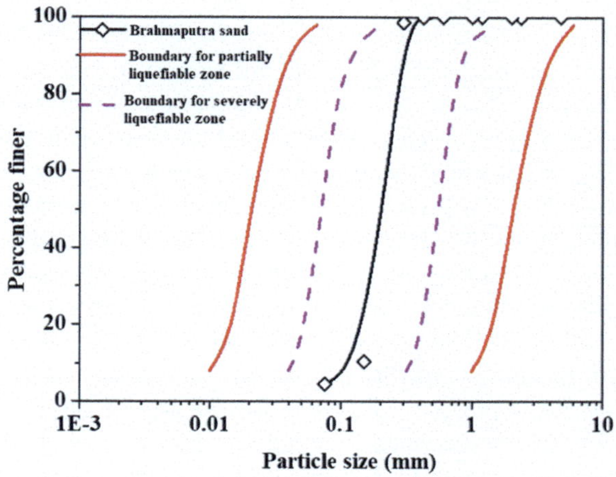

Fig. 6.1 Particle size distribution of the Brahmaputra sand along with the limiting boundaries for the liquefiable zones [as per 47]

6.3 Experimental Investigations

6.3.1 Properties of Material Used

In the present study, the dynamic properties of the cohesionless Brahmaputra Sand (BS), collected from the Guwahati region (Assam, India), have been evaluated under varying frequency content. Figure 6.1 depicts the particle size distribution of BS [45], which has approximately 5% fine content. According to the standard literature [46, 47], the particle size distribution (PSD) falls within the zone of 'severely liquefiable soil'. The index and physical properties of BS were determined as per the relevant standards [48–51], which are listed in Table 6.2. As per the Unified Soil Classification System [52], BS is classified as poorly graded sand (SP).

6.3.2 Sample Preparation for Triaxial Tests

Several sample preparation techniques, such as dry pluviation, moist tamping, and water sedimentation, are commonly in practice for the preparation of reconstituted

Table 6.2 Physical properties of Brahmaputra Sand

Physical properties	Values
Mean grain size, D_{50} (mm)	0.21
Min. dry unit weight (kN/m^3)	13.85
Max. dry unit weight (kN/m^3)	16.84
Max. void ratio (e_{max})	0.91
Min. void ratio (e_{min})	0.57
Specific gravity (G_s)	2.7

sand specimens [53–56]. The different available specimen preparation techniques provide different characteristics at the same relative density of the specimen owing to the varying orientation of reconstituted soil matrix and fabric structures [57]. For the present study, the dry pluviation technique, along with the use of CO_2 flushing, was adopted to prepare all the specimens for experimental testing [58]. Reconstituted solid cylindrical specimens of diameter 70 mm and height 140 mm were prepared for all the tests [53]. The required amount of sand, based on the relative density, was poured in three layers through a cone-shaped funnel into the specimen-forming mould [59]. The prepared specimen was then subjected to flushing with CO_2 for 10–15 min to expedite the saturation process, at a pressure lower than the applied initial cell pressure [59, 60]. Subsequently, de-aired water was allowed through the CO_2 flushed specimen. Thereafter, the cell pressure (CP) and backpressure (BP) were gradually increased in stages, while maintaining a constant differential pressure of 10 kPa [61]. After each increment of CP, Skempton's B-value was estimated to check the status of saturation. The specimen was assumed to be completely saturated when the achieved B-value was greater than 0.96. Subsequently, the saturated specimens were consolidated to achieve the desired effective stress and then they were subjected to cyclic loading.

6.3.3 *Triaxial Testing Apparatus and Testing Procedure*

An automated pneumatic controlled cyclic triaxial testing apparatus was used to perform the experimental investigations at different testing conditions. The dynamic loading on the soil specimen was applied using a dynamic actuator, having a displacement capacity of 30 mm and an operational frequency range of 0.01–10 Hz. The load and displacement on the soil specimens were measured from a submersible load cell and linear variable displacement transducer, respectively. Cyclic triaxial tests have been performed under strain-controlled and stress-controlled loading. For strain-controlled cyclic triaxial test, firstly, the desired single amplitude shear strain (γ_{SA}) was selected, as shown in Table 6.3, and then, required axial strains (ε_{SA}) were calculated using Eq. 6.1. The resulting single amplitude deformation employed on the soil specimen was calculated using Eq. 6.2.

Table 6.3 Investigation parameters for strain-controlled and stress-controlled loadings

Parameters	Strain-controlled	Stress-controlled
D_r (%)	60	60
σ'_c (kPa)	50, 100, 150	100
f (Hz)	0.5, 1, 2, 3, 4	0.1, 0.5, 1, 2, 4
γ (%)	0.15, 0.6, 1.0, 1.5, 3.0, 4.5	NA
CSR	NA	0.1, 0.2, 0.3, 0.4

$$\gamma_{SA} = \varepsilon_{SA}(1 + \nu) \Rightarrow \varepsilon_{SA} = \frac{\gamma_{SA}}{(1+\nu)} \tag{6.1}$$

$$\varepsilon_{SA} = L_{SA}/L_S \Rightarrow L_{SA} = \varepsilon_{SA} \times L_S \tag{6.2}$$

where ν = Poisson ratio, L_{SA} = single amplitude deformation (mm), L_s = length of test specimen (mm) and ε_{SA} = single amplitude axial strain (dimensionless). For this study, $\nu = 0.5$ is considered [62]. For the stress-controlled cyclic triaxial test, the magnitude of cyclic load (shown in Table 6.3) was estimated based on the desired cyclic stress ratio (CSR_{desired}). CSR_{desired} is the ratio of the desired deviator stress to twice the effective stress. The magnitude of the cyclic load to be employed on the soil specimens was estimated using Eq. 6.3.

$$P_c = 2 \times \sigma'_c \times CSR_{\text{desired}} \times A_c \tag{6.3}$$

where P_c = estimated cyclic load to be applied to the specimen, σ'_c = targeted effective stress, CSR_{desired} = desired cyclic stress ratio ($\pm\sigma_d/2\sigma'_c$) and A_c = area of the specimen after consolidation.

6.4 Strain and Stress-controlled Cyclic Loading: Input and Output

Strain-controlled as well as stress-controlled cyclic shear tests were performed on isotropically consolidated sand specimens. A typical set of plots of the input and output of the strain-controlled test, carried out on sandy soils, are presented in Fig. 6.2. For the strain-controlled loading input (axial strain = 0.10%, i.e., shear strain γ = 0.15%, f = 0.5 Hz, Number of cycles = 40, σ'_c = 100 kPa, D_r = 60%) shown in Fig. 6.2a, Fig. 6.2b presents the stress–strain variations during cyclic loading, Fig. 6.2c represents the exponential decay of deviatoric stress due to the generation of excess PWP as shown in Fig. 6.2d. It is to be noted in this case that Fig. 6.2d is a typical representation of the generated excess PWP, which in this case, has not reached the value of 1.0 (or, that the soil has not liquefied) solely due to the low strain amplitude considered for this typical representation. Similar to the strain-controlled

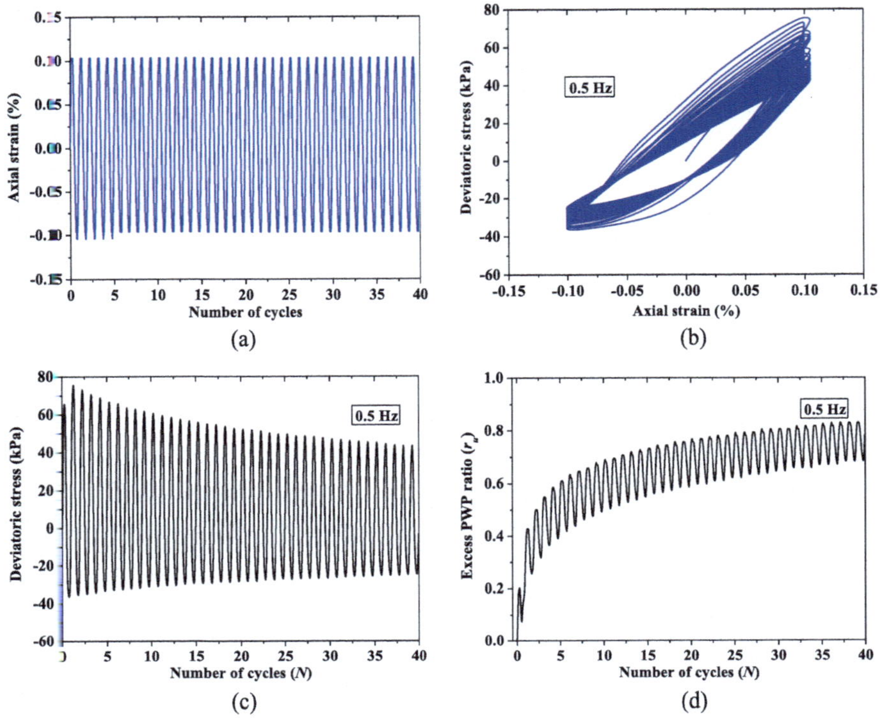

Fig. 6.2 **a–d** A typical set of plots representing the input and output of the strain-controlled loading tests conducted at $\gamma = 0.15\%$, $\sigma'_c = 100$ kPa, $D_r = 60\%$, $f = 0.5$ Hz

tests, stress-controlled tests were also performed at different investigating parameters listed in Table 6.3.

A typical set of plots of input and output of stress-controlled test are depicted in Fig. 6.3. For the controlled stress input (CSR = 0.2, $\sigma'_c = 100$ kPa, $f = 0.5$ Hz, Number of cycles = 40, $D_r = 60\%$) as shown in Fig. 6.3a, Fig. 6.3b presents the stress–strain variations during cyclic loading. Figure 6.3c represents the variations in shear strain, whereas Fig. 6.3d shows the generation of excess PWP. Both the strain-controlled and stress-controlled cyclic tests were performed at varying frequencies, ranging from 0.5–4 Hz. However, for the sake of brevity, only typical plots are presented in Figs. 6.2 and 6.3, corresponding to only one chosen frequency. For the calculation of the dynamic soil properties (shear modulus and damping ratio), the asymmetric hysteretic stress–strain plot (shown in Fig. 6.4) as obtained from the first cycle of the loading was used. The detailed methodology for evaluating the shear modulus and damping ratio out of the asymmetric hysteretic plot is described in Kumar et al. [59, 63].

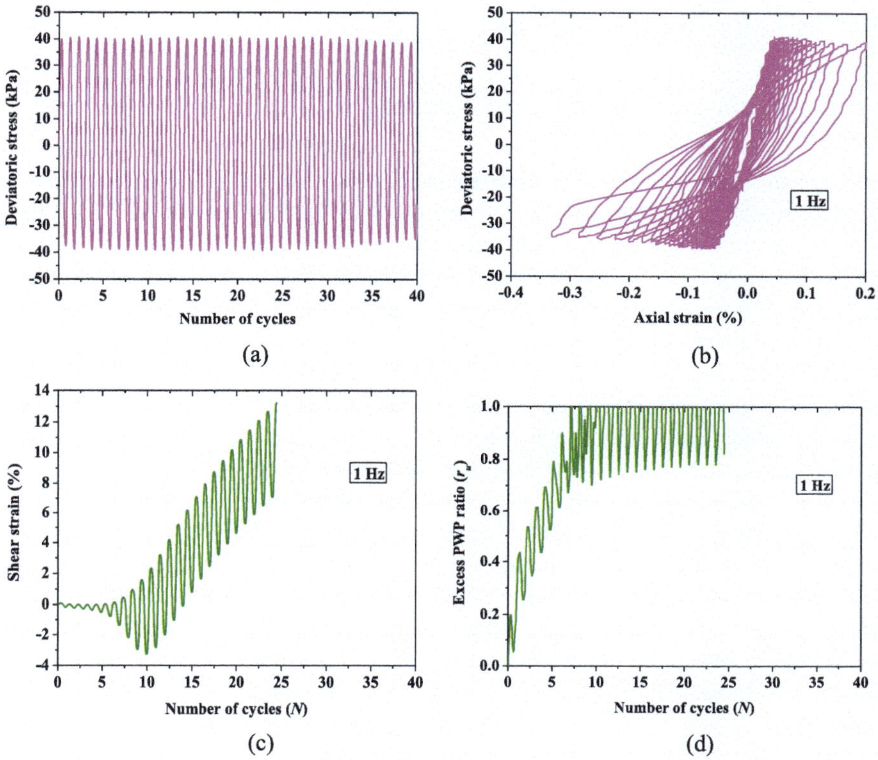

Fig. 6.3 a–d A typical set of plots representing the input and output of the stress-controlled loading tests conducted at CSR = 0.2, $\sigma'_c = 100$ kPa, $D_r = 60\%, f = 1$ Hz

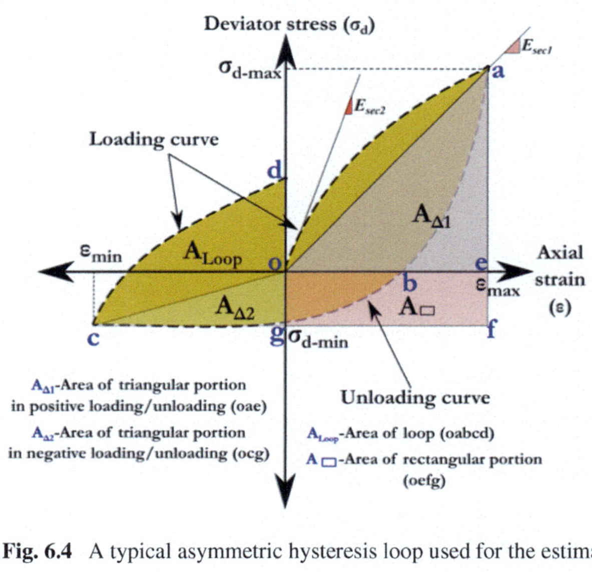

$$E_{\text{sec},a} = \frac{E_{\text{sec}1} + E_{\text{sec}2}}{2}$$

$$G_a = E_{\text{sec},a} / [2(1+\upsilon)]$$

$$\gamma = (1+\upsilon)\varepsilon$$

$$D^{\#} = \frac{1}{\pi} \times \frac{A_{L(o-a-b-c-d)}}{A_{\Delta 1} + A_{\Delta 2} + A_{\text{X}}}$$

Fig. 6.4 A typical asymmetric hysteresis loop used for the estimation of dynamic properties

6.5 Results and Discussions

6.5.1 Strain-controlled Cyclic Loading

Based on the strain-controlled cyclic loading tests, the effects of loading frequency on the shear modulus and damping ratio are presented in Fig. 6.5. Figure 6.5a depicts that at higher confining pressures, as obvious, a higher shear modulus is obtained, while the loading frequency is not observed to cast a prominent influence on the magnitude of dynamic shear modulus. Since these shear modulus values are presented for $N = 1$, there might be a possibility of negligible variations in the pore pressure at $N = 1$ for all frequencies. Figure 6.5b, however, presents a significant effect of frequency on the magnitude of damping ratio. For a particular peak shear strain ($\gamma = 0.15\%$), from the trend line plot, it can be observed that the increase in frequency results in a noticeable decrement in the damping ratio; however, the scatter data shows an increase in the damping ratio up to 1 Hz followed by a decrease with increasing loading frequency. This decrease in the damping ratio is attributed to the increasing intensity of the inter-particle interaction with increasing loading frequency, which leads to the higher dissipation of energy within a faster time interval, thereby causing less amount of PWP generation. Thus, it can be noted that BS specimens when subjected to a sinusoidal stress history up to $f = 1$ Hz, the maximum possible viscous effect is exhibited on the dynamic properties, which arises due to pore water pressure generation.

Figure 6.6 presents the variations in excess PWP ratio (r_u) at different frequencies along with the number of cycles (N). Figure 6.6a illustrates the variations in r_u at different loading frequencies with a continuous increase of loading cycles (i.e., $N = 0$–40). It is exhibited that for all magnitudes of confining pressure (σ'_c), the rate of increase in r_u increases up to a loading frequency of 2 Hz, beyond which the rate of generation of r_u decreases. Figure 6.6b presents the obtained r_u in the first

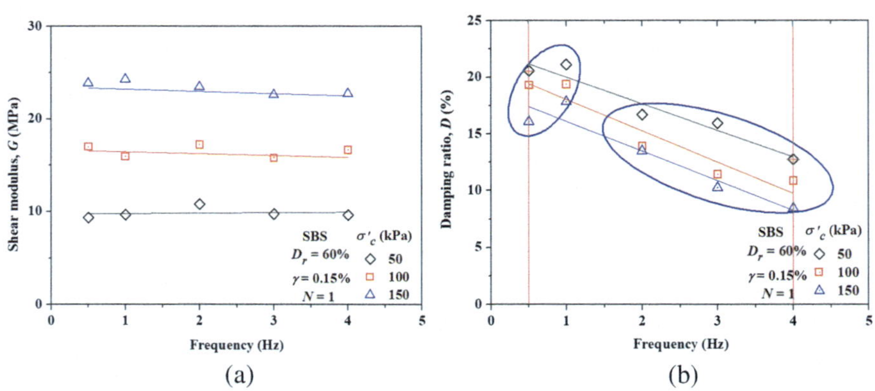

Fig. 6.5 Typical variations in **a** shear modulus **b** damping ratio with loading frequency

cycle ($N = 1$) at different loading frequencies, which reflects that with the increase of loading frequency, there is a distinct reduction in r_u, thereby influencing the dynamic properties of the saturated soil.

Figure 6.7 illustrates the variations in shear modulus and damping ratio with shear strain for the saturated BS subjected to an effective confining stress $\sigma'_c = 100$ kPa and a loading frequency ranging between 0.5–4.0 Hz. It can be observed that the effect of loading frequency on the shear modulus is insignificant for a particular σ'_c (Fig. 6.7a), whereas the damping ratio is found to be recognizably decreased with the increase in frequency (Fig. 6.7b). This significant change in damping ratio is attributed to the relatively lower value of r_u, at $N = 1$ for higher frequency in comparison to the same obtained for lesser frequencies. Figure 6.7b also reflects that the damping ratio beyond 1% of shear strain decreases for all loading frequencies at all effective confining stresses.

Figure 6.8 presents the shear modulus and damping ratio with shear strain for $f = 0.5$–4.0 Hz for $\sigma'_c = 50$–100 kPa. It is found that the shear modulus is negligibly

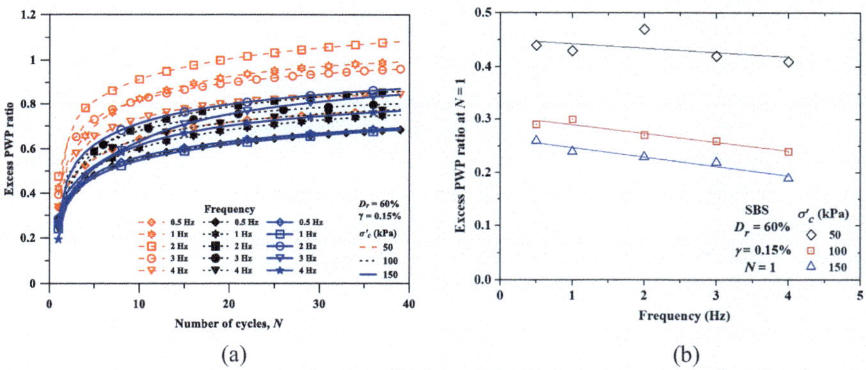

Fig. 6.6 Variation of r_u **a** with N at different f **b** with f at $N = 1$

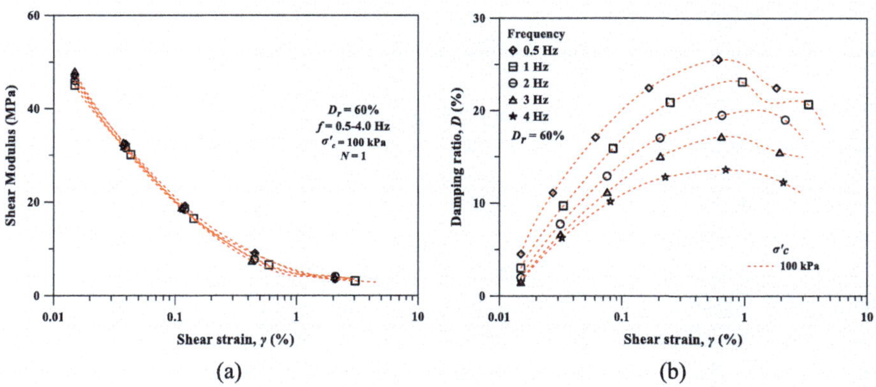

Fig. 6.7 Variation of **a** G **b** D with γ for $f = 0.5$–4.0 Hz and $\sigma'_c = 100$ kPa

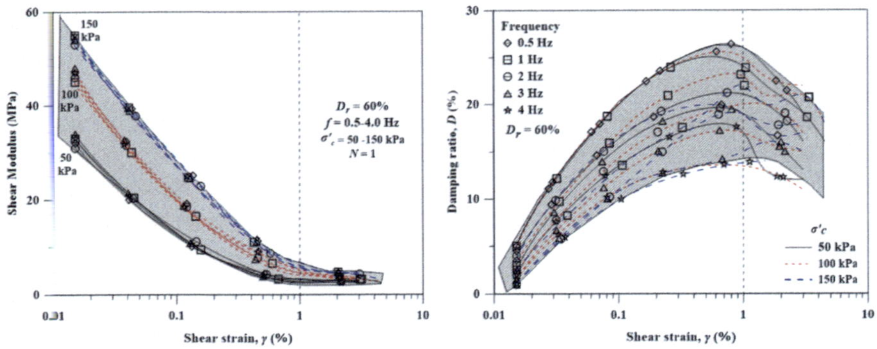

(a) (b)

Fig. 6.8 Variation of **a** G **b** D with γ for $f = 0.5$–4.0 Hz and $\sigma'_c = 50$–150 kPa

affected by loading frequencies at all the tested ranges of σ'_c (Fig. 6.8a), whereas the damping ratio is significantly affected by the loading frequencies at all the tested ranges of σ'_c (Fig. 6.8b). Considering all the above results, the range of strain-dependent shear modulus and damping ratio for the Brahmaputra Sand, examined at various confining pressures and loading frequencies, can be recognized with the aid of a limiting band. It can be noted that for shear strains beyond 1%, the shear modulus remains primarily unaffected for any magnitudes of confining pressure and loading frequency, while the damping ratio offers a significant dip from its peak value under all circumstances of loading frequencies and confining pressure, thereby indicating a general trend in the frequency-dependent response of the BS soil.

6.5.2 *Stress-controlled Cyclic Loading*

Stress-controlled cyclic triaxial tests have also been conducted to evaluate the effects of loading frequency on the shear modulus and damping ratio. Figure 6.9 depicts the results obtained at different loading frequencies, ranging from 0.1–4 Hz, for the specimens tested at $CSR = 0.2$, $D_r = 60\%$ and $\sigma'_c = 100$ kPa. Figure 6.9a illustrates that the shear modulus and damping ratio are significantly affected by loading frequency. It is also seen that the shear modulus increases up to a frequency of 1 Hz and decreases thereafter, whereas the damping ratio followed a decreasing trend. Figure 6.9b presents the variations of r_u and γ at $N = 1$ for different loading frequencies, which helps to indicatively illustrate the effect of r_u on the shear modulus as well as on the damping ratio. It can be clearly seen that r_u and γ for $f = 0.1$ Hz are significantly higher than that observed for other higher frequencies, thereby resulting in a lower shear modulus and higher damping ratio (as shown in Fig. 6.9a). The results reported in Fig. 6.9a shows contradiction to the results reported by Dash et al. [43], while they are in parity with those reported by Rascol [30]. The shear modulus at $f = 1$ Hz is higher, however, the corresponding r_u and γ values are lower in comparison to that obtained for other loading frequencies. Figure 6.9c describes that the rate of generation of excess PWP is significantly affected by the frequency of cyclic loading. It was found that in comparison to various loading frequencies, at the loading frequency of 1 Hz, a lesser excess PWP generation is manifested up to 10 cycles. It may be noted that the generation of excess pore water pressure ratio depends on the generation of shear strain at the first loading cycle (see Fig. 6.9b). It may be noted that the development of shear strain at first loading cycles corresponding to $f = 1$ Hz is relatively lesser than the other frequencies. Furthermore, the continuous accumulation of shear strain and pore water pressure development within the soil specimens depends on the initial value of shear strain and excess pore water pressure ratio, respectively. Consequently, the developed excess PWP for $f = 1$ Hz was found to be lesser compared to the lower and higher frequency ranges.

Figure 6.10a presents the variations in r_u with N at different CSR. At $CSR = 0.1$, the magnitude of r_u is nearly 0.1 at the end of 40 cycles, which is attributed to the development of very low shear strain (almost constant $\gamma = 0.02\%$). In the stress-controlled tests, a value of cyclic stress ratio ($CSR = \sigma_d/2\sigma'_c$) is first decided for the test and then based on the chosen confining stress, the deviatoric stress (σ_d)

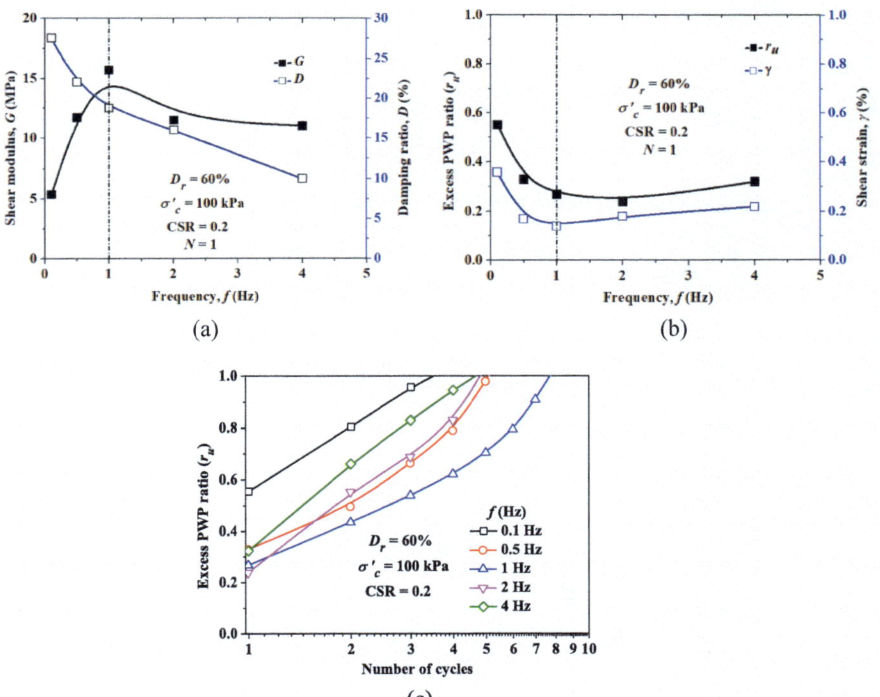

Fig. 6.9 Variations of **a** shear modulus and damping ratio with frequency at $N = 1$ **b** r_u and shear strain with frequency at $N = 1$ **c** r_u with number of cycles at different frequencies

to be employed on the specimen is estimated. For a constant σ'_c (=100 kPa), σ_d on the specimens increases with the increasing *CSR*. As an outcome of the test, the accumulation of both shear strains and excess PWP is observed with the number of cycles in the stress-controlled tests; however, the accumulation of shear strain is absent in strain-controlled tests although there is an accumulation in excess PWP. At $CSR \geq 0.2$, significant variation in r_u was observed during cyclic loading. The excess PWP ratio (r_u) reached 1, at 4th, 7th and 28th number of loading cycles when γ was nearly ~1.0%, 1.75% and 2.4% for *CSR* values of 0.2, 0.3 and 0.4, respectively. Figure 6.10b presents the variations in γ with N at different *CSR*. It is seen that at $CSR = 0.1$, the amount of shear strain until 40 cycles are substantially less, i.e., less than or equal to 0.02%, whereas γ was observed nearly 1.0%, 1.75% and 2.4% for *CSR* values of 0.2, 0.3 and 0.4, respectively. Figure 6.10c presents the development of r_u along with the accumulation of shear strain. It shows that r_u increases with the increase in γ, and for a constant D_r and σ'_c, higher shear strain results in a higher value of r_u. It is worth mentioning that the accumulation of γ and development of r_u depends on the soil type, stress applied or tests conditions.

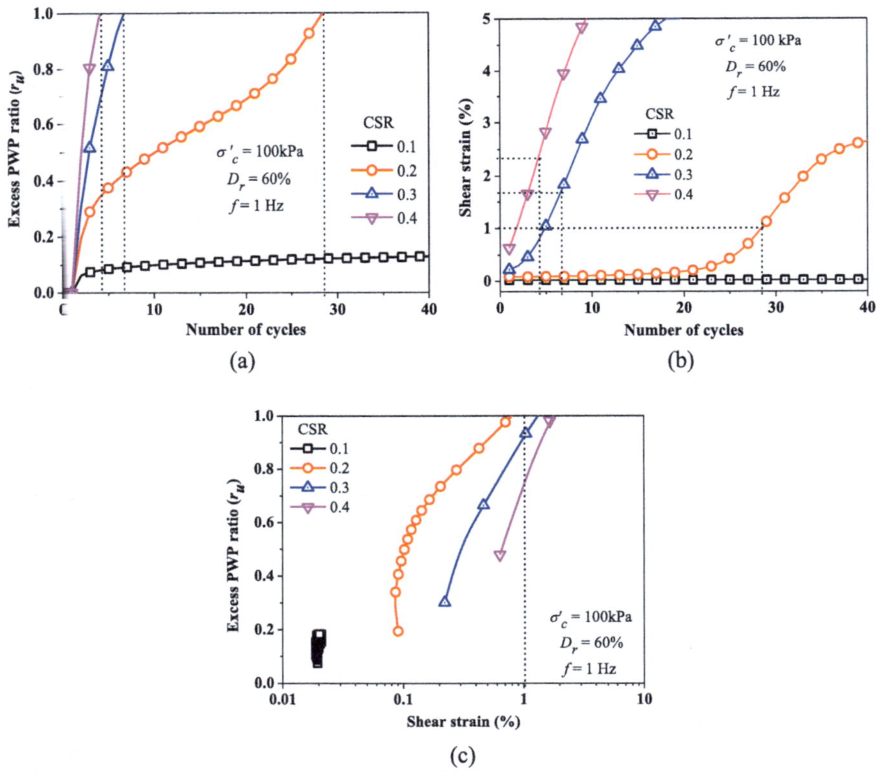

Fig. 6.10 Variation of **a** r_u with N **b** γ with N **c** r_u with γ, at different *CSR*

6.6 Summary and Conclusions

Although literature is available on the dependency of frequency over the response of soils, the comparative study on the responses of soil under both strain-controlled and stress-controlled loads are scanty. Previous literature available on the frequency-dependent dynamic response of soil shows contradictory observations. This paper presents the dynamic behaviour of saturated sandy soil (Brahmaputra sand) subjected to both strain-controlled and stress-controlled cyclic loading at different frequency contents.

For the strain-controlled tests, when the samples are subjected to the same effective confining pressure, the deviatoric stress (σ_d) was found to increased with the increase in the shear strains at the first loading cycle. Furthermore, a continuous decrease of σ_d is noted with an increasing number of cycles. Therefore, it is concluded that the excess pore water pressure ratio, obtained at the first cycle, depends on both γ for the same σ'_c. In the stress-controlled tests, for a constant *CSR*, σ_d applied on the specimens increases with the increase of σ'_c. Therefore, corresponding to a particular σ'_c, r_u depends on the shear strain generated at the first cycle. For the stress-controlled tests, the developed shear strain mainly depends on *CSR* or σ'_c, and it may not be necessarily the same as that of strain-controlled tests. The accumulation of shear strains with the number of cycles was also observed in stress-controlled approach. In comparison to the strain-controlled tests, the information of the shear strain accumulation with a number of loading cycles obtained for a stress-controlled approach is a contributory factor to understand the initiation of soil liquefaction.

From both the strain-controlled and stress-controlled methodologies, it can be concluded that the effect of loading frequency on the response of soil depends largely on the loading conditions since the dynamic properties of saturated soil depend intricately on the generation of excess PWP ratio. In case of strain-controlled tests, the shear modulus is meagrely affected by the loading frequency, whereas the damping ratio is significantly affected in the tested range of shear strain. In case of stress-controlled tests, both the shear modulus, as well as damping ratio are significantly affected by the loading frequency. In strain-controlled loading conditions, the damping ratio increases with frequencies of loading up to 1 Hz followed by a decrement for higher frequencies; while for a stress-controlled loading condition, the shear modulus increases up to a loading frequency of 1 Hz followed by a decrement for higher frequencies. It is understood that the two different types of loading conditions (strain-controlled and stress-controlled) would generate variability in the failure mechanisms of the soil specimen. Hence, the choice of any of these loading processes to evaluate the frequency-dependent dynamic properties or responses of the cohesionless soil should be solely based on the site conditions and the design characteristics of geotechnical systems.

References

1. Sandoval E, Bobet A (2017) Effect of frequency and flexibility ratio on the seismic response of deep tunnels. Undergr Space 2:125–133
2. Pap ZB, Kollar LP (2018) Effect of resonance in soil-structure interaction for Finite soil layers. Periodica Polytech Civil Eng. https://doi.org/10.3311/PPci.11960
3. Wu H (2015) Effect of ground motion frequency on non-structural seismic damage. PhD Thesis, Department of Civil and Environmental Engineering, Massachusetts Institute of Technology, USA
4. Vucetic M, Dobry R (1991) Effect of soil plasticity on cyclic response. J Geotech Eng ASCE 117(1):89–107
5. Shibuya S, Mitachi T, Fukuda F, Degoshi T (1995) Strain rate effects on shear modulus and damping of normally consolidated clay. Geotech Testing J ASTM 18(3):365–375
6. Malagnini L (1996) Velocity and attenuation structure of very shallow soils: evidence for a frequency-dependent Q. Bull Seismol Soc Am 86(5):1471–1486
7. Lo Presti DCF, Jamiolkowski M, Pallara O, Cavallaro A (1996) "Rate and creep effect on the stiffness of soils. ASCE Geotech Spec Publ 61:166–180
8. Stoll RD (1978) Damping in saturated soil. In from Volume-I of Earthquake engineering and soil dynamics. In: Proceedings ASCE geotechnical engineering division specialty conference, Pasadena, California
9. Bhattacharya S (2007) Design and foundations in seismic areas: principles and applications. National Information Centre of Earthquake Engineering, IIT Kanpur, India
10. Lee KL, Fitton JA (1969) Factors affecting the cyclic loading strength of soil. In: Proceedings of vibration effects of earthquakes on soils and foundations, ASTM, vol 450, pp 71–95
11. Wang MS (1972) Liquefaction of triaxial sand samples under different frequencies of cyclic loading. ME Thesis, University of Western Ontario, Canada
12. Wong RT, Seed HB, Chan CK (1975) Cyclic loading liquefaction of gravelly soils. J Geotech Eng ASCE 101(6):571–583
13. Lin ML, Huang TH, You JC (1996) The effects of frequency on damping properties of sand. Soil Dyn Earthq Eng 15(4):269–278
14. Lin ML, Ni SH, Wright SG, Stokoe II KH (1988) Characterization of material damping in soil. In: Proceedings of 9th world conference on earthquake engineering, vol 3, pp 5–10
15. D'Onofrio A, Silvestri F, Vinale F (1999) Strain rate dependent behavior of a natural stiff clay. Soils Found 39(2):69–82
16. Iwasaki T, Tatsuoka F, Takagi Y (1978) Shear modulus of sands under cyclic torsional shear loading. Soils Found 18(1):39–56
17. Bolton MD, Wilson JMR (1989) An experimental and theoretical comparison between static and dynamic torsional soil tests. Geotechnique 39(4):585–599
18. Kim DS, Stokoe KH (1995) Deformational characteristics of soils at small to medium strains. Earthquake Geotechnical Engineering, Tokyo, Japan, pp 89–94
19. Li XS, Yang WL, Shen CK, Wang WC (1998) Energy injecting virtual mass resonant column system. J Geotech Geoenviron Eng ACSE 124(5):428–438
20. Teachavorasinskun S, Thongchim P, Lukkunaprasit P (2002) Shear modulus and damping of soft Bangkok clays. Can Geotech J 39:1201–1208
21. Ravishankar BV, Sitharam TG, Govinda Raju L (2005) Dynamic properties of Ahmedabad sands at large strains. In: Proceedings of Indian geotechnical conference, Ahmedabad, pp 17–19
22. Maheshwari B, Kale S, Kaynia A (2012) Dynamic properties of Solani sand at large strains: a parametric study. Int J Geotech Eng 6(3):353–358
23. Yilmaz MT, Pekcan O, Bakir BS (2004) Undrained cyclic shear and deformation behaviour of silt-clay mixtures of Adapazari, Turkey. Soil Dyn Earthq Eng 24:497–507
24. Wichtmann T, Niemunis A, Triantafyllidis T (2005) Strain accumulation in sand due to cyclic loading: drained triaxial tests. Soil Dyn Earthq Eng 25:967–979

25. Govindaraju L (2005) Liquefaction and dynamic properties of sandy soils. Doctoral dissertation, Indian Institute of Science Bangalore, India
26. Wang JN, Kavazanjian E (1989) Pore pressure development during non-uniform cyclic loading. Soils Found 29(2):1–14
27. Finn WD, Pickering DJ, Bransby PL (1971) Sand liquefaction in triaxial and simple shear tests. J Soil Mech Found Div ASCE SM4: 639–659
28. Boulanger RB, Seed RB, Chan CK, Seed HB, Sousa J (1991) Liquefaction behavior of saturated sands under uni-directional and bi-directional monotonic and cyclic simple shear loading. Research Report No. UCB/GT-91–08, Department of Civil Engineering, University of California at Berkeley, USA
29. Polito CP (1999) The effects of non-plastic and plastic fines on the liquefaction of sandy soils. Doctoral dissertation, Virginia Institute of Technology, USA
30. Rascol E (2009) Cyclic properties of sand: dynamic behaviour for seismic applications. Thesis No. 4546. École Polytechnique Fédérale De Lausanne (EPFL), Switzerland
31. Choudhary SS, Maheshwari BK, Kaynia AM (2010) Liquefaction resistance of Solani sands under cyclic loads. In: Proceeding of Indian geotechnical conference, IIT Bombay, India, pp 14–17
32. Matsui T, Ohara H, Ito T (1980) Cyclic stress-strain history and shear characteristics of clay. J Geotech Eng Div ASCE 106(GT10):1101–1120
33. Zhou J, Gong XN (2001) Strain degradation of saturated clay under cyclic loading. Can Geotech J 38(2):208–212
34. Hyde AF, Yasuhara K, Hirao K (1993) Stability criteria for marine clay under one-way cyclic loading. J Geotech Eng ASCE 119(11):1771–1789
35. Chen YM, Ji MX, Huang B (2004) Effect of cyclic loading frequency on undrained behaviors of undisturbed marine clay. China Ocean Eng 18(4):643–651
36. Zhang R, Tu YJ, Fei WP, Zhao ZH (2006) Effect of vibration frequency on dynamic properties of saturated cohesive soil. J Rock Soil Mech 27(5):699–703
37. Dash HK, Sitharam TG (2009) Undrained cyclic pore pressure response of sand–silt mixtures: effect of non-plastic fines and other parameters. Geotech Geol Eng 27:501–517
38. Dash HK, Sitharam TG (2011) Undrained cyclic and monotonic strength of sand-silt mixtures. Geotech Geol Eng 29:555–570
39. Jiang M, Cai Z, Cao P, Liu D (2010) Effect of cyclic loading frequency on dynamic properties of Marine clay. In: International conference on soil dynamics and earthquake engineering, GeoShanghai, pp 240–245
40. Khan Z, El Naggar MH, Cascante G (2011) Frequency dependent dynamic properties from resonant column and cyclic triaxial tests. J Franklin Inst 348:1363–1376
41. Lombardi D, Bhattacharya S, Hyodo M, Kaneko T (2014) Undrained behaviour of two silica sands and practical implications for modelling SSI in liquefiable soils. Soil Dyn Earthq Eng 66:293–304
42. Chattaraj R, Sengupta A (2016) Liquefaction potential and strain dependent dynamic properties of Kasai River sand. Soil Dyn Earthq Eng 90:467–475
43. Dash HK, Sitharam TG (2016) Effect of frequency of cyclic loading on liquefaction and dynamic properties of saturated sand. Geotech Geol Eng 10(5):487–492
44. Rees S (2014) Introduction to triaxial testing—Part 3: Dynamic triaxial testing GDS Instruments. https://www.gdsinstruments.com/information/white-paper-dynamic-triaxial-testing. Accessed Sept 07, 2021
45. IS: 2720_Part-IV (1975) Grain size analysis. Bureau of Indian Standard, New Delhi
46. Tsuchida H (1970) Prediction and counter measure against the liquefaction in sand deposits. Seminar in the Port and Harbour Research Institute, Ministry of Transport, pp 1–33
47. Xenaki VC, Athanasopoulos GA (2003) Liquefaction resistance of sand-mixtures: an experimental investigation of the effect of fines. Soil Dyn Earthq Eng 23:183–194
48. IS: 2720_Part-III (1981) Determination of specific gravity-fine, medium and coarse grained soils. Bureau of Indian Standard, New Delhi

49. IS: 2720_Part-XIV (1983) Determination of density index of cohesionless soils. Bureau of Indian Standard, New Delhi
50. ASTM D4253 (2016) Standard test methods for maximum index density and unit weight of soils using a vibratory table. ASTM International, West Conshohocken, PA.https://doi.org/10.1520/D4253-16
51. ASTM D4254 (2016) Standard test methods for minimum index density and unit weight of soils and calculation of relative density. ASTM International, West Conshohocken, PA.https://doi.org/10.1520/D4254-16
52. ASTM D2487 (2011) Standard practice for classification of soils for engineering purposes (Unified Soil Classification System). ASTM International, West Conshohocken, PA.https://doi.org/10.1520/D2487-11
53. ASTM D3999/D3999M (2011) Standard test methods for the determination of the modulus and damping properties of soils using the cyclic triaxial apparatus. ASTM International, West Conshohocken, PA.https://doi.org/10.1520/D3999_D3999M
54. ASTM D5311/D5311M (2013) Standard test method for load controlled cyclic triaxial strength of soil. ASTM International, West Conshohocken, PA.https://doi.org/10.1520/D5311_D5311M
55. Ishihara K (1993) Liquefaction and flow failure during earthquakes. Geotechnique 43(3):351–415
56. Kurar B, Maheshwari BK, Jakka RS (2012) Dynamic properties of Solani sand reinforced with coir fibers. In: Proceedings of 15th world conference on earthquake engineering, September, Lisbon, Portugal
57. Karg C, Haegeman W (2009) Elasto-plastic long-term behavior of granular soils: experimental investigation. Soil Dyn Earthq Eng 29:155–172
58. Kumar SS, Dey A, Krishna AM (2015) Dynamic response of river bed sands using cyclic triaxial tests. In: Proceedings 5th Young Indian geotechnical engineers conference, Vadodara, India
59. Kumar SS, Krishna AM, Dey A (2017) Evaluation of dynamic properties of sandy soil at high cyclic strains. Soil Dyn Earthq Eng 99:157–167
60. Ishihara K, Silver LM, Kitagawa H (1978) Cyclic strength of unsaturated sands obtained by large diameter sampling. Soils Found JSSME 18(4):61–76
61. Head KH (1992) Manual of soil laboratory testing, Vol. 3, Effective stress tests. Wiley, New York
62. Rollins KM, Evans MD, Diehl NB, Daily WD (1998) Shear modulus and damping ratio for gravels. J Geotech Geoenviron Eng ASCE 124:396–405
63. Kumar SS, Krishna AM, Dey A (2018) High strain dynamic properties of perfectly dry and saturated cohesionless soil. Indian Geotech J 48(3):549–557

Chapter 7
Assessment of Reliquefaction Behavior of Solani Sand Specimen Using 1-g Shaking Table Experiments

Gowtham Padmanabhan and B. K. Maheshwari

Abstract Soil liquefaction is a progressive research area in geotechnical earthquake engineering, since the 1964 Niigata, Japan, and 1964 Alaska, USA earthquakes. Researchers worldwide focused on understanding liquefaction mechanism and mitigation system through experimental, numerical, and analytical approaches. In the present study, authors investigated the reliquefaction behavior of Solani sand specimen by varying the acceleration amplitude and shaking duration subjected to repeated shaking events with two different shaking patterns. A rigid tank of dimension 1.05 m × 0.6 m × 0.6 m was employed and the experiments were performed with saturated sand specimen at 25% relative density. Repeated sinusoidal shaking was imparted at an acceleration amplitude in the range (0.1–0.4 g) with a constant excitation frequency of 3.5 Hz and with varying shaking durations (20, 40, and 60 s). Shaking was applied in both the incremental and decremental patterns and results are compared. The experimental results demonstrated that the preshaking effect and shaking pattern are critical in influencing the reliquefaction potential of sand deposits. The analysis of the pore pressure results indicates that the tendency of liquefaction increases even at lower acceleration with higher shaking duration.

Keywords Liquefaction · Shaking duration · Acceleration amplitude · Shaking Table · Preshaking effect

7.1 Introduction

Soil liquefaction in saturated sandy ground is a common phenomenon when subjected to seismic events. Soil is said to be liquefied if the excess pore water pressure exceeds the shear strength of the soil. The occurrence of earthquake swarms (2011 Tohoku,

G. Padmanabhan · B. K. Maheshwari (✉)
Department of Earthquake Engineering, Indian Institute of Technology Roorkee, Roorkee, Uttarakhand 247667, India
e-mail: bk.maheshwari@eq.iitr.ac.in

G. Padmanabhan
e-mail: gowtham@eq.iitr.ac.in

© Indian Society of Earthquake Technology 2023
R. S. Jakka et al. (eds.), *Earthquake Engineering and Disaster Mitigation*, Springer Tracts in Civil Engineering, https://doi.org/10.1007/978-981-99-0081-7_7

2016 Kumamoto and 2019 Vancouver earthquakes) in recent years induced soil reliquefaction and caused severe damage to the built environment. The reliquefaction behavior of the sand deposit was examined in the present study using a shaking table apparatus. By definition, reliquefaction is said to be the occurrence of liquefaction in a sand deposit which is originally liquefied earlier due to some prior seismic shaking events.

Oda et al. [1] was the first to investigate the microstructural investigation on the reliquefaction of sand deposits subjected to cyclic loading events. The experimental results indicate that the reliquefaction resistance significantly increased due to the initial inherent anisotropy of prepared sand deposits. In addition to that, the size and shape of voids are critical in developing liquefaction resistance. Repeated shaking table experiments and numerical analysis were carried out by Ye et al. [2] to understand the factors influencing the liquefaction behavior of sand deposits. The authors observed that the ground densified significantly during the repeated liquefaction process, however, the increased sand density not contributed much to the reliquefaction resistance when subjected to subsequent strong motion events. Yamada et al. [3] studied the behavior of sand deposits subjected to reliquefaction through triaxial experiments and reported a sudden decrease in the reliquefaction resistance despite a significant increase in the relative density of the sand deposits. In addition to sand density, anisotropy was also critical in influencing the resistance, as the higher the rate of induced anisotropy, the lower the resistance to soil reliquefaction. Ha et al. [4] conducted extensive shaking table experiments to evaluate the liquefaction and reliquefaction resistance by varying the sand density and grain size characteristics. The authors observed a substantial decrease in the resistance to reliquefaction, despite significant improvement in the density of the sand specimen. It is concluded that sand density or grain characteristics are not sufficient to examine the reliquefaction behavior/mechanism. The reliquefaction potential of sand specimen was examined by Ecemis et al. [5] through four shake table experiments with different shaking intensities. Authors reported that sand density is not only the critical factor influencing the reliquefaction resistance, but also depends on the sand consolidation characteristics. The authors concluded that the resistance to reliquefaction was found lesser than the liquefaction resistance.

Through centrifugation experiments, Ye et al. [6] attempted to verify the tendency of reduced reliquefaction resistance of sand specimen compared to the first liquefaction. The detailed study concluded that the speed of generation of excess pore water pressure gradually increases after each shaking event despite the increase in sand density. Apart from sand density, factors such as grain size characteristics and seismic input motions are crucial in influencing the resistance to liquefaction subjected to repeated shaking events. By experimenting multiple shaking events through 1-g shake table Wang et al. [7] studied the effect of shaking history on improving the resistance to liquefaction. Some interesting observations were reported by the authors in regard to reliquefaction, sand density was increased as a result of multiple shaking events, thus the soil structure was significantly disturbed, which in turn improves the resistance to reliquefaction to a greater extent. The past published research works reported a reduction in reliquefaction resistance with the increase in sand density. However, this study reported a contrasting nature to the past published research works.

With the help of 1-g shaking table experiments, Padmanabhan and Shanmugam [8] presented the reliquefaction behavior of Solani sand subjected to repeated incremental shaking events and different sand densities for high-frequency excitation of 5 Hz. For such high frequency, authors concluded that the reliquefaction phenomenon mainly depends on applied input motion characteristics, and the shaking-induced densification was not much contributed to the reliquefaction resistance subjected to incremental shaking events. However, this needs to be examined for lower frequencies. To mitigate the reliquefaction-induced damages and ground deformation, Padmanabhan and Shanmugam [9] proposed sand compaction piles as an improvement system. Four compaction piles were installed in a square pattern and subjected to repeated incremental shaking events and test results demonstrated that the installed improvement system was effective in improving the sand resistance to reliquefaction. The authors concluded that the induced sand densification is a viable technique to mitigate reliquefaction. Motivated by the 2016 Kumamoto, Japan sequential earthquakes, Padmanabhan and Maheshwari [10] carried out a case study on the damages induced by liquefaction and subsequent reliquefaction. The study was also compared with the 2005 Fukuoka-ken Seiho-oki independent earthquake in Japan. It is reported that the damage potential was lesser in case of the 2016 sequential earthquakes as the beneficial effect of preshaking was the difference between the two events. Similar observations were also reported by other researchers [11–13].

Padmanabhan and Maheshwari [14] carried out uniaxial shaking table experiments to examine the effect of sequential acceleration on Solani sand specimen and compared it with independent shaking events. A significant difference in the generation of excess pore water pressure was observed in the independent and sequential shaking events, and it was concluded that the preshaking effect was the major reason behind the increased reliquefaction resistance subjected to sequential shaking events. Many questions are still left unanswered in understanding the reliquefaction mechanism, as several researchers reported contradictory observations as some events increased and others decreased the resistance to reliquefaction when subjected to intricate shaking pattern and intensity [15–18].

The objective of the present study is to understand the intricacies in the reliquefaction behavior using uniaxial shaking table experiments. Influence of acceleration amplitude (0.1, 0.2, 0.3, and 0.4 g), shaking duration (20, 40, and 60 s) and pattern of shaking events (incremental and decremental pattern) are examined in this study. Sand specimen was prepared at 25% relative density and the excitation frequency was kept constant as 3.5 Hz for all the experiments. Pore pressure piezometers was embedded at three different depths to measure the response of excess pore water pressure during shaking. Ground surface was leveled after each event and the soil settlement was measured and the corresponding increase in sand density was estimated. Reliquefaction behavior was examined on the basis of generated pore pressure ratio, time taken to attain maximum pore pressure ratio, and the estimated sand density.

7.2 Why Reliquefaction Needs to Be Considered?

Since the 1964 Niigata earthquake, Japan, and the 1964 Good Friday earthquake, USA, the liquefaction research was focused mainly on the independent shaking events. Initial time researchers investigated the liquefaction mechanism and behavior through experimental, numerical, analytical, and case study approaches. Further research paid attention to various ground improvement systems to mitigate liquefaction and associated deformations. Nonetheless, research related to soil liquefaction and associated deformations on sand deposits subjected to sequential and repeated shaking events are limited. Recent instances of soil liquefaction are mainly associated due to earthquake swarms in the form of foreshocks followed by mainshock or the mainshock followed by a series of aftershocks. Historic evidence witnessed that these successive earthquakes are more catastrophic in nature. For instance, the extensive damages observed during the earthquakes such as 2010–2011 Canterbury, New Zealand; 2011 Tohoku, Japan; 2016 Kumamoto, Japan; 2018 Lombok, Indonesia; 2019 Vancouver, Canada; and 2021 Assam, India, bears testimony to successive shaking events [10, 19–21]. This motivated the researchers to work on this interesting and complicated mechanism behind reliquefaction and the effect of preshaking through experimental and case study approaches. From the limited studies, it is conclusive that the reliquefaction mechanism and behavior were different from the liquefaction induced by the independent shaking events. To explore the reliquefaction behavior and to propose a mitigation system, much attention needs to be paid in the future to reliquefaction research.

7.3 Experimental Setup and Test Procedure

7.3.1 *Experimental Setup*

To conduct the liquefaction experiments, a uniaxial 1-g shaking table available in Soil Dynamics Laboratory, Earthquake Engineering Department, IIT Roorkee, India, was used. A watertight rigid tank of dimension 1.05 m × 0.6 m × 0.6 m attached to the shake table was used for the preparation of the sand specimen. Figure 7.1a shows the experimental setup used in the study. The specimen of thickness 570 mm at 25% relative density was prepared. Loose sand density (25%) is selected because, when the specimen is subjected to repeated shaking events, a further increase in relative density could be expected. Hence, the selection of higher relative density may result in exceeding the threshold relative density. Density higher than 25% may have resulted in a higher increment in sand density which might be more resistant to liquefaction and subsequent reliquefaction. Hence, the authors selected 25% as the specimen density for the experimental study.

Acceleration amplitude and excitation frequency can be varied as desired in the range of (0.05–1 g) and 2–5 Hz, respectively. To monitor the pore water pressure

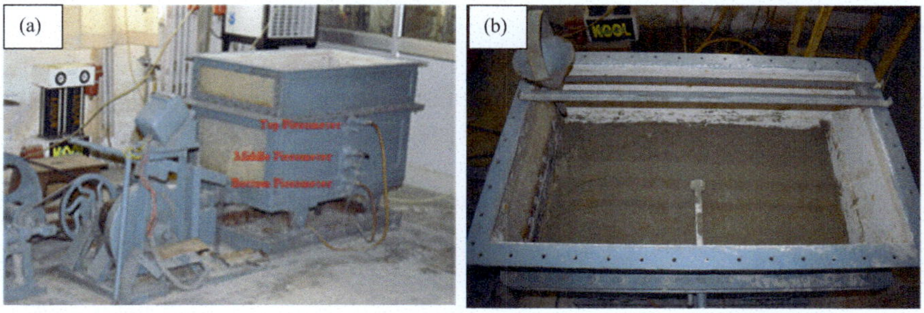

Fig. 7.1 **a** Experimental setup with instrumentation; **b** At the middle of the sample preparation and location of piezometer at the center of the tank in the lateral horizontal direction

at different depths, three piezometers were embedded at heights 40 mm, 200 mm, and 360 mm from the bottom of the tank, and named as bottom, middle, and top piezometers, respectively. To avoid the effect of boundary, the measurements of pore water pressure are performed at the center of the tank (in the lateral direction). The measuring range of the pore pressure transducers is (0–250 kPa) which is sufficient to measure the generated excess pore water pressure. The piezometers can measure the pore water pressure in the interval of 0.25 kPa. As shown in Fig. 7.1a, pore pressure instrumentations are rigidly connected to the tank at a particular height and are resistant to dynamic loading conditions.

The sample was prepared using the water pluviation method, so as to obtain maximum uniformity in sand density [4, 22, 23]. Figure 7.1b shows the sample preparation in the halfway stage and the location of piezometer. The total amount of sand and water required to prepare the saturated sand specimen at a particular sand density and depth was estimated using the procedure followed by Maheshwari et al. [24].

To counter the rigid boundary effects and constraint on horizontal movement, the pore pressure response was measured at the center of the tank. Considering the dimension of the tank, the authors believe that the measured pore water pressure is free from the boundary effects and seismic wave reflections.

The sinusoidal harmonic motion was applied at the base of the tank to impart seismic motion under constant excitation frequency. Considering the capacity of the shaking table and the dimension of the prototype, the similarity ratio of the geometry type was determined to be 1:50 [25]. The similarity law was proposed based on Buckingham π theory with the assumption that the soil obeys equivalent viscoelastic criteria under dynamic conditions [26]. Geometry size, gravity acceleration, and mass density are considered as control variables. In the shaking table experiments, acceleration amplitude and shaking duration were set as input control variables.

7.3.2 Test Sand

Sand used in the present study was procured from the Solani riverbed near Roorkee, Uttarakhand, India. According to the soil classification system [27], Solani sand was reported as poorly graded (SP) with little fines. Past published research works reported that the Solani sand was vulnerable to soil liquefaction subjected to independent shaking as well as repeated shaking events [24, 28, 29].

7.3.3 Experimental Procedure

A total of 24 shaking events (12 with incremental and 12 with decremental patterns) were performed to study the reliquefaction behavior of the sand specimen. Acceleration amplitude, shaking duration, and pattern of the shaking events were varied and experimented using a shaking table. Table 7.1 shows the test matrix used in the study. Varying duration of the shaking events was represented as S1, S2, and S3 for 20 s, 40 s, and 60 s, respectively. In an incremental shake pattern, 0.1 g loading was applied initially to the specimen followed by 0.2 g, 0.3 g, and ended with 0.4 g loading. Whereas in the case of the decremental shaking pattern, reverse application of loading (0.4 g followed by 0.3 g, 0.2 g, and finally 0.1 g) was experimented.

Soil is said to be liquefied, if the maximum pore pressure ratio $r_{u,\max}$ reaches unity. The $r_{u,\max}$ was estimated using Eq. 7.1.

$$r_{u,\max} = \frac{U_{\max}}{\sigma'_{vo}} \tag{7.1}$$

where $U_{\max}$ is the maximum generated excess pore water pressure in kPa and σ'_{vo} is the effective overburden pressure in kPa observed at that particular depth. The settlements were measured only after ensuring the complete dissipation of excess pore water pressure. The surface was then leveled and the settlement was measured

Table 7.1 Test matrix of the study

Shaking pattern	Shaking duration (s)	Acceleration amplitude (g)
Incremental	20 (S1)	0.1, 0.2, 0.3, and 0.4
	40 (S2)	
	60 (S3)	
Decremental	20 (S1)	0.4, 0.3, 0.2, and 0.1
	40 (S2)	
	60 (S3)	

Note Excitation frequency and relative density of the specimen were kept constant at 3.5 Hz and 25%, respectively, for all the events from the top of the tank using a measuring scale. From the observed soil settlement, the relative density of the specimen was estimated using the procedure followed by Maheshwari et al. [24].

7.4 Results and Discussions

7.4.1 Effect of Shaking Duration and Pattern on Pore Pressure Generation

Figure 7.2 presents the generated $r_{u,\max}$ at different acceleration amplitudes and shaking duration subjected to incremental and decremental shaking patterns. As expected, the $r_{u,\max}$ increases with an increase in shaking duration and acceleration amplitude. Interestingly, it can be observed that for a particular acceleration, the effect of shaking duration is initially considerable (from S1 to S2), however, it is not so much later (from S2 to S3). For example, in case of incremental shaking pattern for top piezometer (Fig. 7.2a$_1$) for varying shaking duration from (20–40 s) maximum pore pressure ratio increment rate was reported as (47%, 53%, 47%, and 47%) at 0.1 g, 0.2 g, 0.3 g, and 0.4 g, respectively. However, when the shaking duration is varied from 40 to 60 s, this increment is reduced to 17, 14, 11, and 9% only. Similarly, for the decremental shaking pattern for the top piezometer (Fig. 7.2a$_2$), the reported increment in max. pore pressure ratio was (34%, 21%, 24%, and 24%) and (10%, 13%, 14%, and 13%), respectively, for these two increments in shaking duration. A similar trend in results was also observed for both the middle and bottom piezometers as shown in Fig. 7.2b$_1$, b$_2$ and c$_1$, c$_2$ respectively. The decrease in the rate of increment of $r_{u,\max}$ was because of shaking densification and reorientation of sand particles. During long-duration shaking events, the ground specimen is subjected to larger disturbance and inevitably results in high sand density compared to short- and medium-duration shaking events. This could be the probable explanation for the decreased rate of pore pressure generation when the ground is subjected to long-duration repeated shaking events.

From Fig. 7.2, it can be seen that the maximum pore pressure ratio did not reach the value of 1.0, which means that the soil is not completely liquefied. However, from Fig. 7.2a$_2$, it can be seen that max. pore pressure ratio reached 0.90 for 0.4 g acceleration loading and S3 shaking duration (60 s). It means that the soil is almost liquefied. However, due to the effect of repeated shaking and preshaking of the sand specimen, it resulted in significant resistance to the generation of excess pore water pressure. The occurrence of partial liquefaction was reported in the manuscript. Further, for middle and bottom piezometers, the overburden pressure is higher than that for the top piezometer, thus reducing the pore water pressure ratio further.

Figure 7.2 also compares the generated maximum pore pressure ratio subjected to incremental and decremental shaking patterns at different depths. Figure 7.2a$_1$, a$_2$ shows that for the 20 s shaking duration and 0.1 g loading, $r_{u,\max}$ are 0.43 and 0.51

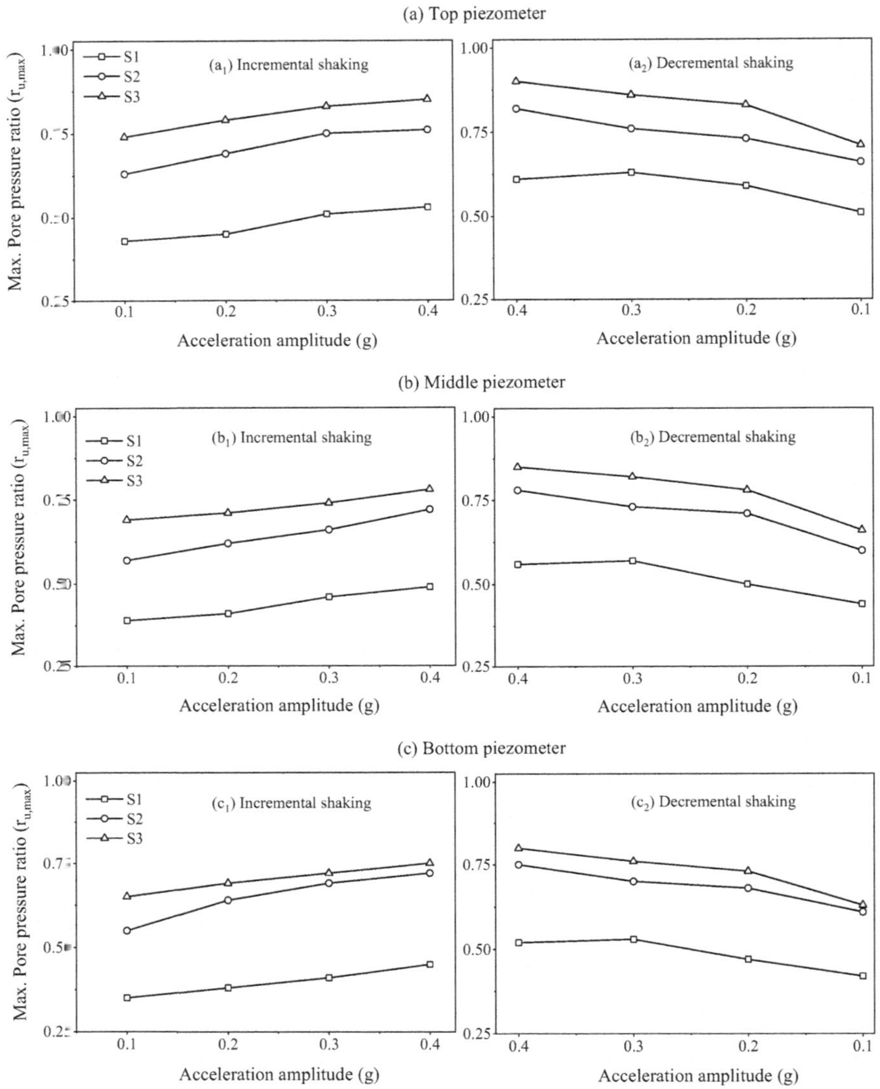

Fig. 7.2 Maximum pore pressure ratio $r_{u,\max}$ at varying acceleration amplitude for short (S1), medium (S2), and long (S3) duration shaking events

and these numbers for 0.4 g are 0.53 and 0.61, in case of incremental and decremental shaking patterns, respectively. Thus, test results indicate that the incremental shaking pattern was effective in improving the resistance to reliquefaction subjected to repeated shaking events. In a decremental shaking pattern, the application of high intense shaking event (0.4 g) drastically alters/collapses the inherent anisotropy of sand and the reorientation of sand particles changes the void structure of the specimen.

Then the effect of successive application of medium and low intense shaking events (0.3, 0.2, and 0.1 g) is not much different from the main event (0.4 g). Whereas, in case of incremental shaking pattern, the application of low intense shaking event (0.1 g) initially was not much significant in inducing the excess pore water pressure, which in turn slightly alters the soil orientation. Upon further application of medium to high intense shaking (0.2, 0.3, and 0.4 g), the ground subsidence increased gradually after each event and resulted in reduced generation of excess pore water pressure. The beneficial effect of preshaking events was more pronounced in case of incremental shaking compared to the decremental shaking patterns. From Fig. 7.2b_1, b_2 and c_1, c_2, it is clear that a similar trend was observed for both middle and bottom piezometer repeated events. The quantitative assessment of preshaking effect is complicated, as many researchers reported differences in resistance to liquefaction when subjected to intricate shaking patterns and combinations. This phenomenon demands further investigation and research in the future.

The present study also focused on understanding the generation of excess pore water pressure during repeated shaking events. In case of incremental shaking pattern, it was observed that irrespective of shaking duration, high intense shaking event (0.4 g) was not much contributed in the generation of pore water pressure. It also justifies the beneficial effect obtained from the previously applied low to medium-intense shaking events. Padmanabhan and Maheshwari [14] reported a higher generation of excess pore water pressure for 0.4 g compared to 0.3 g loading in case of an independent shaking event, where the effect of preshaking is not applicable. In the decremental shaking pattern, despite the reorientation of sand particles during the initial high-intense shaking event (0.4 g), the resistance to liquefaction was not much pronounced during the successive shaking events. Though partial liquefaction was reported in all the events, neither the shaking duration nor the acceleration amplitude contributed to the resistance to reliquefaction under repeated shaking events, in the decremental shaking pattern.

Figure 7.3 shows the time history of generated pore pressure ratio (r_u) for 60 s shaking duration (S3) for both incremental and decremental shaking patterns. Partial liquefaction was observed during all the shaking events. Long-duration shaking event (S3) was selected for reference, as the generation of excess pore water pressure was expected to be higher. A slight difference in pore pressure ratio and reliquefaction behavior was observed in both shaking patterns even though the shaking duration and acceleration amplitude remains the same. The generated excess pore pressure started dissipating quickly after attaining the maximum value and the stay time of peak value of pore pressure is insignificant.

From both Figs. 7.2 and 7.3, it can be observed that the generated pore pressure ratio was higher in the top and smaller in the bottom piezometer, and the values lie in the median for the middle piezometer. This is due to the effect of overburden pressure, minimum at the top and maximum at the bottom piezometer. A comparison of Figs. 7.3a_1 and a_2 indicate that the peak value of (r_u) is greater for the decremental shaking pattern for all accelerations at (0.2, 0.3, and 0.4 g) but not so at 0.1 g. This holds for middle and bottom piezometers too. Thus, for higher amplitudes (>0.1 g), the decremental pattern is more critical.

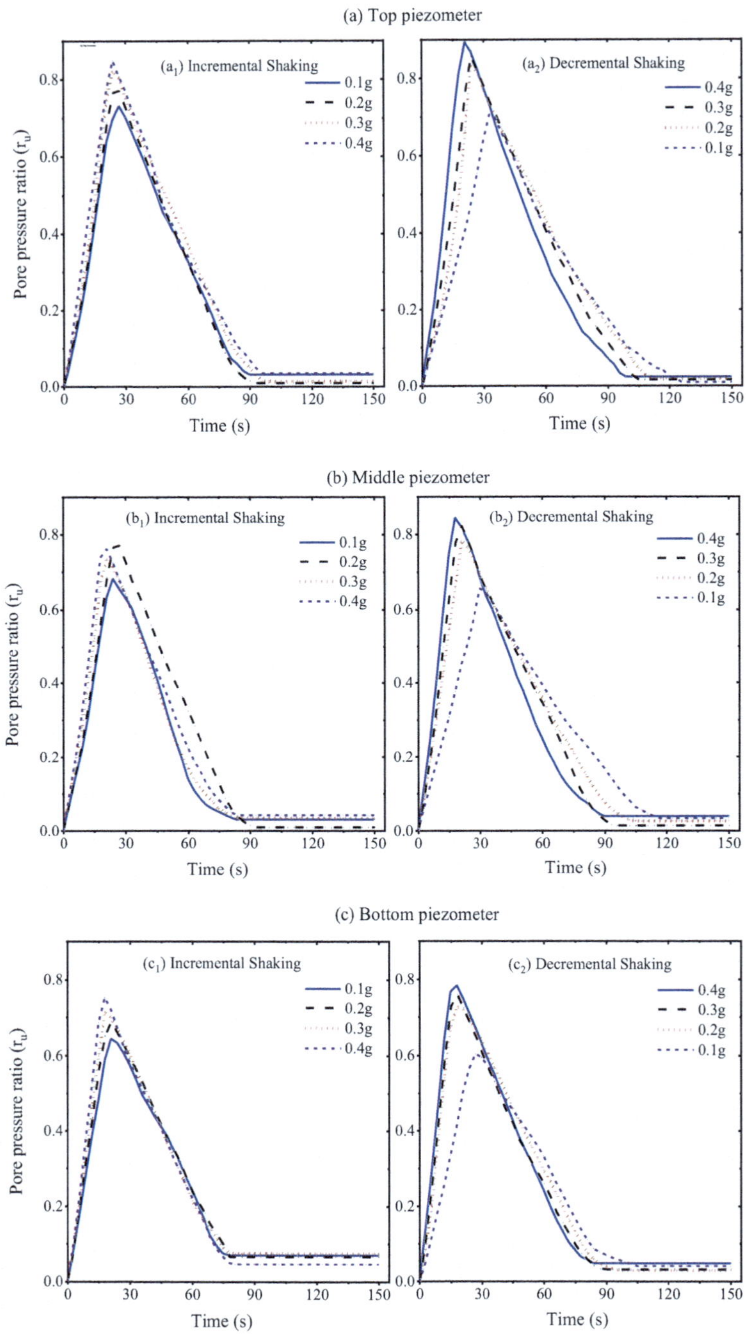

Fig. 7.3 Time history of pore pressure ratio (r_u) for 60 s shaking subjected to repeated shaking events

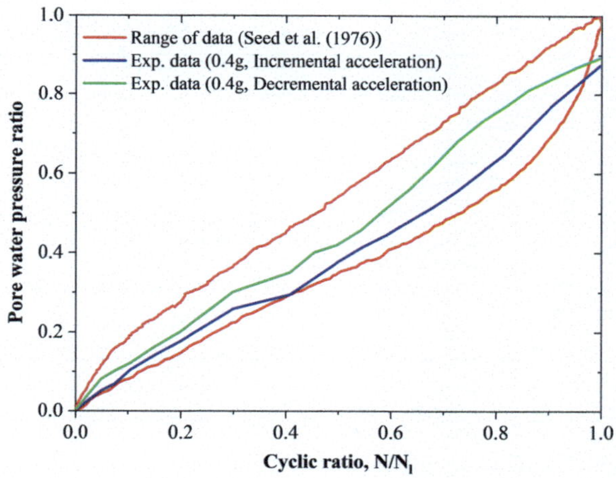

Fig. 7.4 Ratio of pore water pressure generation during the initial and final shaking events with 3.5 Hz and 1 min shaking

The present experimental data was compared with the typical band presented by Seed et al. [30] to validate the accuracy of the results for both the liquefaction and reliquefaction of the sand specimen. The cyclic ratio (N/N_l) was plotted against the variation of the pore pressure ratio, where N is the total number of cycles and N_l is the number of cycles to attain the maximum pore pressure ratio. Figure 7.4 shows the variations in (r_u) for initial shaking of 0.4 g (for decremental pattern, which can also be treated as an independent shaking) and for final shaking of 0.4 g (for incremental pattern) with a frequency of 3.5 Hz and 1 min shaking duration. It can be observed that both these curves lie within the range of data proposed by Seed et al. [30]. Further, for all values of cyclic ratio (N/N_l), the values of r_u for incremental shaking are smaller than that for independent shaking, clearly indicating the effect of shaking pattern. Moreover, the difference in r_u values for two shaking patterns is increasing with the number of cycles.

7.4.2 *Effect on Time Taken to Attain Maximum Pore Pressure*

Figure 7.5 shows the time taken to attain the maximum pore pressure ratio for varying shaking duration and different shaking patterns corresponding to the top piezometer. The reason for the selection of the top piezometer is that shallow-depth liquefaction causes more damage to the built infrastructure and shallow foundations are more vulnerable to damage during liquefaction. Some interesting observations are made from Fig. 7.5. The time taken to attain maximum pore pressure ratio depends on the shaking duration and also on the acceleration amplitude for both shaking patterns. For incremental shaking, at 0.1 g acceleration loading, the time taken to attain $r_{u,\max}$ was reported as 21 s, 24 s, and 26 s for short (S1), medium (S2), and long (S3) duration shaking events, respectively. However, this difference in time is insignificant as compared to the significant increase in shaking duration. The shaking pattern is an

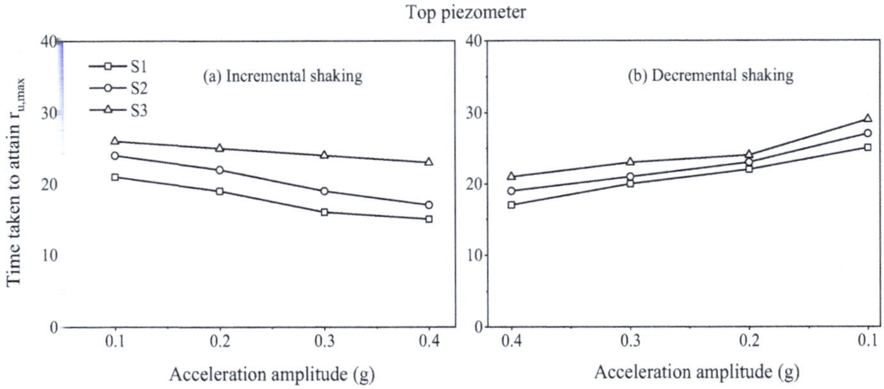

Fig. 7.5 Time taken to attain the maximum pore pressure ratio $r_{u,max}$ for short (S1), medium (S2), and long (S3) duration shaking events

important parameter in deciding the time taken to attain the maximum pore pressure ratio. As a slight contrasting nature was observed in comparing both the incremental and decremental shaking pattern.

It can be observed that the effect of shaking duration is smaller for the decremental shaking pattern as compared to the incremental shaking pattern. For instance, in case of medium duration (S2) shaking event, the time taken to attain $r_{u,max}$ was reported as (24 s, 22 s, 19 s, and 17 s) and (16 s, 21 s, 25 s, and 29 s) for incremental and decremental shaking patterns, respectively. In the previous section, it was discussed that the preshaking effect was much more beneficial in case of incremental shaking as compared to the decremental shaking pattern. Though the rate of generation of pore water pressure was limited with the incremental shaking pattern, the time taken to attain $r_{u,max}$ was delayed in case of a decremental shaking pattern.

For instance, in short-duration shaking events, the time taken to attain $r_{u,max}$ was reported as 21 s and 25 s for 0.1 g loading and 15 s and 18 s for 0.4 g loading in case of incremental and decremental shaking patterns, respectively. Increment in time taken was achieved in a decremental pattern despite the observed sand densification and reorientation of sand particles. Both pattern shows a slight contrast nature in resisting reliquefaction when subjected to repeated shaking events.

7.4.3 Effect of Repeated Shaking on Soil Displacement

Figure 7.6 presents the observed soil displacement (settlement) for varying acceleration amplitude and shaking duration for both the shaking patterns. Soil displacement gradually increases with the increase in acceleration amplitude. However, the first/initial shaking event (irrespective of the intensity of shaking) contributes to the majority of the soil displacement, irrespective of acceleration amplitude and shaking

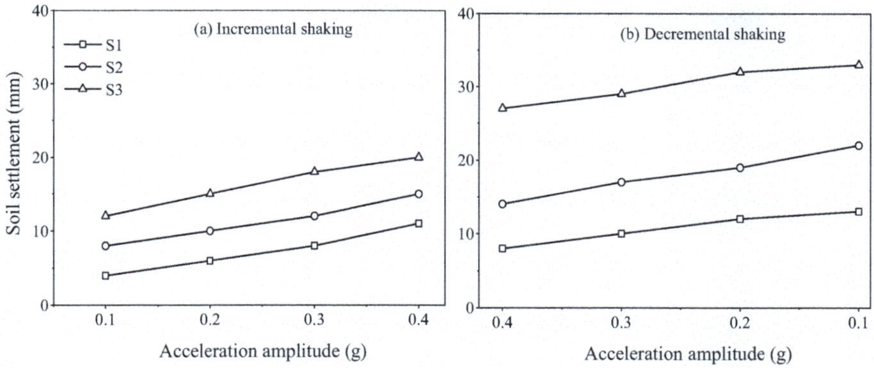

Fig. 7.6 Soil settlement observed after each shaking event for short (S1), medium (S2), and long (S3) duration shaking events

duration. It substantiates that the prepared sand specimen was vulnerable to soil liquefaction and associated deformations. As repeated shaking was applied only after ensuring complete dissipation of pore water pressure, low-density samples are subjected to greater settlement during the shaking event. Settlement resulted during shaking as well as during dissipation of pore water pressure. This also resulted in the increased sand densification of the specimen.

Percentage increment in soil displacement with varying acceleration was found higher in the incremental shaking pattern compared to the decremental shaking pattern. For instance, in short-duration (S1) shaking events, percentage increment was reported as (50%, 33%, and 38%) and (25%, 25%, and 8%) for incremental and decremental shaking patterns, respectively. A similar trend was followed in both the medium (S2)- and long (S3)-duration shaking events. The reason is the application of high intense shaking event (0.4 g) as an initial event in the decremental shaking pattern, which contributes the majority of the soil displacement and made soil specimens resistant to the reorientation phenomenon during the successive shaking events.

In case of longer-duration shaking events, the increment rate was found to decrease compared to short- and medium-duration shaking events. For instance, in case of decremental shaking pattern, the percentage increment in soil displacement was reported as (21%, 12%, and 16%) and (7%, 10%, and 3%) for both the medium (S2)- and long (S3)-duration shaking events, respectively.

7.4.4 Influence of Shaking Duration on Sand Density

The increment in sand density obtained after each shaking event was estimated using the soil settlement values and presented in Fig. 7.7. Initial sand specimen was prepared at 25% relative density and tested under repeated shaking conditions. Sand

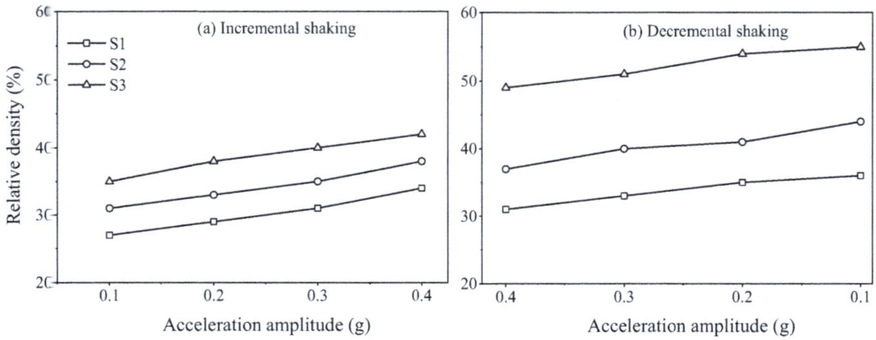

Fig. 7.7 Estimated sand density after each shaking event for short (S1), medium (S2), and long (S3) duration shaking events

density is an important parameter in soil liquefaction studies, as sand densification is commonly used for ground improvement in mitigating liquefaction. However, in case of repeated shaking events, the influence of sand density was associated with the shaking pattern and initially applied acceleration loading. For instance, in case of long-duration (S3) shaking event, the maximum pore pressure ratio and sand density was reported as (0.83 and 40%) and (0.86 and 51%) for 0.3 g loading for incremental and decremental shaking patterns, respectively. Another interesting observation is that the application of 0.4 g acceleration loading resulted in the $r_{u,\max}$ and sand density as (0.85 and 42%) and (0.90 and 49%) for incremental and decremental shaking patterns, respectively. These results indicate that the preshaking effect and shaking pattern govern the influence of sand density in increasing the resistance to reliquefaction subjected to repeated shaking events. The increment in sand density was greatly influenced by the acceleration amplitude, shaking duration, and shaking pattern. Interestingly, the rate of increment in sand density was not pronounced higher for a particular loading condition and pattern. Despite the increment in sand density during the initial shaking event, the percentage increment was found similar for all the cases. For instance, in case of incremental shaking pattern, the increment in sand density was reported as (7%, 7%, and 10%) and (6%, 6%, and 9%) for short- and medium-duration shaking events, respectively.

7.5 Conclusions

The following conclusions can be derived based on the experimental test results:

1. Acceleration amplitude and shaking duration are critical in influencing the reliquefaction phenomenon under both incremental and decremental shaking patterns. All shaking events, generate significant excess pore water pressure (PWP) and the magnitude of PWP increases as the shaking duration increases. However, the

rate of increment in PWP is not uniform with shaking duration and decreases gradually.
2. The time taken to attain the maximum pore pressure ratio was slightly influenced by acceleration amplitude and shaking duration. It also depends on the shaking pattern. A decremental shaking pattern delays the time taken to attain the maximum pore pressure ratio compared to the incremental pattern.
3. When the sand specimen is subjected to repeated shaking events, the first/initial event contributes to the majority of the soil displacement and sand density. Though factors such as acceleration amplitude and shaking duration also influences the outcome. The beneficial effect of preshaking and the associated reliquefaction resistance also depends on the initial shaking event. This effect of preshaking in terms of an increase in liquefaction resistance was more pronounced in case of incremental shaking when compared to the decremental shaking pattern.

The appropriate ground improvement technique needs to be designed and experimented to mitigate the catastrophic effects of reliquefaction.

Acknowledgements Financial Assistance to the first author by the Ministry of Education (MoE), Govt. of India for this research work is gratefully acknowledged.

References

1. Oda M, Kawamoto K, Suzuki K, Fujimori H, Sato M (2001) Microstructural interpretation on reliquefaction of saturated granular soils under cyclic loading. J Geotech Geoenviron Eng 127(5):416–423
2. Ye B, Ye G, Zhang F, Yashima A (2007) Experiment and numerical simulation of repeated liquefaction-consolidation of sand. Soils Found 47(3):547–558
3. Yamada S, Takamori T, Sato K (2010) Effects on reliquefaction resistance produced by changes in anisotropy during liquefaction. Soils Found 50(1):9–25
4. Ha IS, Olson SM, Seo MW, Kim MM (2011) Evaluation of reliquefaction resistance using shaking table tests. Soil Dyn Earthq Eng 31(4):682–691
5. Ecemis N, Demirci HE, Karaman M (2015) Influence of consolidation properties on the cyclic re-liquefaction potential of sands. Bull Earthq Eng 13(6):1655–1673
6. Ye B, Zhang L, Wang H, Zhang X, Lu P, Ren F (2019) Centrifuge model testing on reliquefaction characteristics of sand. Bull Earthq Eng 17(1):141–157
7. Wang J, Salam S, Xiao M (2020) Evaluation of the effects of shaking history on liquefaction and cone penetration resistance using shake table tests. Soil Dyn Earthq Eng 131:106025
8. Padmanabhan G, Shanmugam GK (2022) Reliquefaction assessment studies on saturated sand deposits under repeated acceleration loading using 1-g shaking table experiments. J Earthq Eng 26(6):2888–2910
9. Padmanabhan G, Shanmugam GK (2021) Liquefaction and reliquefaction resistance of saturated sand deposits treated with sand compaction piles. Bull Earthq Eng 19(4):1–25
10. Padmanabhan G, Maheshwari BK (2021) Case studies on preshaking and reliquefaction potential for different earthquakes in Japan. In: Local site effects and ground failures. Springer, Singapore, pp 137–144
11. Kiyota T, Ikeda T, Konagai K, Shiga M (2017) Geotechnical damage caused by the 2016 Kumamoto earthquake, Japan. Int J Geoeng Case Hist 78–95

12. Bhattacharya S, Hyodo M, Nikitas G, Ismael B, Suzuki H, Lombardi D, Egami S, Watanabe G, Goda K (2018) Geotechnical and infrastructural damage due to the 2016 Kumamoto earthquake sequence. Soil Dyn Earthq Eng 104:390–394
13. Anderson DJ (2019) Understanding soil liquefaction of the 2016 Kumamoto earthquake. Master thesis, Brigham Young University, United States
14. Padmanabhan G, Maheshwari BK (2022) Liquefaction resistance of solani sand under normal and sequential shaking events. In: Proceedings of the 4th international conference on performance based design in earthquake geotechnical engineering (Beijing 2022). Springer International Publishing, Cham, pp 1656–1663
15. Andrus RD, Hayati H, Mohanan NP (2009) Correcting liquefaction resistance for aged sands using measured to estimated velocity ratio. J Geotech Geoenviron Eng 135(6):735–744
16. Dobry R, Abdoun T, Stokoe KH, Moss RES, Hatton M, El Ganainy H (2015) Liquefaction potential of recent fills versus natural sands located in high-seismicity regions using shear-wave velocity. J Geotech Geoenviron Eng 141(3):04014112
17. Darby KM, Boulanger RW, DeJong JT (2017) Effect of multiple shaking events on cone penetration resistances in saturated sand. In: Taiebat M et al (eds) Proceedings 3rd international conference on performance-based design in earthquake geotechnical engineering, PBDIII. Vancouver, BC, Canada
18. Darby KM, Boulanger RW, DeJong JT, Bronner JD (2019) Progressive changes in liquefaction and cone penetration resistance across multiple shaking events in centrifuge tests. J Geotech Geoenviron Eng 145(3):04018112
19. Bhattacharya S, Hyodo M, Goda K, Tazoh T, Taylor CA (2011) Liquefaction of soil in the Tokyo Bay area from the 2011 Tohoku (Japan) earthquake. Soil Dyn Earthq Eng 31(11):1618–1628
20. Cubrinovski M, Henderson D, Bradley BA (2012) Liquefaction impacts in residential areas in the 2010–2011 Christchurch earthquakes. Tokyo, Japan: One Year after 2011 Great East Japan Earthquake. In: International symposium on engineering lessons learned from the giant earthquake, 3–4 Mar 2012
21. Cubrinovski M, Hughes M, Bradley BA, Noonan J, Hopkins R, McNeill S, English G (2014) Report on performance of horizontal infrastructure in Christchurch city through the 2010–2011 Canterbury earthquake sequence. University of Canterbury. Civil and Natural Resources Engineering, 139pp
22. Varghese RM, Latha GM (2014) Shaking table tests to investigate the influence of various factors on the liquefaction resistance of sands. Nat Hazards 73(3):1337–1351
23. Banerjee R, Konai S, Sengupta A, Deb K (2017) Shake table tests and numerical modeling of liquefaction of Kasai River sand. Geotech Geol Eng 35(4):1327–1340
24. Maheshwari BK, Singh HP, Saran S (2012) Effects of reinforcement on liquefaction resistance of Solani sand. J Geotech Geoenviron Eng 138(7):831–840
25. Zhou Z, Lei J, Shi S, Liu T (2019) Seismic response of aeolian sand high embankment slopes in shaking table tests. Appl Sci 9(8):1677
26. Wang J, Yao L, Hussain A (2010) Analysis of earthquake-triggered failure mechanisms of slopes and sliding surfaces. J Mt Sci 7(3):282–290
27. IS 2720, Part IV (1985) Methods of test for soils: Part 4 Grain size analysis. Bureau of Indian Standards, New Delhi, India, pp 1–38
28. Maheshwari BK, Patel AK (2010) Effects of non-plastic silts on liquefaction potential of Solani sand. Geotech Geol Eng 28(5):559–566
29. Singh HP, Maheshwari BK, Saran S, Paul DK (2010) Improvement in liquefaction resistance of pond ash using stone-sand columns. Int J Geotech Eng 4(1):23–30
30. Seed HB, Martin PP, Lysmer J (1976) Pore-water pressure changes during soil liquefaction. J Geotech Eng Div 102(4):323–346

Chapter 8
Behaviour of Batter Piles Under Machine Induced Vibrations

M. Bharathi, R. N. Dubey, and S. K. Shukla

Abstract The batter piles have several applications in infrastructure projects, such as offshore structures, transmission towers, bridges, and high-speed turbines involving machine-induced vibrations. This paper presents the details of a series of dynamic lateral and vertical loading tests as conducted on a vertical pile, B0 ($\beta = 0^{\circ}$) and batter (i.e., inclined) piles, B10 ($\beta = 10^{\circ}$) and B20 ($\beta = 20^{\circ}$) constructed in layered silty sandy soil, where β is the inclination of the pile to the vertical. All the piles had a length of 2.5 m and a diameter of 0.20 m. These piles were subjected to six force levels in lateral (X and Z) and vertical (Y) directions. Accelerometers were placed on the pile cap to obtain the dynamic response of these piles. The results show that the resonant frequency of the soil-pile system decreases by 37–50% along the lateral direction and 43–50% along the vertical direction, with increasing force level. When the force level is increased, the maximum peak displacements in the X and Z directions increase by 11 and 4 times in the case of B0 and B20 piles, respectively. However, the peak displacement increases by 4–5 times in the Y direction with an increase in force level. The increase in lateral strains (in X and Z) in the piles with force level was more prominent in the case of B20 compared to B0 and B10. Further, the increase in axial strain (in Y) with force level was almost four times in all three piles. In both lateral directions, the rotational stiffness of the soil-pile system decreases nonlinearly, whereas the damping ratio increases nonlinearly with an increase in lateral strain.

Keywords Batter piles · Dynamic load · Machine vibrations

M. Bharathi
Department of Civil Engineering, SKIT, Jaipur 302017, India

R. N. Dubey (✉)
Department of Earthquake Engineering, IIT Roorkee, Roorkee 267667, India
e-mail: rn.dubey@eq.iitr.ac.in

S. K. Shukla
School of Engineering, Edith Cowan University, Joondalup, Perth, WA 6027, Australia

© Indian Society of Earthquake Technology 2023

R. S. Jakka et al. (eds.), *Earthquake Engineering and Disaster Mitigation*,
Springer Tracts in Civil Engineering, https://doi.org/10.1007/978-981-99-0081-7_8

8.1 Introduction

Commonly pile foundations are used in relatively weak soil deposits to support various structures. These structures, along with their foundation systems, are often exposed to dynamic loads in addition to the bending moment, and static vertical and lateral loads. Hence, the pile foundations experience comparatively higher bending moments in addition to low and high strain rate reversals at each cycle, leading to passive pressure mobilization of soil-mass in the surrounding. This phenomenon may result in the collapse of the integrated soil-pile system along with the structural failure of the pile itself. Understanding the dynamic response of pile foundations has been exciting and challenging to researchers in the past few decades. To carry large lateral loads, batter (i.e., inclined) piles are usually selected despite some construction-related complications. When structures experience huge overturning moments due to wind and wave action, using combination of vertical and batter piles is recommended [1, 2].

The underperformance of batter piles were reported during Costa Rica earthquake [3] and Loma Prieta earthquake [4]. Later on, Berrill et al. [5] and Kastranta et al. [6] revealed evidence in support of batter piles during Edgecumbe and Kobe earthquake respectively. Poulos [7] suggests paying extra attention especially when the pile experiences either horizontal or vertical ground movements in addition to applied loads. Thus, the available literature on the seismic performance of batter piles indicated both advantageous and disadvantageous effects, and in some cases, even vertical piles failed abruptly during seismic actions.

Both AFPS [8] and EC8 [9] were restrictive, especially to the use of inclined piles to resist seismic and transmit lateral loads, respectively. Seismic guidelines for ports [10] also advised not to use batter piles at ports because of their poor performance during past earthquakes. AASHTO [11] recommended considering increased foundation stiffness for batter piles used in seismic-prone areas while designing the pile foundations. Later on, researchers and engineers gained confidence after understanding the reason for the failure of batter piles during seismic events, they further developed strategies to improve the seismic performance of batter piles.

Zhang et al. [12] reported that the induced shear force in batter piles subjected to vertical load is less than laterally loaded batter piles in sand. Gerolymos et al. [13] reported that tall structures founded on symmetric batter piles resulted in the most satisfying performance of both foundation and superstructure. Centrifuge tests with repeated horizontal and impact loading on floating and end-bearing pile groups were conducted by Escoffier et al. [14]. Batter piles helped in achieving improved seismicity at almost all frequency ranges, according to Tazoh et al. [15]. Giannakou et al. [16] claimed batter piles could be beneficial or detrimental based on the relation between shear force and bending moment. The soil pile interaction of pile groups consisting vertical and batter piles has also been reported [17, 18]. Deeper insights on inclined piles, focusing on dynamic stiffness functions [19], kinematic bending moment and shear force [20], the effect of structural response [21], kinematic interaction factors [22], and stiffness/damping [23], helped in understanding their behavior much better. In addition, the horizontal impedance function value of inclined single piles decreases as the angle of the pile inclination increases [24].

In recent years, centrifuge experiments on batter piles were carried out through a series of earthquake excitation [25] and sinusoidal dynamic loads [26], where batter piles supporting short superstructures had more significant beneficial effects. Moreover, the dynamic lateral behavior of batter piles embedded in clay was examined through laboratory-scaled pile models by Subramanian and Boominathan [27]. Wang et al. [28] emphasized that an optimum batter angle for piles in a group would significantly influence the dynamic horizontal impedance of a single pile and interaction factors for adjacent piles. Most of the reported studies on the dynamic behavior of batter piles were based on either centrifuge tests/ scaled lab experiments or analytical/numerical solutions.

The dynamic behavior of combined batter and underreamed piles, embedded in silty sand, in horizontal (both along batter and across batter) [29] and vertical [30] directions were reported through detailed experimental investigations conducted on field conditions. Extensive experimental work on the dynamic lateral and vertical response of cast in-situ reinforced concrete, vertical, and batter pile groups in silty sand was reported by Bharathi et al. [31]. 3D finite element (FE) analyses were also carried out to verify the experimental investigations [32]. In this paper, an experimental investigation on vertical and batter piles under dynamic lateral and vertical machine-induced vibrations is reported. Reinforced concrete (RC) single piles were constructed in the field by bored cast in-situ method. Dynamic lateral and vertical response of single piles, B0 (vertical, $\beta = 0°$), piles with batter angle, β (measured from vertical) i.e., B10 ($\beta = 10°$), B20 ($\beta = 20°$), constructed on the same soil conditions were studied. The dynamic response recorded in terms of acceleration at the pile cap was analyzed and presented in the form of peak displacement amplitude, resonant frequency, induced strain, rotational stiffness, and damping ratio.

8.2 Field Investigation

Three single piles considered in the present study were constructed at the geographical location of 29°51′58.0″N and 77°54′02.8″E. All the details of the experimental investigation are reported in the following sections.

8.2.1 Site and Soil Conditions

A borehole of depth 6.0 m was created to explore the soil profile. The groundwater table was encountered in the borehole at a depth of 5.5 m below the ground level. To conduct laboratory tests, disturbed representative soil samples from the borehole were obtained from the test location at every half-meter interval. Different laboratory tests have been performed to obtain the index properties of the collected soil samples

Table 8.1 Properties of soil

Depth (m)	Sand (%)	Silt and clay (%)	C_u	C_c	LL	PL	PI	Soil type
0–3.5	88	12	3.33	1.88	42.50	–	42.50	SP-SM
3.5–5.5	92	8	3.00	1.53	37.87	23.43	14.44	SP-SC
5.5–6	89	11	7.50	4.80	39.23	28.89	10.34	SP-SM

[33–35]. Based on different laboratory test results, the soil profile at the site was divided into three layers as per BIS [36]. The SPT N-value up to a depth of 8 m below the ground surface is reported in [31]. The angle of internal friction, ϕ has been calculated using the relation between ϕ and SPT-N value from BIS [37]. The average specific gravity and mass density of SP-SM layers are 2.67 and 1.61 g/cm^3 whereas, for SP-SC layer, they are 2.72 and 1.34 g/cm^3, respectively. The properties of soil are summarized in Table 8.1.

8.2.2 Construction of Piles with a Cap

The dimension of piles constructed for this experimental study is presented in Fig. 8.1a. All three piles were bored cast in-situ RC (M25 grade) piles built at the test site. Boring at the test site was carried out with the help of auger to bore vertical boreholes of 0.20 m diameter for pile B0. Clear covers were fixed to the reinforcement cage before placing it into the holes. To construct B10 and B20, auger and boring guide were used together to create inclined boreholes. The boring guide used for guiding the auger had arrangements to guide the auger at different inclinations i.e., 5°, 10°, 15° and 20°.

Figure 8.1b shows the setting of the boring guide used while boring pile B20. Nominal clear cover and reinforcement were provided for piles and pile cap according to BIS [38]. To ensure a firm and monolithic connection between the pile and its cap reinforcement bars were welded before concreting. The reinforcement cage of the pile cap was welded with four bolts (Fig. 8.1c) to facilitate mounting of the oscillator motor assembly. The pile cap has a length and width of 0.70 m and a depth of 0.30 m. The borehole was first filled with concrete and then the pile cap was cast. To ensure proper compaction of fresh concrete, needle vibrator was used at several intervals. It was ensured to practice quality workmanship during various stages of construction at the site to avoid any adverse effects. Before the commencement of the experiment, all these cast in-situ reinforced concrete piles were cured for twenty-eight days with the help of temporary mud barriers.

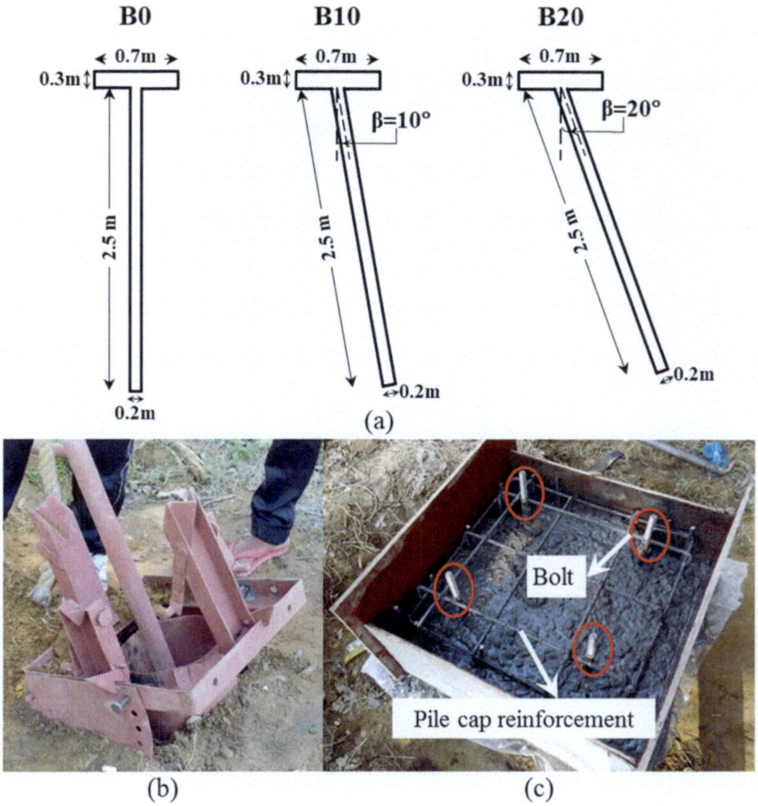

Fig. 8.1 **a** Geometry of piles; **b** boring guide; and **c** pile cap reinforcement cage with bolts

8.2.3 Loading and Response

The sinusoidal dynamic load was generated with the help of an oscillator box which contains two eccentric masses. To achieve desired frequency range, a speed control unit has been used along with a DC motor (of 3 horsepower with the highest working frequency of 45 Hz i.e., 2700 rpm). The oscillator box is mounted on top of the pile cap firmly with the help of a base plate, four connecting bolts, connecting rods, and side frames. A motor base plate placed on top of the oscillator box helps connect the DC motor with the oscillator beneath.

Assembly of the oscillator motor mounted on the pile cap for lateral and vertical excitations is depicted in Fig. 8.2a–b, respectively. During lateral excitation, three accelerometers were placed along the pile cap depth (Fig. 8.2a). Whereas, for vertical excitation, two accelerometers are placed on top of the adjacent faces of the pile cap (Fig. 8.2b). Figure 8.3 presents the complete test setup with piles B0 and B20. The left and right side of Fig. 8.3 presents the setup for lateral and vertical excitations, respectively. In a particular excitation direction, each pile was subjected to six force

Fig. 8.2 Test setup for: **a** lateral vibration; and **b** vertical vibration

levels in increasing order. The dynamic load generated is same as that reported earlier [30]. In this study, dynamic tests have been conducted at six different eccentricities i.e., at $e = 10°, 20°, 30°, 60°, 80°$ and $100°$, with excitation frequency ranging between $f_1 = 1$ to $f_n = 40$ Hz at all considered eccentricities of the oscillator. A pictorial representation of the oscillator with the eccentric masses representing *e*, *r*, and other details of the oscillator is presented in Fig. 8.4a–b.

The experimental data were obtained through LabVIEWSignalExpress [39] software and processed in real-time. PCB accelerometers with sensitivity 100 mVolts/g ($\pm 10\%$); frequency range between 0.5 and 10,000 Hz and measurement range of ± 50 g pk were connected to a multi-channel data acquisition device from National Instruments, NI 9233 (24 bit delta sigma (with analog prefiltering) ADC and 102 dB dynamic range) to record the dynamic response in terms of acceleration during the experiment(Fig. 8.4c–d). This device is connected to the computer with the help of a portable USB carrier NI USB-9162 (Fig. 8.4c). To understand the effect of eccentricity on the generated dynamic load two typical force time history has been presented in Fig. 8.4e–f for $e = 20°$ and $80°$ respectively. At each operating frequency, 12,000 continuous samples of acceleration time history data were recorded with a sampling rate of 2000 Hz in real-time for all considered cases.

A typical acceleration time history record thus obtained during the vertical vibration test is presented in Fig. 8.5a. The two different color bands in Fig. 8.5a represent the record corresponding to two different accelerometers. When the operating frequency reaches f_n, the power supply to the oscillator is withdrawn by switching off the speed control unit and rundown is recorded, which captures the entire test in the reverse order (i.e., from 40 to 0 Hz). A typical rundown record thus obtained is

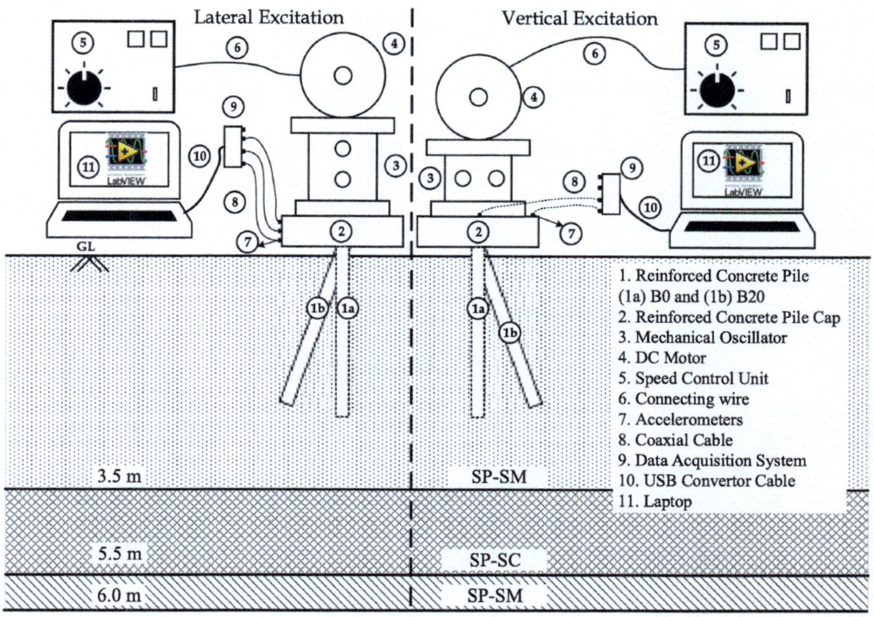

Fig. 8.3 Schematic representation of the entire test setup

presented in Fig. 8.5b where the red encircled region represents the range of resonant frequency. Since this record is obtained after reaching f_n it has a frequency in the reverse order. The rundown record thus obtained has been only used for verifying the resonant frequency obtained.

Figure 8.6a–i present acceleration time histories corresponding to $e = 30^{\circ}$ at f approaching 15 Hz of all three piles excited in X (Fig. 8.6a–c), Z (Fig. 8.6d–f) and Y (Fig. 8.6g–i) directions. Thus, from the acceleration time histories recorded, the response of the pile foundations is obtained in terms of displacement amplitudes, resonant frequencies, induced strain, etc. In addition, the effect of lateral loading direction on these single piles is also examined. By changing the orientation of the entire assembly of the oscillator motor the direction of loading can be changed. Figure 8.7a–b depict both the directions of excitation considered for this experimental investigation. At a particular eccentricity, the piles were first tested in one of the lateral (X) directions, followed by another lateral (Z) direction. Further, the piles were excited in vertical (Y) direction to measure the dynamic response. It was not possible to study the combined interaction between the axial (Y) and lateral (X or Z) behaviour of batter piles with the available experimental setup. However, at a particular forcing level, a minimum duration of 48 h was maintained as the rest period between each test. The tests were conducted on the piles at a higher eccentricity following the same loading direction sequence (i.e., first X, second Z, and third Y). This loading sequence was maintained so that the pile is subjected to the same force level in all three directions before testing it for a higher force level. The experimental

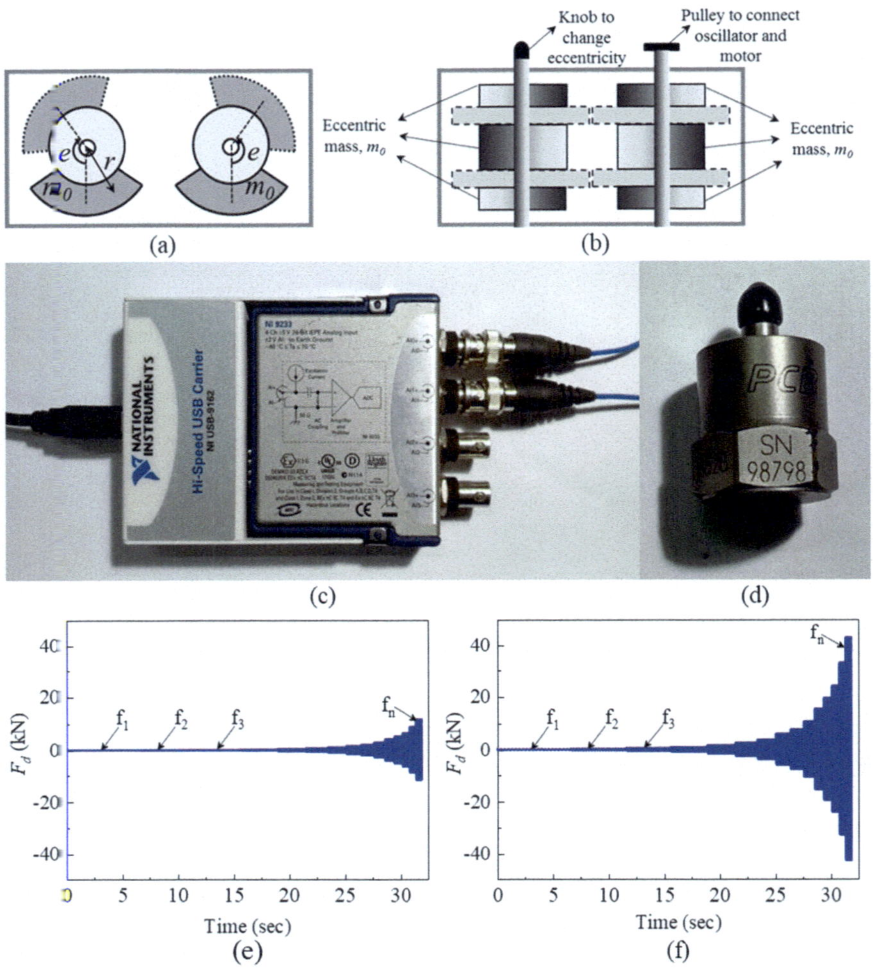

Fig. 8.4 Details of the oscillator: **a** cross section; **b** plan; **c** data acquisition system; **d** accelerometer; **e** load pattern at $e = 20°$; and **f** load pattern at $e = 80°$

investigations reported in this study were carried out in multiple phases. The dynamic lateral response of piles B0, B10, and B20 excited in both the directions and the dynamic vertical response are discussed herein.

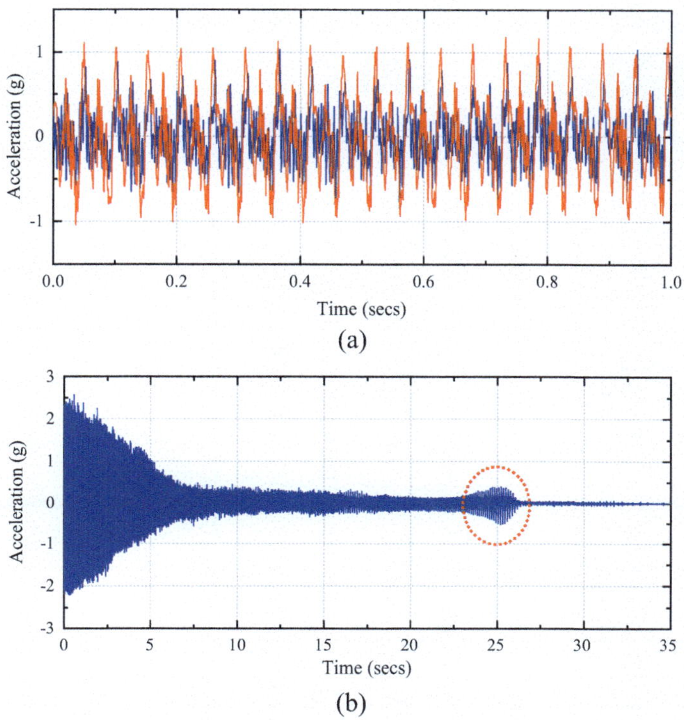

Fig. 8.5 Typical acceleration record of: **a** vertical excitation; and **b** rundown

8.3 Results and Discussion

The dynamic response of piles B0, B10, and B20 in lateral and vertical directions is discussed here, in terms of resonant frequency, displacement amplitude, induced strain level, etc., in detail. In addition, the effect of an underream bulb on the dynamic response of vertical and batter piles is also discussed.

8.3.1 Lateral Response

The dynamic lateral behavior of piles B0, B10, and B20 is presented in terms of lateral displacement, resonant frequency, rotational stiffness, and damping ratio and are discussed herein. The frequency has been obtained from the rotations (i.e., rotations per minute) of the oscillator motor assembly. The reported displacement response of the piles was obtained as the average of the measured acceleration response with the help of three accelerometers.

The dynamic response of piles B0, B10, and B20 excited, at all considered force levels, both in X and Z directions, are presented in terms of frequency (f_x or f_z) against

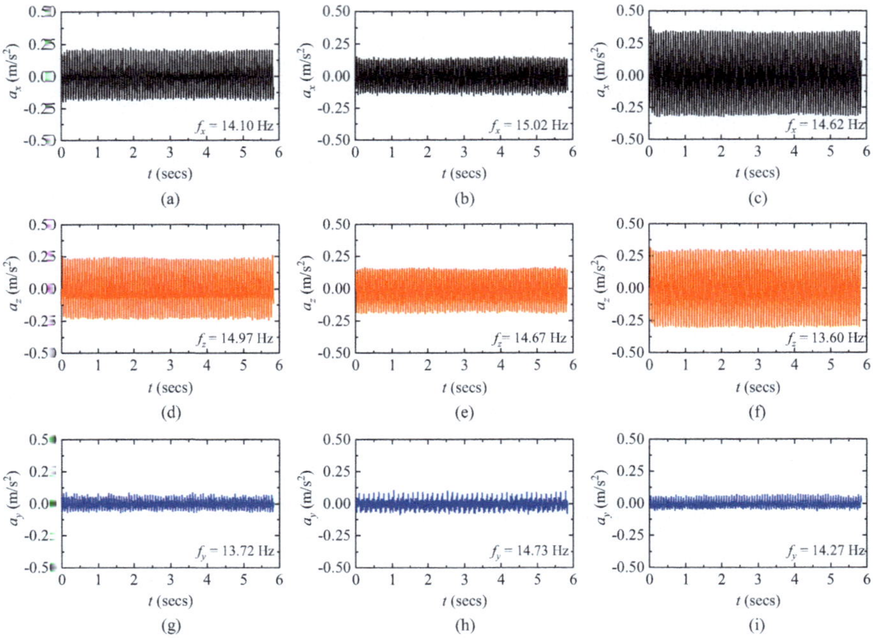

Fig. 8.6 Acceleration time histories at $e = 30°$: **a** B0 in X direction; **b** B10 in X direction; **c** B20 in X direction; **d** B0 in Z direction; **e** B10 in Z direction; **f** B20 in Z direction; **g** B0 in Y direction; **h** B10 in Y direction; and **i** B20 in Y direction

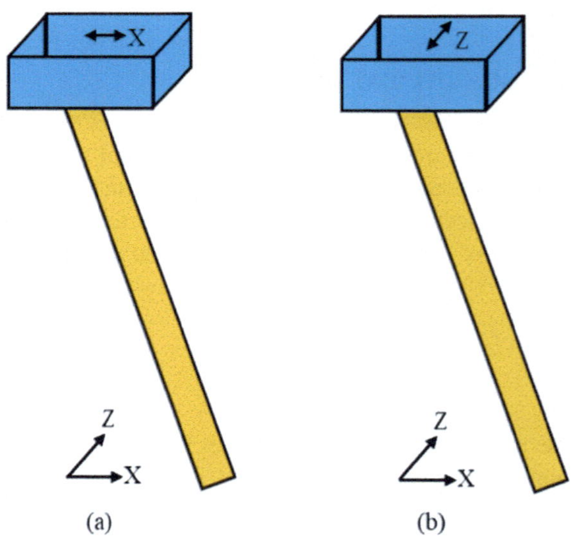

Fig. 8.7 The direction of lateral loading: **a** X direction; and **b** Z direction

lateral displacement (d_x or d_z) in Fig. 8.8a–c and d–f respectively. The displacement is obtained, as a ratio of acceleration and $4\pi^2 f^2$ where f is the frequency in Hz, according to the procedure suggested by BIS [40]. The frequency displacement response consists of only a single peak in all the cases within the excitation range considered, as observed in Fig. 8.8.

The single peak indicates that the soil-pile system is similar to a single degree of freedom system, as also reported in the literature [27, 41, 42]. A minimal gap (less than 50 mm) was created between the ground surface and the bottom of all the pile caps to avoid friction between them. The rotational mode of vibration is more significant for large freestanding length or slender piles, but in this study, the freestanding length is less than 50 mm, and the slenderness ratio is also very less. Hence, only the translation mode is significant, as also observed in the experimental records. Based on the observations from the frequency displacement response, further analysis has been carried out considering the soil-pile system as a single degree of freedom.

The resonant frequency (f_x' or f_z') and peak displacement amplitude (d_x' or d_z') at each force level, in both X and Z directions, were obtained from the frequency displacement amplitude records and are reported in Fig. 8.9a–b and c–d respectively. The resonant frequency of the soil-pile system (f_x' or f_z') significantly reduces (by 37–50%) with the increase in force level (i.e., eccentricity) in both the lateral directions. Possible reasons may be that as the eccentricity increases, force level also increases, and degradation in stiffness of the soil mass also increases. Hence, the stiffness of the soil-pile system participating in vibration decreases which leads to a decrease in the resonant frequency with increasing excitation force. A similar trend in reduction of resonant frequency with increasing excitation force was also observed by Manna and Baidya [43], and Boominathan et al. [44] based on the dynamic horizontal field tests on RC vertical piles. On the other hand, it can also be observed from Fig. 8.13 that the stiffness of the soil pile system increases with the increase in angle β. This increase in the stiffness has resulted in the increase of resonant frequency with increase in angle β. The peak displacement (d_x' or d_z') increases with an increase in the force level in both directions (Fig. 8.9c–d).

The increase in d_x' or d_z' at the maximum forcing level was about 11 times that of the minimum forcing level for the B0 pile and about four times for the B20 pile. It is worth mentioning that at all lower eccentricities (i.e., $\leq 80°$ eccentricity), there is no significant difference in the displacement amplitude in the X and Z direction of B0. When pile B0 was excited at $e = 100°$ in the Z direction, the displacement amplitude was much higher. This may be attributed to the permanent deformation of the soil pile system, which was evident through gap formation along the circumference of the piles, as shown in Fig. 8.10.

From the acceleration response of piles B0, B10 and B20 exited in X and Z directions, the displacement response was obtained corresponding to all the three accelerometer positions. The top, middle, and bottom displacement values in the X direction are denoted by d_{x1}, d_{x2}, and d_{x3}, respectively. The displacement response revealed linear variation in displacement from top to bottom, d_{x1} the maximum and d_{x3} the minimum in all the cases. When d_{x1}, d_{x2}, and d_{x3} values extend to meet the

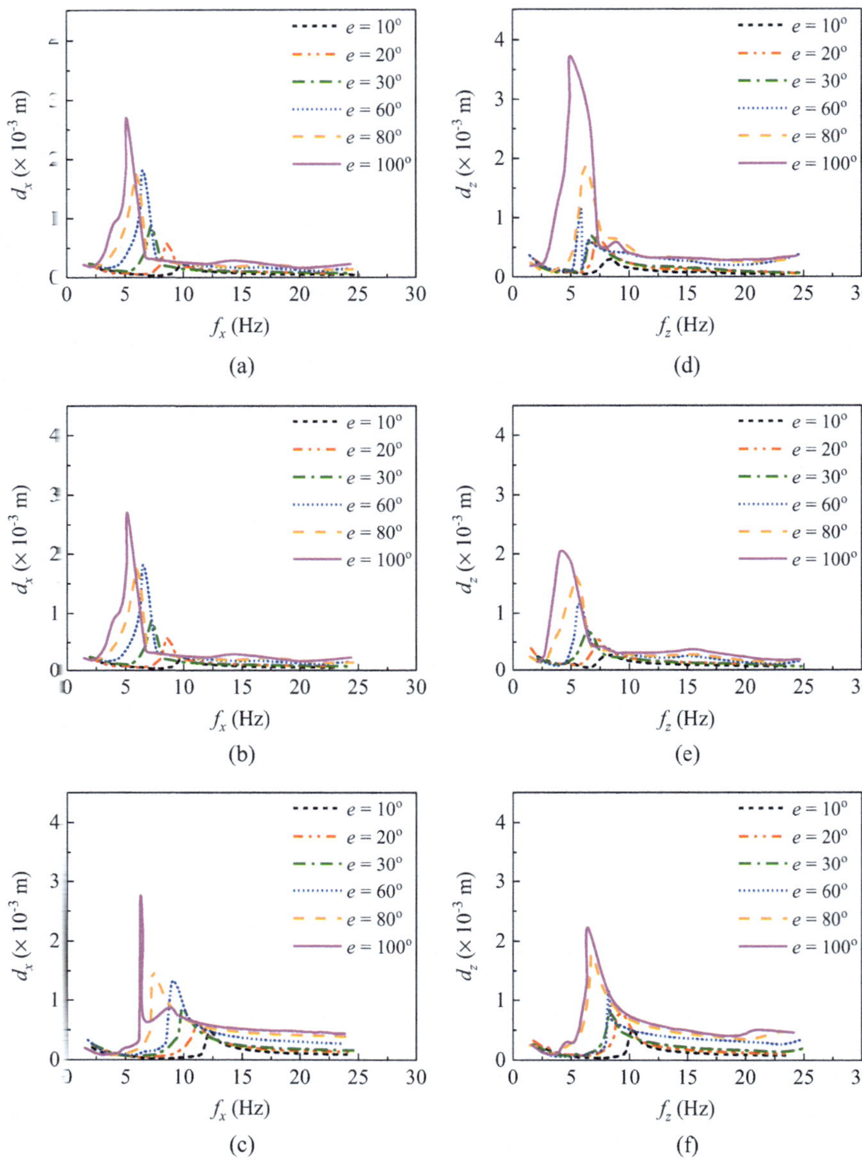

Fig. 8.8 Frequency variation with lateral displacement amplitudes for: **a** B0 in X direction; **b** B10 in X direction; **c** B20 in X direction; **d** B0 in Z direction; **e** B10 in Z direction; and **f** B20 in Z direction

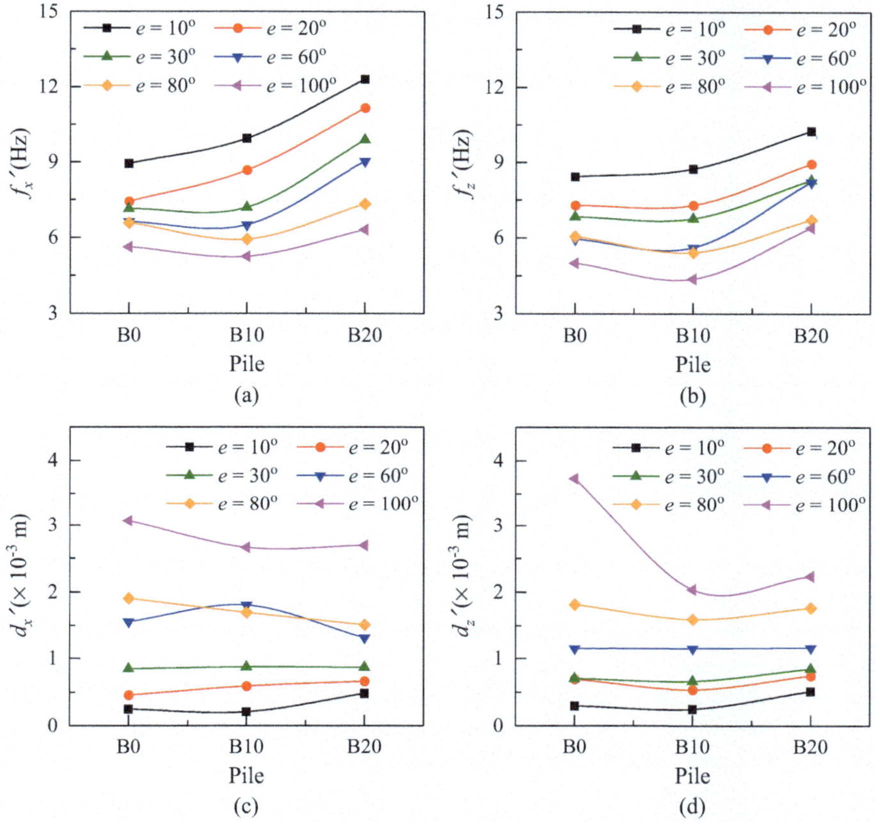

Fig. 8.9 Experimental observations: **a** resonant frequency in X direction; **b** resonant frequency in Z direction; **c** peak displacement amplitude in X direction; and **d** peak displacement amplitude in Z direction

Fig. 8.10 Gap developed along circumference of B0 in Z direction after $e = 100°$

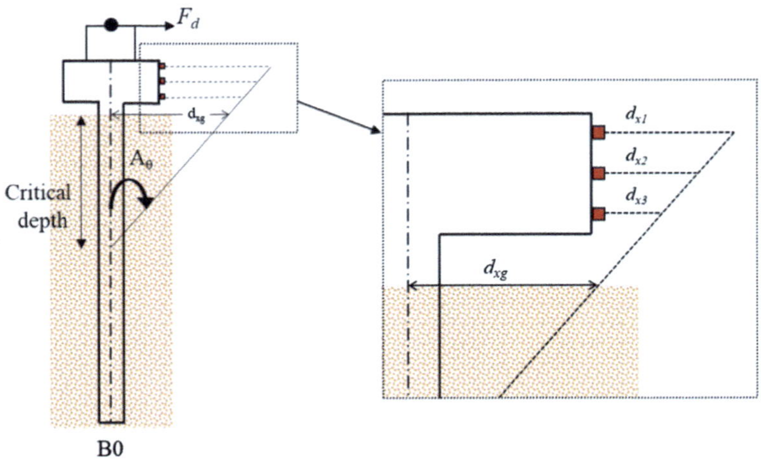

Fig. 8.11 Critical depth

pile, it intersects the pile's center line at a critical depth, as shown in Fig. 8.11. The rotational amplitude, A_θ is obtained from the critical depth and d_{x1}.

Knowing the rotational amplitude and distance between the pile cap and ground level and following the assumption that the lateral displacement at critical depth is zero, the displacement at ground level (d_{xg} or d_{zg}) is attained. The ratio of d_{xg} or d_{zg} and pile diameter is defined as the induced lateral strain (ε_x or ε_z) in this study. The variation of ε_x or ε_z with force level is reported in the following sections. The resonant frequency obtained from the experimental frequency displacement response is assumed as the damped natural frequency of the system (f_{ndx} or f_{ndz}). Then assuming a damping ratio (ξ), undamped natural frequency (f_{nx} or f_{nz}), magnification factor (μ) and static displacement (δ_{stat}) were attained. The exciting moment M_d is obtained at the C.G. of pile cap as the product of its height from C.G. of oscillator mass (0.235 m in this case) to F_d . Then, the slope of M_d against δ_{stat} gives the stiffness of the soil-pile system for an assumed ξ. Now, ξ is recalculated from the stiffness value obtained. This process is repeated until the difference between both assumed and calculated ξ becomes small enough to be neglected. The equations provided in Sect. 5 of BIS [40] has been used for this purpose. Thus the value obtained is the ξ_x or ξ_z, and the corresponding stiffness is $K_{\theta x}$ or $K_{\theta z}$, which are reported in the following sections.

Effect on induced lateral strain

The variation in lateral strain (ε_x or ε_z) against force level (F_{dx}' or F_{dz}') for piles B0, B10, and B20 in X and Z directions are presented in Fig. 8.12a–c and d–f, respectively. To evaluate the force levels (F_{dx}' or F_{dz}') at an eccentricity for a pile, the corresponding resonant frequency has been considered. It can be observed that lateral strain increases with increasing force level in all the cases. The increase in lateral strain with force level was more prominent in the case of B20 compared to B0 and B10. It can be observed that lateral strain has a linear variation depending upon exciting force (F_{dx}' or F_{dz}') and complex nonlinear variation with the batter angle (β).

Effect on the rotational stiffness of the soil-pile system
The rotational stiffness of the soil-pile system ($K_{\theta x}$ or $K_{\theta z}$) in X and Z directions for piles B0, B10, and B20 decreases nonlinearly with an increase in the strain level, as

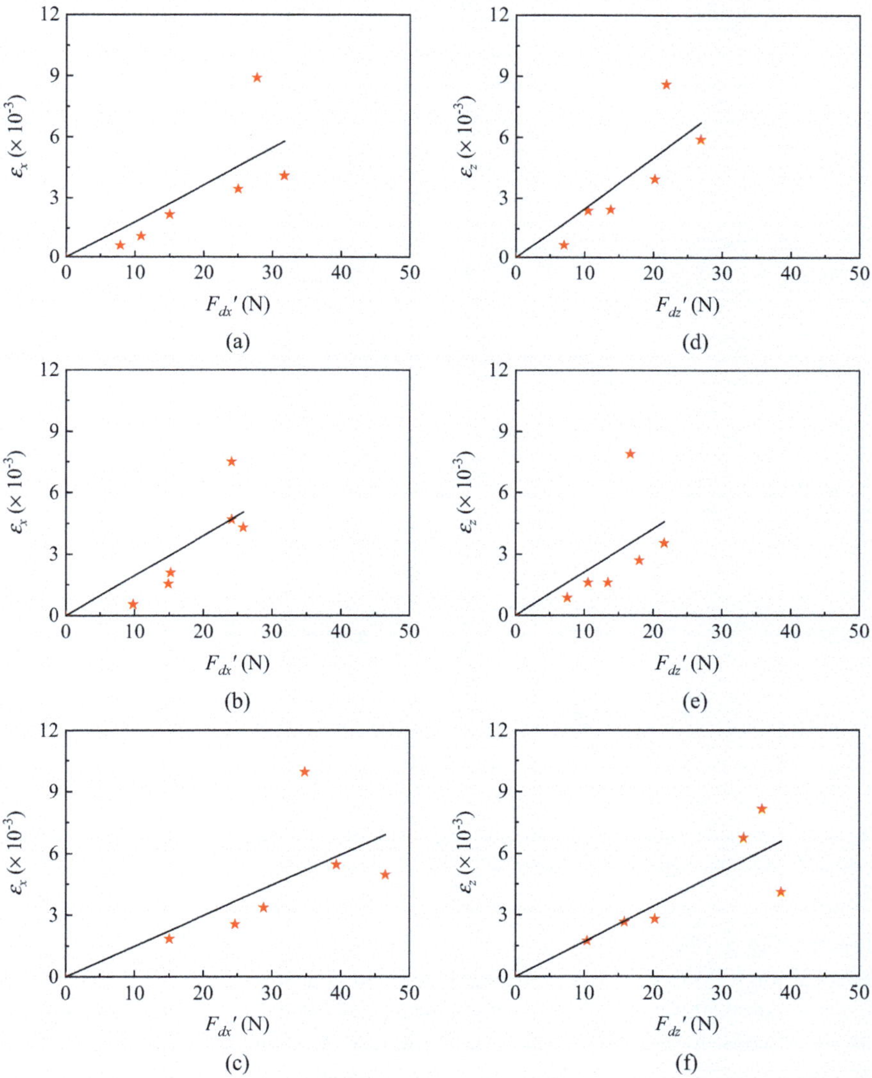

Fig. 8.12 Lateral strain variation for: **a** B0 in X direction; **b** B10 in X direction; **c** B20 in X direction; **d** B0 in Z direction; **e** B10 in Z direction; and **f** B20 in Z direction

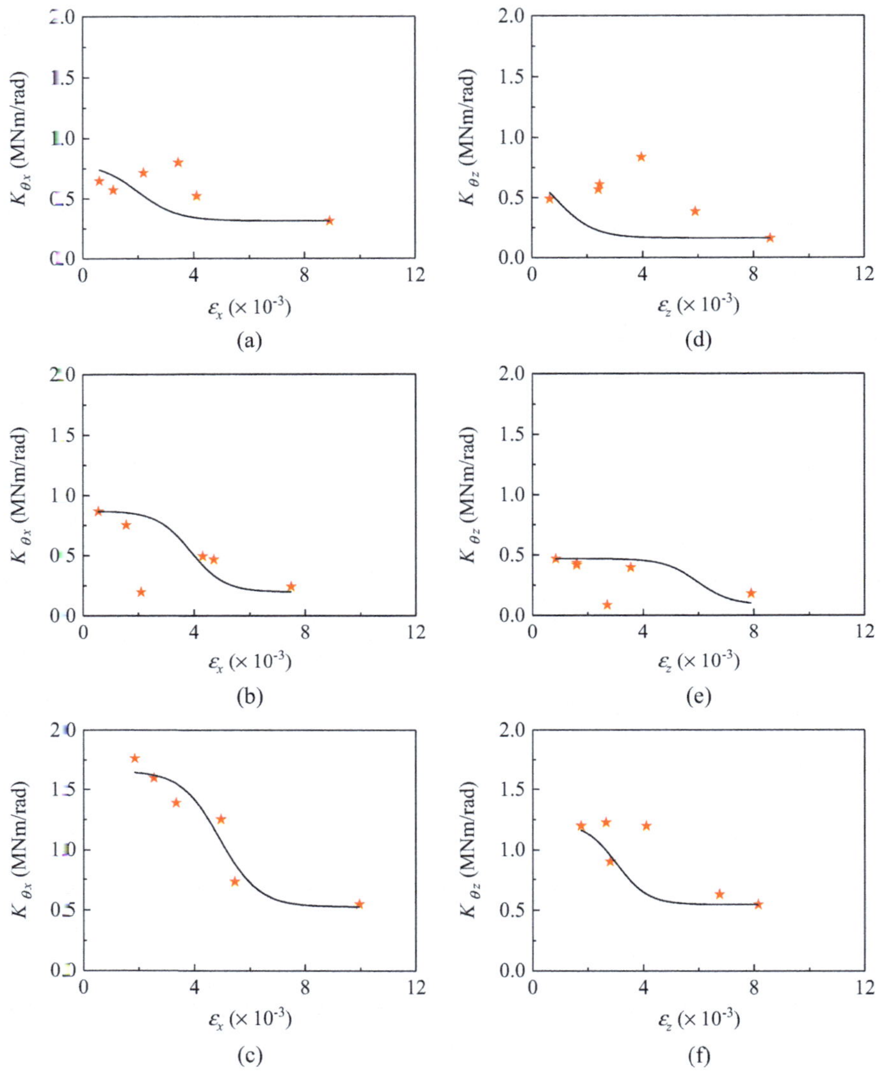

Fig. 8.13 Rotational stiffness for: **a** B0 in X direction; **b** B10 in X direction; **c** B20 in X direction; **d** B0 in Z direction; **e** B10 in Z direction; and **f** B20 in Z direction

shown in Fig. 8.13a–c and d–f, respectively. The variation of rotational stiffness with lateral strain follows a similar nonlinear fashion in almost all cases. The variation of $K_{\theta x}$ is within the range of 1 MNm/rad for both piles B0 and B10, whereas this range further extends up to 2 MNm/rad in case pile B20. It is interesting to note that the nonlinear variation of $K_{\theta z}$ yields a lower value than $K_{\theta x}$, which may be attributed to the fact that in all the piles, the initial test at an eccentricity was carried out in the X direction, and then in the Z direction.

Effect on damping ratio of the soil-pile system
The damping ratios of the soil-pile system (ξ_x or ξ_z) were found to increase nonlinearly with an increase in the strain level, as presented in Fig. 8.14a–c and d–f for piles B0, B10, and B20 excited in both directions (X and Z).

ξ_x and ξ_z vary between 5 and 15% for B0 and B20 in both directions and for B10 in the X direction. In the case of pile B10 in the X direction, the variation is in the range 5–10% but in Z direction, abruptly the damping ratio increased to 30%. This

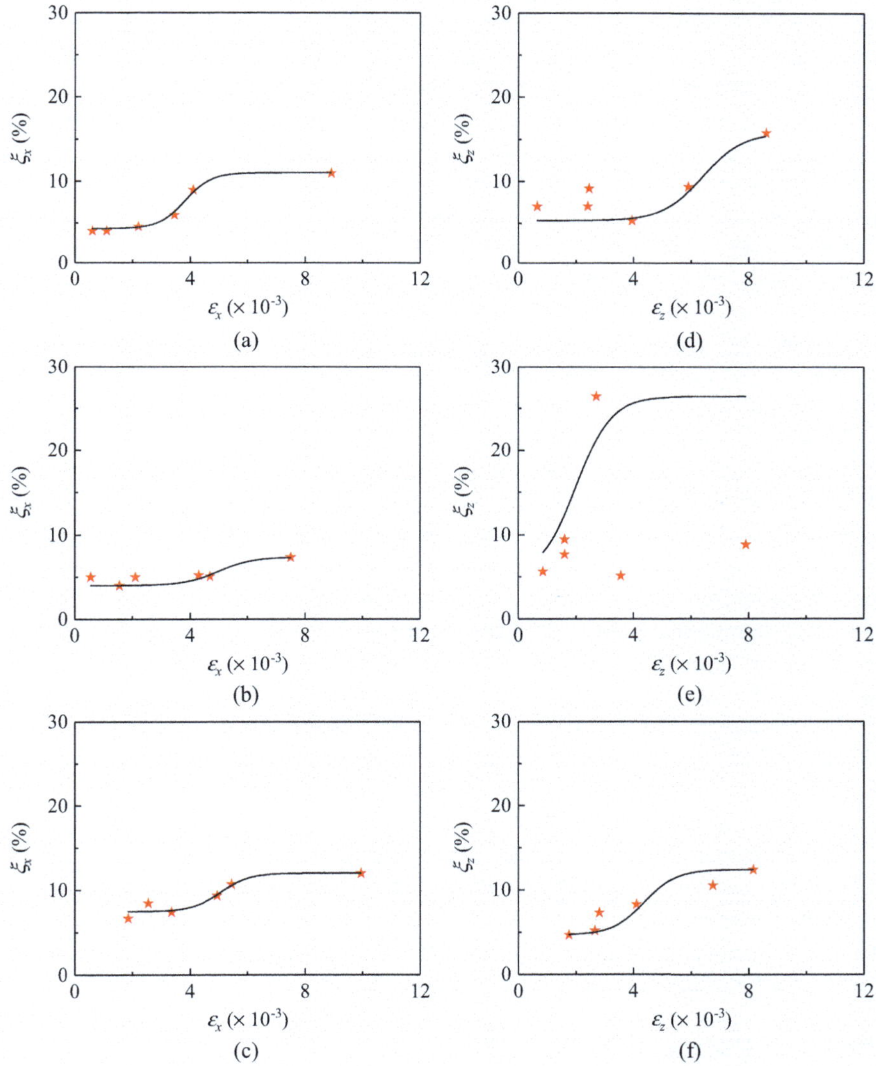

Fig. 8.14 Damping ratio for: **a** B0 in X direction; **b** B10 in X direction; **c** B20 in X direction; **d** B0 in Z direction; **e** B10 in Z direction; and **f** B20 in Z direction

abrupt change in damping ratio might be due to gap formation between the soil and pile surface, which was visible at the ground surface in the form of cracks around the circumference of the pile B10 observed by the authors during the experiment.

8.3.2 Vertical Response

The dynamic vertical response of all three piles is discussed in terms of resonant frequency (f_y'), peak vertical displacement amplitude (d_y'), and axial strain (ε_y). At an eccentricity, the tests in the Y direction were conducted after the Z direction. Then the eccentricity was increased to a higher value and followed the same loading direction cycle (i.e., first X, Z, and then Y). However, after excitation in the Z direction a duration of at least 48 h was maintained as the rest period before testing in Y direction. The variations in vertical displacement amplitude with operating frequency are presented in Fig. 8.15a–c for all considered cases.

It was observed that the difference in vertical displacement obtained from two accelerometers placed on top of the pile cap at different operating frequencies was insignificant in all cases. However, the average of the two was considered for further discussion in this paper. It is obvious from Fig. 8.15a–c that there is only one peak in the frequency displacement response of each considered case confirming the behavior of the soil-pile system in its first fundamental frequency.

The variation in f_y' of piles B0, B10, and B20 at different force levels considered is shown in Fig. 8.16a. It is apparent that the resonant frequency (frequency corresponding to peak displacement) decreases with an increase in force level for each individual pile. With the increase in force level, f_y' decreases by 43–50%. This may be attributed to the fact that the increase in force level leads to the increase in the degradation of the soil mass and hence a decrease in resonant frequency. It is also interesting to note that the f_y' of B0 at a particular force level was the minimum in almost all the cases. The variation of d_y', at all considered force levels is shown in Fig. 8.16b. It is worth noticing that d_y' increases with an increase in force level for each individual pile. At $e = 100°$, d_y' of a pile is almost 4–5 times d_y' of the same pile at $e = 10°$. Similar trend showing an increase in d_y' and a decrease in f_y' with the increase in excitation force on the dynamic vertical vibration field tests on RC vertical piles was reported [45].

ε_y is obtained as the ratio of d_y' and the effective length of pile, L_e. The definition of L_e is depicted in Fig. 8.17. The variation of axial strain induced in these model piles thus obtained is depicted in Fig. 8.18. It can be observed that the axial strain induced is linearly varying with the force level. In all cases, at $e = 100°$ the ε_y is almost four times higher than ε_y of the same pile at $e = 10°$.

Effect of an Underream Bulb

To understand the effect of an underream bulb on inclined piles, the experimental results of B0, B10 and B20 were compared to similar experimental work carried out at the same site, by the authors on vertical (B0U1) and batter piles (B10U1 and

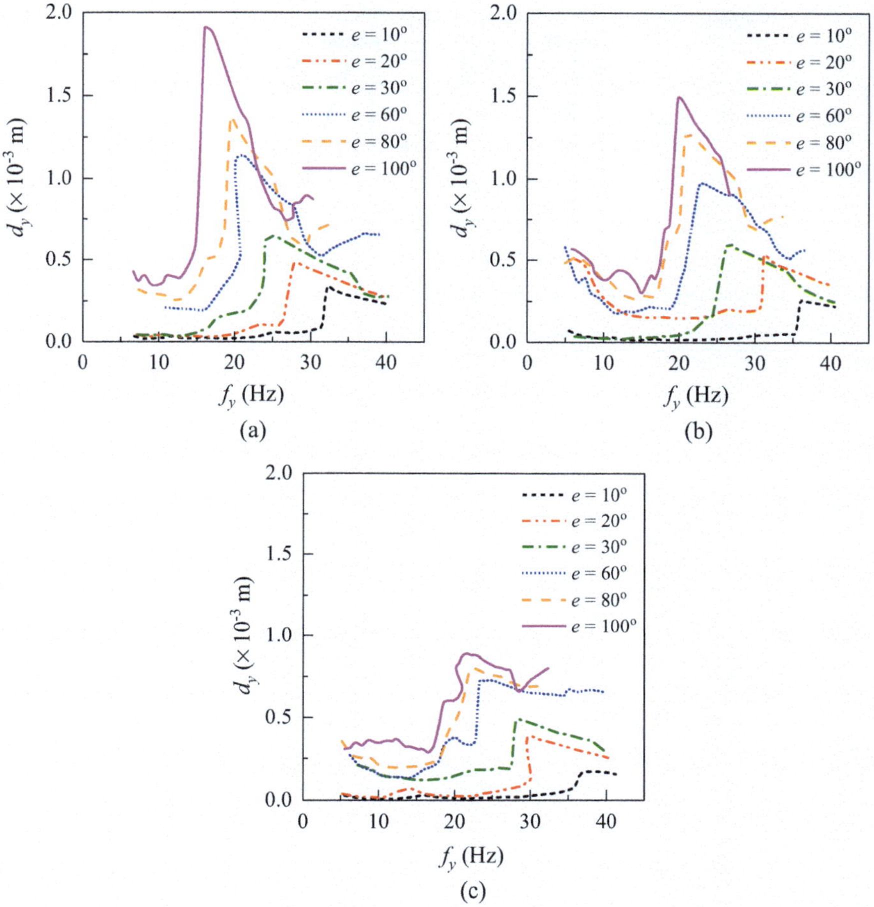

Fig. 8.15 Experimental vertical response of piles: **a** B0; **b** B10; and **c** B20

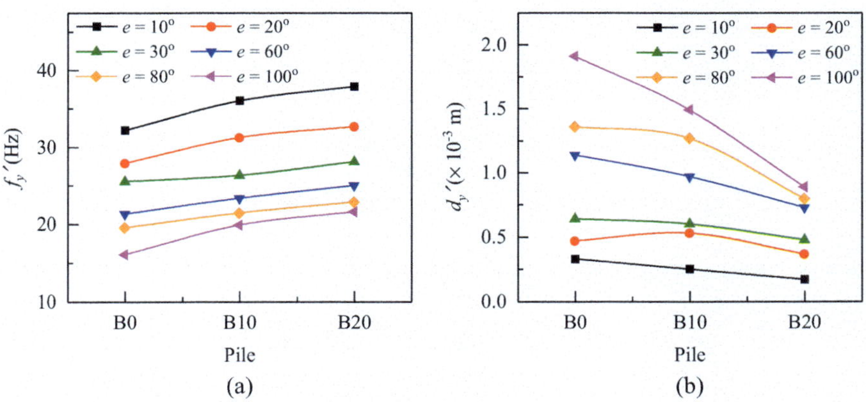

Fig. 8.16 Variation of: **a** f_y; and **b** d_y

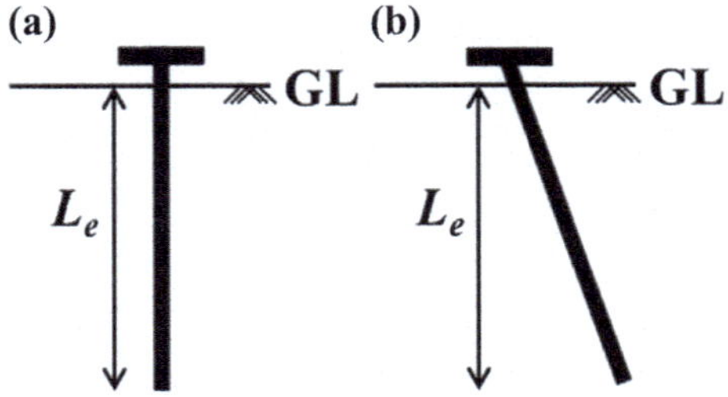

Fig. 8.17 Critical length, L_e for: **a** B0; and **b** B10 and B20

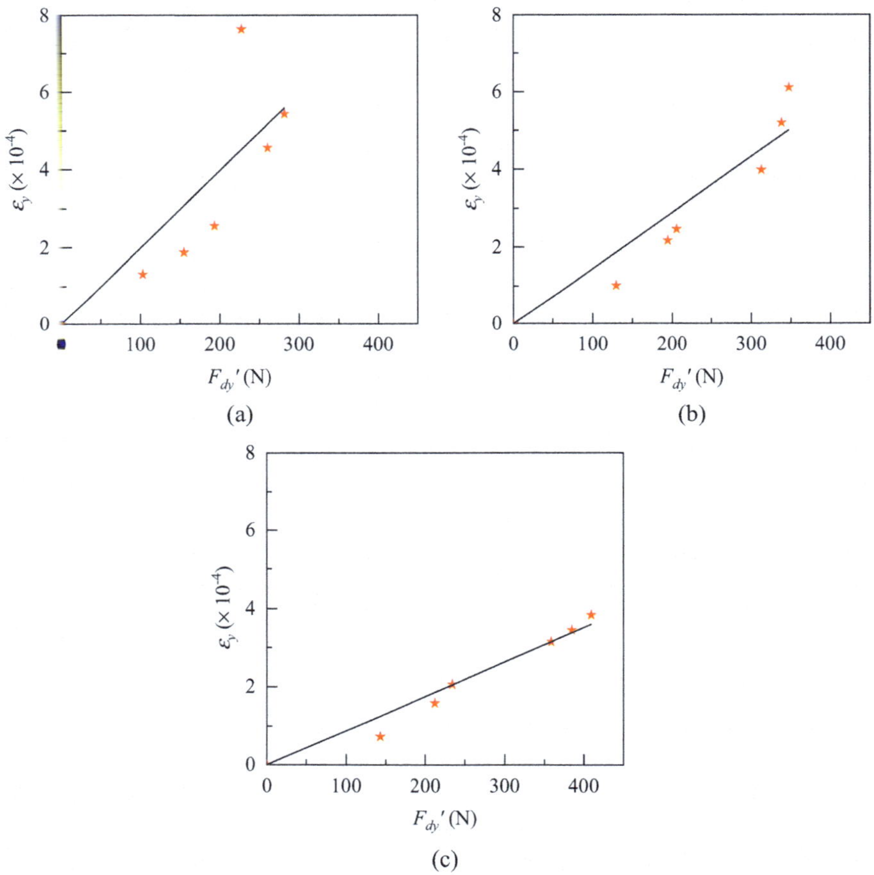

Fig. 8.18 Variation of axial strain: **a** B0; **b** B10; and **c** B20

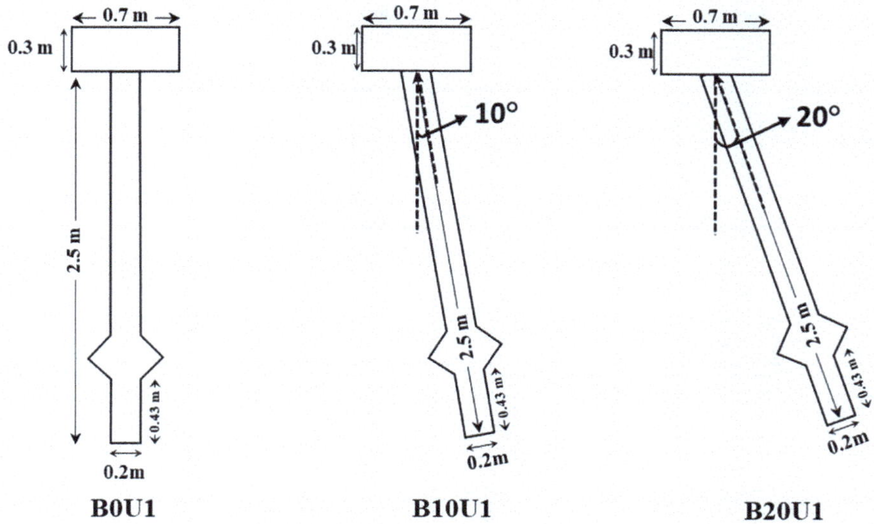

Fig. 8.19 Geometry of piles: **a** B0U1; **b** B10U1; and **c** B20U1

B20U1) subjected to dynamic lateral [29] and vertical [30] loads. The diameter of the bulb in B0U1, B10U1 and B20U1 was 2.5 times the diameter of the pile. The geometry of piles B0U1, B10U1 and B20U1 are presented in Fig. 8.19. The results are compared in terms of d_x', d_z', and d_y' for all the above-mentioned piles and presented in Fig. 8.19a–c, respectively.

It can be observed from Fig. 8.20a–b that the variation in d_x' and d_z' is different for batter piles with varying inclinations (i.e., 10° and 20°). B20U1 has lower d_x' when compared to B20 (till $e = 30°$), beyond which the trend reverses, whereas B10 has lower d_x' in comparison to B10U1. B20U1 experienced lower d_z' when compared to B20, and B10 had lower d_z' when compared to B10U1. The effect of underream bulb on d_y' (Fig. 8.20c) appears marginal till $e = 30°$ and comparatively prominent beyond $e = 30°$, especially for piles with $\beta = 20°$ (B20U0 and B20U1). However, it is interesting to note that B10 and B20 experienced the least vertical displacement when compared to B10U1 and B20U1, respectively. In the X direction, the presence of an underream bulb has insignificant effect on peak displacement at eccentricities less than equal to 30. Whereas, in the Z direction, presence of an underream bulb in batter piles slightly reduces the peak displacement almost at all considered eccentricities. In Y direction, the presence of an underream bulb in a batter pile does not help in reducing the peak vertical displacement.

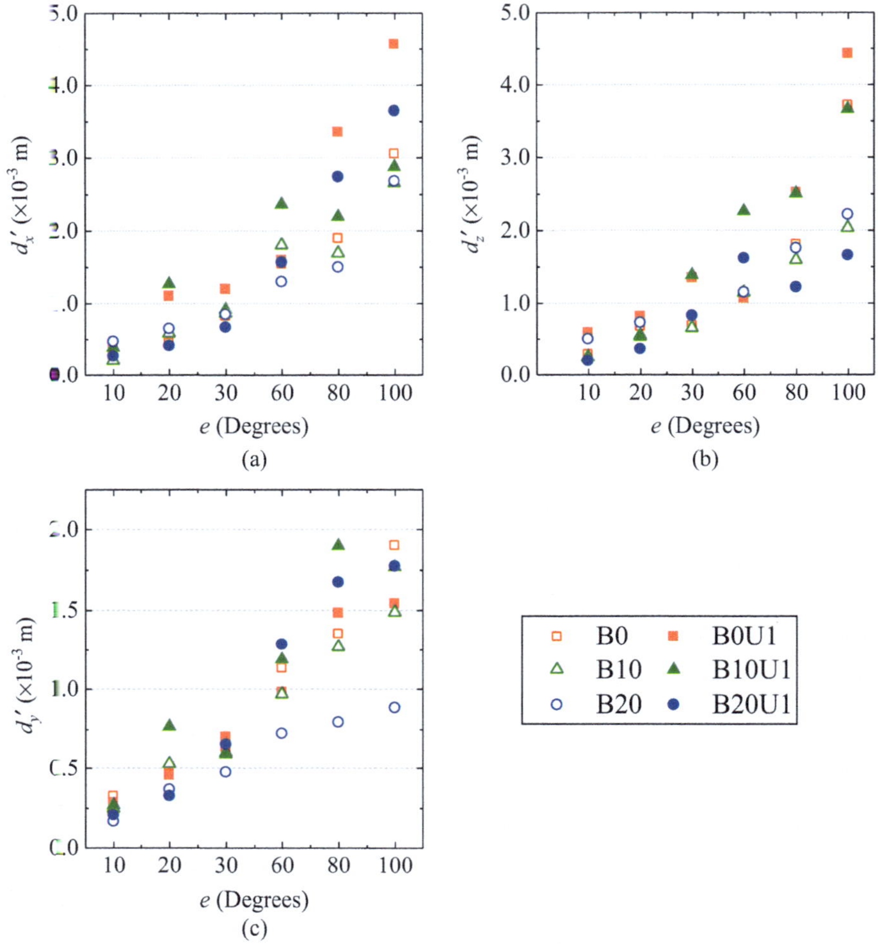

Fig. 8.20 Effect of an underream bulb on dynamic response of piles: **a** in X direction; **b** in Z direction; and **c** in Y direction

8.4 Limitations of the Present Study

The observations reported in the present work are limited to the field experimental results obtained for vertical/batter piles subjected to dynamic lateral and vertical vibration tests in dry conditions. In real situations, the pile dimensions are much larger when compared to the dimensions considered in this study. However, the outcomes of this experimental study can be used to understand the dynamic behavior of single batter piles built in the dry sand. When compared to the pile dimension, the pile cap is comparatively massive. But the pile cap dimension has been selected such that it provides enough plan area to facilitate mounting of the oscillator motor assembly on the top and has appropriate depth to fix the accelerometers for lateral vibration measurements. All the observations reported in this experimental study are based on the acceleration measurements obtained at the pile cap. The effect of varying static vertical load, the gap between the soil and pile interface, residual deformation, and the interaction of pile behavior in lateral and vertical directions have not been considered herein. The effect of residual deformation at lower eccentricities might be insignificant, whereas, at higher eccentricities, residual deformation might be significant.

8.5 Conclusions

Several field tests were conducted on cast in-situ vertical (B0) and batter piles (B10 and B20), installed in silty sand, subjected to dynamic lateral and vertical vibrations individually. The following conclusions can be drawn from the present experimental study as presented above:

- Resonant frequency (f_x' or f_y' or f_z') of the soil-pile system decreases with an increase in the force level (e) in both lateral (X and Z) and vertical (Y) directions. With an increase in force level, the degradation of stiffness in soil mass also increases. Thus, at the particular force level, when the mass participation increases, the resonant frequency (f_x' or f_y' or f_z') of the system automatically decreases. In all the cases, the highest resonant frequency was observed for the 20° batter pile (B20), regardless of the force level and direction of excitation, and this signifies the importance of the batter pile subjected to dynamic lateral load over the vertical pile.
- The peak displacement ($d_x'/d_y'/d_z'$) and lateral strain ($\varepsilon_x/\varepsilon_z$) or axial strain ($\varepsilon_y$) increase linearly with an increase in the force level (e).
- The rotational stiffness ($K_{\theta x}/K_{\theta z}$) decreases nonlinearly, while the damping ratio (ξ_x / ξ_z) increases nonlinearly with an increase in lateral strain ($\varepsilon_x/\varepsilon_z$). It means that with increasing force level (e), $K_{\theta x}$ or $K_{\theta z}$ decreases and ξ_x or ξ_z increases nonlinearly.
- The direction of loading (X or Z) has a significant influence on the dynamic performance of batter piles. This may be attributed to the soil mass participation, stiffness degradation, and the pressure distribution in the two directions of loading in the horizontal plane.
- In the Z direction, presence of an underream bulb in batter piles slightly reduces the peak displacement almost at all considered eccentricities. Whereas, in Y direction, the presence of an underream bulb in a batter pile does not help in reducing the peak vertical displacement.

Acknowledgements Authors acknowledge the Department of Earthquake Engineering, IIT Roorkee for providing all the necessary facilities for this experimental investigation. The first author would like to acknowledge the MoE (formerly MHRD) for providing the Institute Fellowship during the course of her Ph.D. at the Department of Earthquake Engineering, IIT Roorkee.

References

1. Tomlinson M, Woodward J (2015) Pile design and construction practice. CRC Press
2. Juneja A, Mohammed-Aslam AK (2018) Application of a cyclic degradation model for pore pressure accumulation in loose sands and silts subjected to dynamic loading. J Earthq Tsunami 12(05). https://doi.org/10.1142/s1793431118500148
3. Priestley N, Singh J, Youd T, Rollins K (1991) Costa Rica earthquake of April 22, 1991, reconnaissance report. vol 91-02. Earthquake Engineering Research Institute
4. Mitchell D, Tinawi R, Sexsmith RG (1991) Performance of bridges in the 1989 Loma Prieta earthquake—lessons for Canadian designers. Can J Civ Eng 18(4):711–734. https://doi.org/10.1139/l91-085
5. Berrill JB, Christensen SA, Keenan RP, Okada W, Pettinga JR (2001) Case study of lateral spreading forces on a piled foundation. Géotechnique 51(6):501–517. https://doi.org/10.1680/geot.51.6.501.40462
6. Kastranta G, Gazetas G, Tazoh T (1998) Performance of three quay walls in Maya Wharf: Kobe 1995. In: 11th European conference on earthquake engineering, Paris, France
7. Poulos HG (2006) Raked piles virtues and drawbacks. J Geotech Geoenviron Eng 132(6):795–803. https://doi.org/10.1061/(ASCE)1090-0241(2006)132:6(795)
8. AFPS (1990) Association française du génie parasismique, Presses des Ponts et et Chaussees [French Association for Earthquake Engineering, Presses of Bridges and Engineering]
9. EC8 (2003) Eurocode 8 Part 5. Design provisions for earthquake resistance of structures: foundations, retaining structures, and geotechnical aspects. British Standards Institution, London
10. Werner SD (1998) Seismic guidelines for ports. ASCE Publications, Reston, VA
11. AASHTO (2014) LRFD bridge design specifications. Seventh edn. American Association of State Highway and Transportation Officials ©2014, Washington, D.C.
12. Zhang L, McVay M, Lai P (1998) Centrifuge testing of vertically loaded battered pile groups in sand. Geotech Test J 21(4):281–288. https://doi.org/10.1520/GTJ11367J
13. Gerolymos N, Giannakou A, Anastasopoulos I, Gazetas G (2008) Evidence of beneficial role of inclined piles: observations and summary of numerical analyses. Bull Earthq Eng 6(4):705–722. https://doi.org/10.1007/s10518-008-9085-2
14. Escoffier S, Chazelas J-L, Garnier J (2008) Centrifuge modelling of raked piles. Bull Earthq Eng 6(4):689–704. https://doi.org/10.1007/s10518-008-9094-1
15. Tazoh T, Sato J, Jang J, Taji Y, Gazetas G, Anastaspoulos I (2010) Kinematic response of batter pile foundation. In: Soil-Foundation-Structure Interaction. Taylor & Francis Group, London. https://doi.org/10.1201/b10568-8
16. Giannakou A, Gerolymos N, Gazetas G, Tazoh T, Anastasopoulos I (2010) Seismic behavior of batter piles: elastic response. J Geotech Geoenviron Eng 136(9):1187–1199. https://doi.org/10.1061/(ASCE)GT.1943-5606.0000337
17. Ghasemzadeh H, Alibeikloo M (2011) Pile–soil–pile interaction in pile groups with batter piles under dynamic loads. Soil Dyn Earthq Eng 31(8):1159–1170. https://doi.org/10.1016/j.soildyn.2011.04.005
18. Ghazavi M, Ravanshenas P, El Naggar MH (2013) Interaction between inclined pile groups subjected to harmonic vibrations. Soils Found 53(6):789–803. https://doi.org/10.1016/j.sandf.2013.08.009
19. Padrón LA, Aznárez JJ, Maeso O, Santana A (2010) Dynamic stiffness of deep foundations with inclined piles. Earthq Eng Struct Dyn 39(12):1343–1367. https://doi.org/10.1002/eqe.1000

20. Padrón LA, Suárez A, Aznárez JJ, Maeso O (2015) Kinematic internal forces in deep foundations with inclined piles. Earthq Eng Struct Dyn 44(12):2129–2135. https://doi.org/10.1002/eqe.2559
21. Medina C, Padrón LA, Aznárez JJ, Maeso O (2015) Influence of pile inclination angle on the dynamic properties and seismic response of piled structures. Soil Dyn Earthq Eng 69:196–206. https://doi.org/10.1016/j.soildyn.2014.10.027
22. Medina C, Padrón LA, Aznárez JJ, Santana A, Maeso O (2014) Kinematic interaction factors of deep foundations with inclined piles. Earthq Eng Struct Dyn 43(13):2035–2050. https://doi.org/10.1002/eqe.2435
23. Padrón LA, Aznárez JJ, Maeso O, Saitoh M (2012) Impedance functions of end-bearing inclined piles. Soil Dyn Earthq Eng 38:97–108. https://doi.org/10.1016/j.soildyn.2012.01.010
24. Goit CS, Saitoh M (2013) Model tests and numerical analyses on horizontal impedance functions of inclined single piles embedded in cohesionless soil. Earthq Eng Eng Vib 12(1):143–154. https://doi.org/10.1007/s11803-013-0158-0
25. Li Z, Escoffier S, Kotronis P (2016) Centrifuge modeling of batter pile foundations under earthquake excitation. Soil Dyn Earthq Eng 88:176–190. https://doi.org/10.1016/j.soildyn.2016.05.013
26. Li Z, Escoffier S, Kotronis P (2016) Centrifuge modeling of batter pile foundations under sinusoidal dynamic excitation. Bull Earthq Eng 14(3):673–697. https://doi.org/10.1007/s10518-015-9859-2
27. Subramanian RM, Boominathan A (2016) Dynamic experimental studies on lateral behaviour of batter piles in soft clay. Int J Geotech Eng 10(4):317–327. https://doi.org/10.1080/19386362.2016.1150006
28. Wang J, Zhou D, Ji T, Wang S (2017) Horizontal dynamic stiffness and interaction factors of inclined piles. Int J Geomech 17(9):04017075. https://doi.org/10.1061/(ASCE)GM.1943-5622.0000966
29. Bharathi M, Dubey RN (2018) Dynamic lateral response of under-reamed vertical and batter piles. Constr Build Mater 158(Supplement C): 910–920. https://doi.org/10.1016/j.conbuildmat.2017.10.042
30. Bharathi M, Dubey RN, Shukla SK (2022) Dynamic response of underreamed batter piles subjected to vertical vibration. Int J Geotech Eng 16(8):991–999. https://doi.org/10.1080/19386362.2021.2025304
31. Bharathi M, Dubey RN, Shukla SK (2019) Experimental investigation of vertical and batter pile groups subjected to dynamic loads. Soil Dyn Earthq Eng 116:107–119. https://doi.org/10.1016/j.soildyn.2018.10.012
32. Bharathi M, Dubey RN, Shukla SK (2022) Numerical simulation of the dynamic response of batter piles and pile groups. Bull Earthq Eng 20(7):3239–3263. https://doi.org/10.1007/s10518-022-01362-7
33. BIS (1973) IS2720 (Part2). Methods of test for soils: determination of water content. Bureau of Indian Standards, New Delhi, India
34. BIS (1985b) IS2720 (Part5). Methods of test for soils: determination of liquid and plastic limit. Bureau of Indian Standards, New Delhi, India
35. BIS (1985a) IS2720 (Part4). Methods of test for soils: grain size analysis. Bureau of Indian Standards, New Delhi, India
36. BIS (1970) IS1498. Classification and identification of soils for general engineering purposes. Bureau of Indian Standards, New Delhi, India
37. BIS (1981b) IS 6403. Code of practice for determination ofbbearing capacity of shallow foundations. Bureau of Indian Standards, New Delhi, India
38. BIS (2010) IS 2911(Part1/Sec2). Design and construction of pile foundations, cocrete piles, bored cast in-situ concrete piles. Bureau of Indian Standards, New Delhi, India
39. LabVIEWSignalExpress (2009) Getting started with LabVIEW signal express. National Instruments Corporation Austin, Texas, USA
40. BIS (1981a) IS9716. Guide for lateral dynamic load test on piles. Bureau of Indian Standards, New Delhi, India

41. Boominathan A, Ayothiraman R (2005) Dynamic behaviour of laterally loaded model piles in clay. Proc Inst Civil Eng Geotech Eng 158(4):207–215. https://doi.org/10.1680/geng.2005.158.4.207
42. Boominathan A, Ayothiraman R (2007) An experimental study on static and dynamic bending behaviour of piles in soft clay. Geotech Geol Eng 25(2):177. https://doi.org/10.1007/s10706-006-9102-7
43. Manna B, Baidya DK (2010) Nonlinear dynamic response of piles under horizontal excitation. J Geotech Geoenviron Eng 136(12):1600–1609. https://doi.org/10.1061/(ASCE)GT.1943-5606.0000388
44. Boominathan A, Krishna Kumar S, Subramanian R (2015) Lateral dynamic response and effect of weakzone on the stiffness of full scale single piles. Ind Geotech J 45(1):43–50. https://doi.org/10.1007/s40098-014-0106-6
45. Manna B, Baidya DK (2009) Vertical vibration of full-scale pile—analytical and experimental study. J Geotech Geoenviron Eng 135(10):1452–1461. https://doi.org/10.1061/(ASCE)GT.1943-5606.0000110

Chapter 9
A Comparative Study on Seismic Response of Pile in Liquefiable Soils Considering Level and Sloping Ground

G. M. Basavanagowda, S. V. Dinesh, R. Ramesh Babu, and L. Govindaraju

Abstract The catastrophic failure of pile foundations used for the support of bridges and other structures in liquefiable soils is still observed after every major earthquake. This paper presents the results of 3D finite difference analyses conducted to evaluate the influence of a non-liquefiable crust overlaid by a liquefiable crust and the pile is embedded into non-liquefiable soil layer below the liquefiable soil in level and sloping ground. The ground having a slope of 2° with the horizontal consists of three layers such as slightly cemented sand at the top and bottom, sand in the middle and water table at the ground level. Analysis was carried out for different cases of pile tip embedment into the liquefiable and non-liquefiable soil subjected to the 2001 Bhuj Earthquake motion. Results demonstrate that the maximum bending moment occurred at the interface of the liquefiable and non-liquefiable layers. Seismic displacement and bending moment values in the sloping ground are more compared to level ground.

Keywords Pile · Liquefaction · Displacement · Bending moment · Sloping ground

9.1 Introduction

Pile foundations are usually recommended where the soil strata are weak at shallow depth and soil has a very low bearing capacity to transfer the load from the superstructure. Centrifuge modelling of pile response to lateral spreading on different pile foundation models such as end-bearing piles, floating piles, single pile and pile groups with and without pile cap in two- and three-layer liquefiable soil has been reported

G. M. Basavanagowda (✉)
Department of Civil Engineering, RIT, Bangalore 560054, India
e-mail: basavanagowdagm@gmail.com

S. V. Dinesh
Department of Civil Engineering, SIT, Tumkur 572103, India

R. R. Babu
CPRI, Bangalore 560080, India

L. Govindaraju
Department of Civil Engineering, UVCE, Bangalore 560056, India

© Indian Society of Earthquake Technology 2023

R. S. Jakka et al. (eds.), *Earthquake Engineering and Disaster Mitigation*,
Springer Tracts in Civil Engineering, https://doi.org/10.1007/978-981-99-0081-7_9

by Abdoun et al. [1, 2], and the results showed that the bending moment depends on the depth of liquefiable soil and shallow non-liquefiable layers. The maximum bending moment in the pile is due to the presence of a shallow non-liquefiable layer showing passive mode failure. The centrifuge data also shows increased moment due to the presence of pile cap by soil densification around the piles and the maximum bending moment occurs at the bottom of the liquefiable layer at the end of shaking in a three-layered soil for end-bearing piles. The behaviour of piles under earthquake loading in liquefiable soil is complex because of the gradual buildup of pore pressure and the corresponding decrease in strength and stiffness.

Basavanagowda et al. [3] have reported that for a pile group subjected to different earthquake loading, the largest value of bending moment at any time during shaking has been observed at the interface of the liquefied and non-liquefied layers as the soil loses its confining pressure. In the case of sloping ground where lateral spreading can be a major concern for pile foundation, downward movement of the non-liquefied crust has the potential to develop a large bending moment leading to substantial damage due to lateral spreading and also due to inertial loads transmitted to piles [4–7]. Lateral spreading is recognized to be a major concern for pile foundations in sloping grounds when a thick non-liquefied soil layer overlies a liquefied soil layer, and piles are embedded in competent non-liquefiable soil layer below the liquefied soil [8].

Rao et al. [9] using 3D FD analysis investigated the behaviour of pile foundations in loose soil and reported that PGA and deflection at the top of the pile in liquefied soil are more compared to non-liquefiable soils. Hazzar et al. [10] used 3D FD analysis to investigate the influence of vertical load on the lateral response of piles in sand and clay, and they reported that the lateral response of piles does not vary considerably with the vertical load in sandy soils, especially in the loose state. In fact, the scope of previous attempts examining the problem of pile foundation is limited to the behaviour of piles installed in homogenous sandy or clayey soils. Little work has been devoted to the behaviour of piles in homogenous or layered soils which are often encountered in many geotechnical sites. Moreover, mechanisms regarding the influence of liquefiable soils in layered soils may be quite different from those of piles in ideal homogenous situations.

Earthquakes impose lateral loads in the piles through lateral spreading. Structurally, piles are long slender columns with lateral support from the surrounding soil. If unsupported, these columns will fail in buckling instability and not due to the crushing of pile material [11]. In addition, piles can also fail by bending or shear and failure of soil (i.e., excessive settlement). Further, these failure mechanisms may interact with each other. In order to understand this phenomenon of failure of piles in liquefiable soils, the present study investigates the possible causes of the failure of pile foundations.

9.2 Numerical Modelling

The 3D finite difference (FD) FLAC3D [4] was used to study the behaviour of piles under the 2001 Bhuj earthquake ground motion. The complete 3D geometric model was used to represent the coupled soil-pile system.

The stress-strain behaviour of soil-strata is considered as elasto-plastic in nature, and the most extensively used Mohr–Coulomb constitutive model was selected to model non-liquefiable layer, and the Byrne [12] constitutive model is chosen to model liquefiable sand layer. The parameters required to effectively define soil behaviour are the elastic bulk modulus (K), elastic shear modulus (G), density (ρ), angle of internal friction (Φ) and cohesion (C).

9.2.1 *Validation Study for Static Loading*

Before analysing the pile behaviour under seismic loading, the applicability of the model was verified by evaluating the pile response under lateral loading from the FD half model of Hazzar et al. [10]. The soil block consists of a single layer of very dense sand. The pile length is 10 m having a diameter of 1 m. Soil thickness is considered as 16 m. The Geotechnical properties of soil strata and the pile used in the analysis are summarized in Table 9.1.

Figure 9.1 shows the general layout and meshing of the FLAC 3D model used for the analysis of the soil-pile system. Lateral load was applied gradually at the pile head till the lateral deflection of the pile reaches 100 mm without any vertical load. Figure 9.2 shows the lateral displacement of the pile due to the applied lateral load. It is observed that, as lateral load increases, lateral displacement of the pile increases. The results of the present study are compared with Hazzar et al. [10] and are indicated in Fig. 9.2. The results obtained are in close agreement with the results reported.

Table 9.1 Soil properties used for the static analysis [4]

Properties	Very dense sand	Pile
Cohesion [Cu] (kPa)	0	–
Density (kg/m^3)	2200	2300
Friction [Ø] (°)	42	–
Elastic bulk modulus [K] (MPa)	58.3	–
Elastic modulus [E] (GPa)	–	25
Elastic shear modulus [G] (MPa)	26.9	10.87
Poisson's ratio [μ]	0.2	0.15

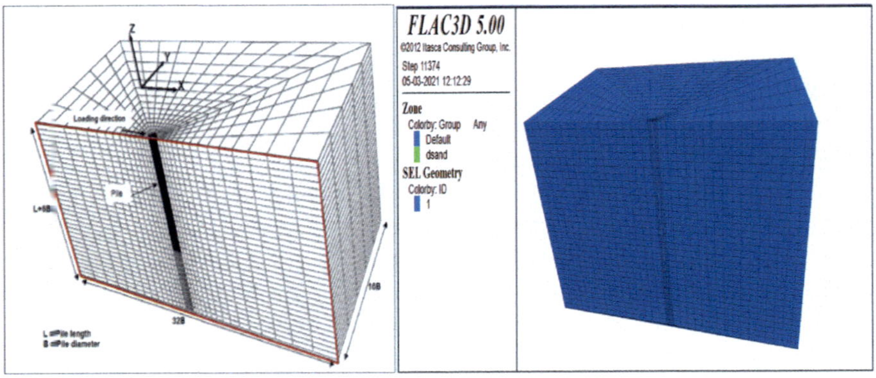

Fig. 9 1 General layout and meshing of FLAC 3D model

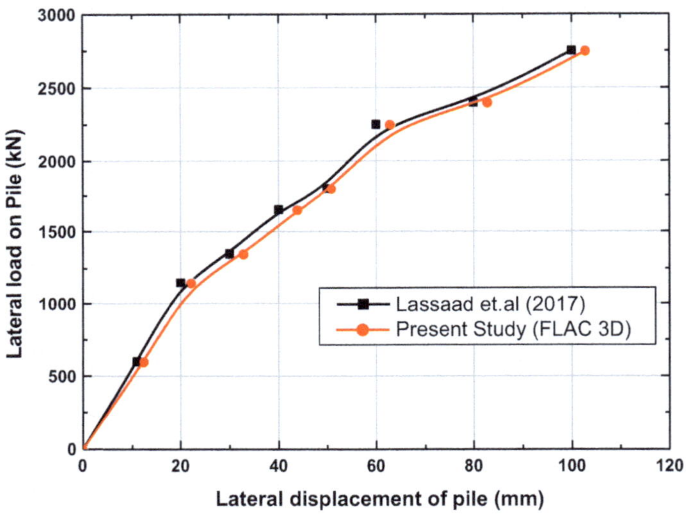

Fig. 9.2 Lateral load versus lateral displacement of pile

9.2.2 Validation Study for Dynamic Loading

The applicability of the adopted FLAC 3-D model was verified for dynamic loading by studying the pile response from the published centrifuge model studies. Abdoun et al. [2] reported the response of a 10 m long and 0.6 m diameter pile embedded in a three-layered strata consisting of non-liquefiable sand (slightly cemented sand) of 2 m thickness at the top and bottom, liquefiable sand (Nevada sand) layer extending to a depth of 6 m in between as shown in Fig. 4a. Soil-pile-pile cap model developed in the present study using FLAC 3D is shown in Fig. 4b. The geotechnical properties of soil layers in the model are summarized in Table 9.2. The dimensions of the pile cap are 2.5 m × 2.5 m made out of aluminium having a thickness of 0.5 m firmly connected to the pile made of polyetherimide rod. The water table is introduced at the ground surface. Byrne (1991) and Mohr–Coulomb constitutive models are adopted to model liquefiable and non-liquefiable sand layers, respectively. The soil model was subjected to harmonic motion of 0.3 g acceleration as shown in Fig. 9.3 at the base of the model. The frequency of harmonic motion was 2Hz. The normal and shear stiffness values calculated using Eqs. (9.1) and (9.2) are 64×10^7 N/m and 57.6×10^7 N/m, respectively. The interaction between soil and the pile cap obtained by Eq. (9.5) is found to be 897.3×10^8 N/m. The variation in bending moment along the depth between the centrifuge study and FLAC 3D modelling is shown in Fig. 9.5. The FLAC 3D results are in accordance with the results reported by Abdoun et al. [1].

Table 9.2 Properties of soil and pile [13, 14]

Properties	Soil types		Pile	Pile cap
	Nevada sand (Liquefiable sand)	Slightly cemented sand (Non-liquefiable sand)	Polyetherimide	Aluminium
Cohesion [Cu] (kPa)	–	5.1	–	–
Density (kg/m^3)	1800	1900	1270	2830
Friction angle [φ] (°)	30	34.5	–	–
Poisson's ratio [μ]	0.25	0.3	0.36	0.35
Relative density [%]	40	–	–	–
Young's modulus [E] (MPa)	10	60	1260	69,000

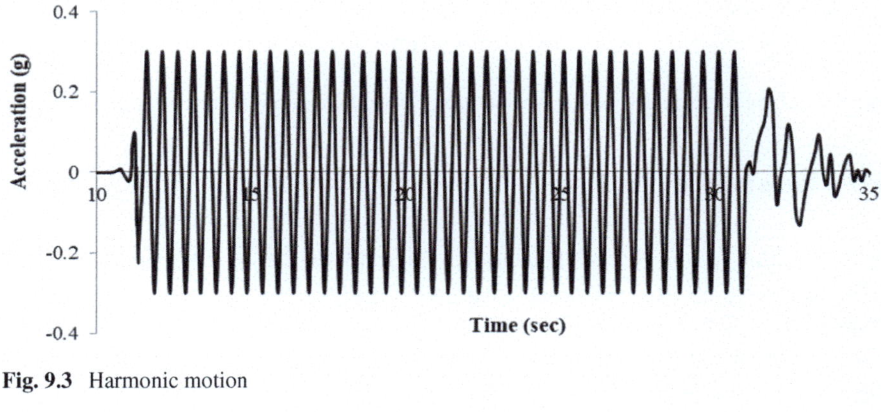

Fig. 9.3 Harmonic motion

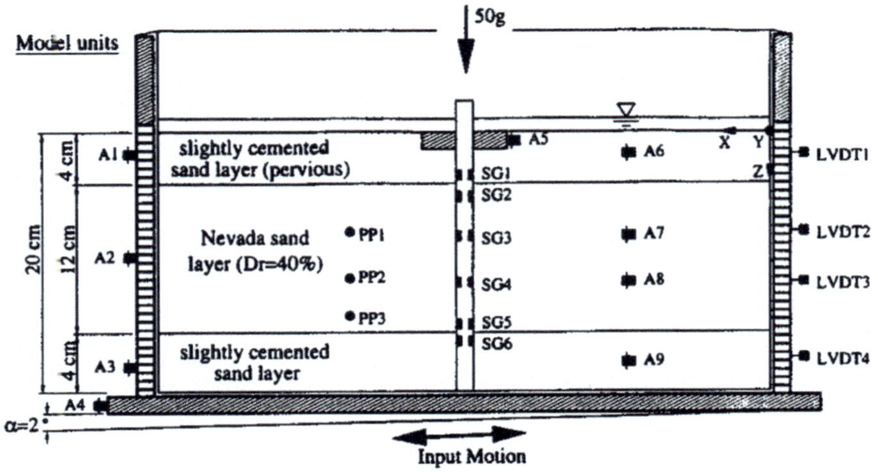

(a)

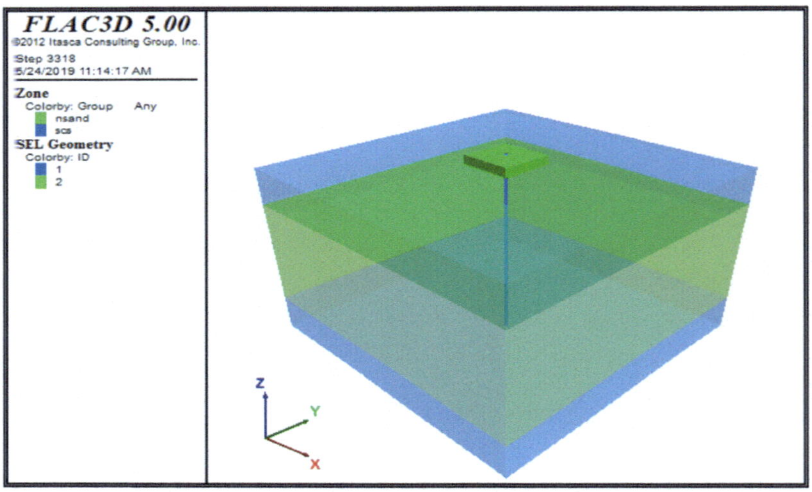

(b)

Fig. 9.4 **a** Centrifuge model used by Abdoun et al. [2] and **b** model developed in the present study

9.3 Parametric Studies

In this section, a pile is modelled in FLAC3D [15] which is a 3D "explicit finite difference program" that performs "Lagrangian analysis" for engineering mechanics computation. A free-headed pile of length 9 m and diameter 0.6 m is modelled in a soil grid of 15 m × 15 m × 10 m dimension consisting of three layers in which top and bottom 2 m thickness is slightly cemented sand (SCS), and in between 6 m Nevada sand (Nsand) layer in level ground condition and ground having a slope of 2°

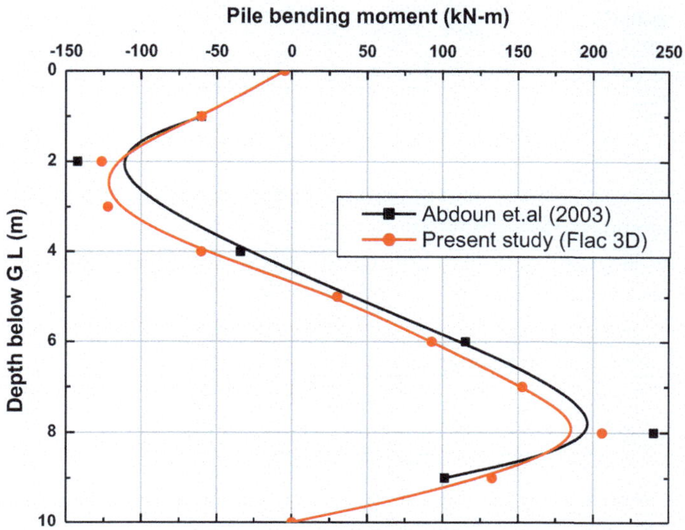

Fig. 9.5 Comparison between present FLAC 3D and centrifuge study data of Abdoun et al. [2]

with the horizontal is considered. The water table is introduced at the ground level. The properties of Nsand, SCS and pile are tabulated in Table 9.2.

9.3.1 Modelling Procedure

The coordinate axes for the FLAC 3D model are located with the origin at the bottom and z-axis oriented along the pile axes and downward. The top of model, at $z = 10$ m, is a free surface. The base of the model, at $z = 0$ m, is fixed in x, y and z directions, and roller boundaries are provided on the sides of the model and further increase in the dimension of the soil block will not affect the results.

Pile was modelled as the collection of pile structural elements (pileSELs). PileSELs are two-noded finite elements having six degrees of freedom per node. The elastic bulk modulus (K_p), elastic shear modulus (G_p) and density of pile (ρ_p) are the three parameters used to define pile material behaviour.

9.3.2 Soil-Pile Interface Model

The pile was divided into 32 segments and as the pile capacity is a function of skin friction resistance along the pile shaft and the end-bearing capacity at the pile tip, skin friction resistance is modelled through normal and shear coupling springs, and these springs transfer forces and motions between the soil grid, pile and pile cap.

Table 9.3 Soil-pile interface properties [IS:2911(Part 1, Sec4)-1985 and Timoshenko and Goodier [16]]

Interface properties	Soil-pile interface
Normal stiffness (K_n) (N/m)	6.4×10^6
Shear stiffness (K_s) (N/m)	5.76×10^6

These springs are non-linear, and the normal and shear behaviour at the interface is both cohesive and frictional (Table 9.3).

The interactions are represented by normal stiffness (K_n) and shear stiffness (K_s) as per Timoshenko and Goodier [16] and are given as follows:

$$\text{Normal stiffness, } K_n = \frac{4\text{Gr}}{1-\mu} \quad (9.1)$$

$$\text{Shear stiffness, } K_s = \frac{32(1-\mu)Gr}{7-8\mu} \quad (9.2)$$

where G = Shear modulus of soil (N/m)

r_0 = Radius of pile (m) and

μ = Poisson's ratio of soil (N/m).

The bulk modulus (K) and shear modulus (G) were calculated using the following equations:

$$K = \frac{E}{3(1-2\mu)} \quad (9.3)$$

$$G = \frac{E}{3(1+\mu)} \quad (9.4)$$

Using Eqs. (9.1) and (9.2), the values of normal stiffness and shear stiffness obtained are 6.4×10^6 N/m and 5.76×10^6 N/m, respectively.

In dynamic analysis, free field boundary conditions are provided to reduce the reflection of input waves thereby simulating the field conditions. The default model of hysteresis damping was chosen.

The interaction between soil and the pile cap is represented through the normal spring stiffness and shear spring stiffness calculated using the equation:

$$K_n = K_s = \frac{K + \frac{4}{3}G}{\Delta Z_{\min}} \quad (9.5)$$

To study the response of the pile in liquefiable soils for level ground and sloping ground, Bhuj ground motion (2001) of 0.106g peak ground acceleration shown in Fig. 9 6 is considered.

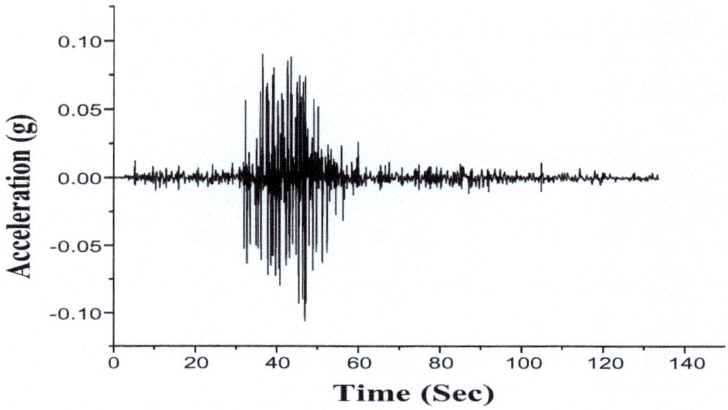

Fig. 9.6 2001 Bhuj earthquake ground motion

9.4 Results and Discussions

The soil-pile model properties considered by Abdoun et al. [2] consist of top and bottom 2 m thick non-liquefiable sand (SCS) and in between 6 m thickness of liquefiable sand (Nsand). Three cases of soil model (Case A, Case B and Case C) were considered both in level ground and sloping ground, and the pile is embedded at the centre of the soil block. The 2001 Bhuj earthquake ground motion (Fig. 9.6) is applied at the base of the model.

9.4.1 Soil-Pile Interaction in Level Ground

9.4.1.1 Case A: Non-liquefiable Crust Overlies Liquefiable Soil, and Pile is Embedded in Non-liquefiable Soil Below the Liquefiable Soil

The typical layered system configuration is shown in Fig. 7a. The pile is supported at the top and bottom non-liquefiable layers which act as fixed support for the pile. Due to the applied dynamic load, pore water pressure builds up in the soil model and as a result pore pressure ratio reaches a value of 1 as shown in Fig. 7b causing the sand (Nsand) to liquefy. The lateral displacement of the pile top reaches a value of 46 mm and tends to zero at the tip of the pile as shown in Fig 7c. The liquefaction of sand (middle layer) resulted in exerting maximum bending moment in the pile at the top and bottom interface. The maximum bending moment at the top interface is 30 kNm and negative moment of 20.5 kNm at the bottom interface as shown in Fig 7d. The vertical settlement of the pile tip is observed to be 0.79 mm due to the self-weight of the pile itself as there is no axial load on the pile under the earthquake loading.

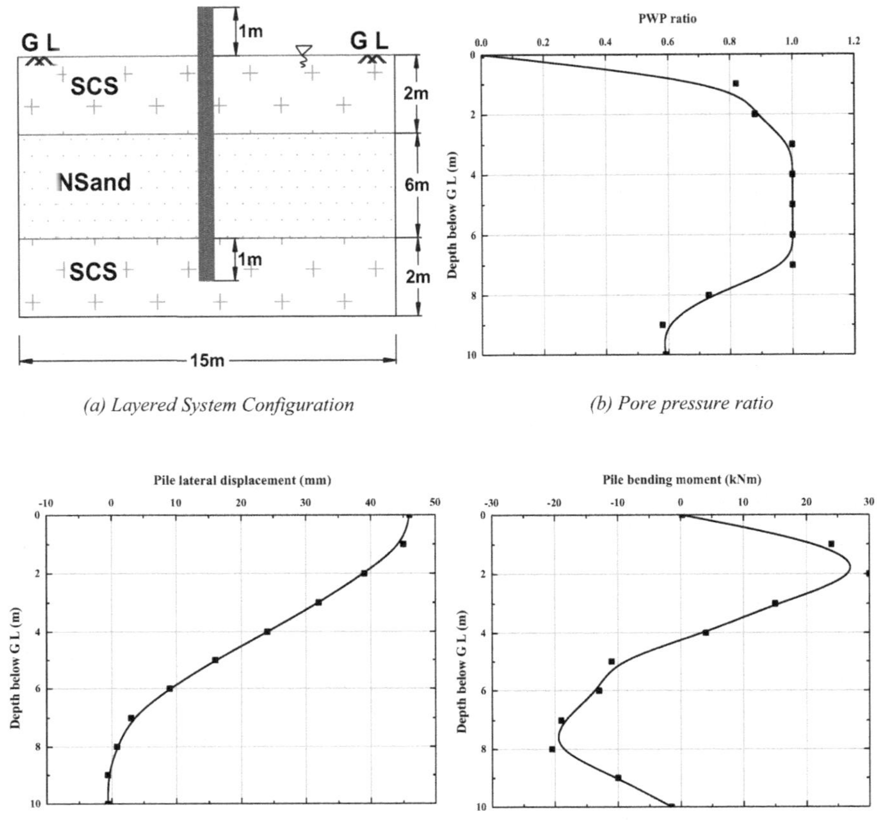

(a) Layered System Configuration

(b) Pore pressure ratio

(c) Pile lateral displacement

(d) Pile bending moment

Fig. 9.7 Profiles of **a** layered configuration, **b** pore pressure ratio, **c** lateral displacement and **d** bending moment for Case A in level ground condition

9.4.1.2 Case B: The Top Non-liquefiable Layer is Absent, and the Pile is Embedded in Non-liquefiable Soil Below the Liquefiable Soil

When the top non-liquefiable layer is absent (Fig. 9.8a), the pile will be unsupported at the top and supported at the bottom (1 m into the non-liquefiable layer), thereby it behaves like a cantilever beam supported by the non-liquefiable soil.

The dynamic load applied at the base of the model develops pore water pressure resulting in a pore pressure ratio of 1 in liquefiable sand (Nsand) (Fig. 9.8b). As a result, the lateral displacement of the pile is found to be 26 mm at the top as shown in Fig. 9.8c, and becomes zero at the tip of the pile. The maximum bending moment induced in the pile is 9.3 kNm at 2 m depth and the fixity of the pile in the non-liquefied crust yields a negative moment of 6.1 kNm at the bottom interface of 8 m below the ground as shown in Fig. 9.8d. Further, in the absence of vertical load, the vertical settlement of the pile tip is 0.64 mm.

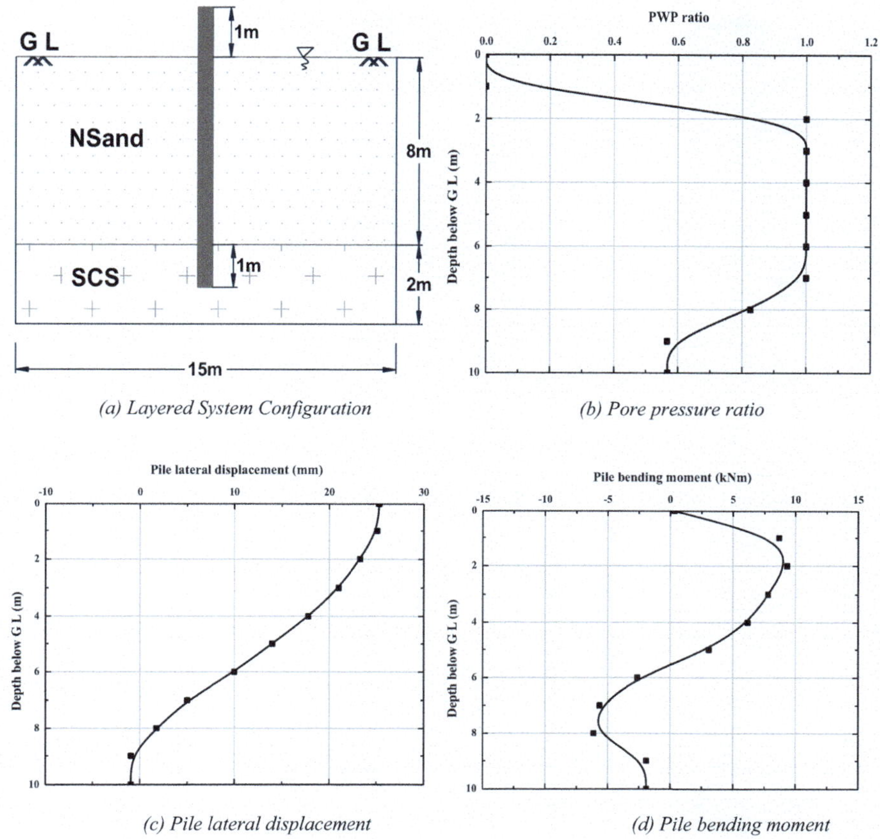

Fig. 9.8 Profiles of **a** layered configuration, **b** pore pressure ratio, **c** pile lateral displacement and **d** pile bending moment for Case B in level ground condition

9.4.1.3 Case C: Thin Non-liquefiable Crust is Overlaid by Liquefiable Soil, and the Pile Tip Rests on Liquefiable Soil

In this case, the pile is embedded completely in liquefiable soil (Fig. 9.9a), and it is simply supported by the soil surrounding the pile as there is no support at the top and bottom of the pile. The pore pressure develops in the soil due to dynamic load, the pore pressure ratio attains a value of 1 and the lateral displacement of the pile top is found to be 12 mm as shown in Fig. 9.9c. Figure 9.9d shows the bending moment induced in the pile due to the liquefied soil exerting lateral pressure similar to the laterally loaded pile. Further, due to self-weight of the pile during shaking, it has shown a vertical settlement of 1.05 mm.

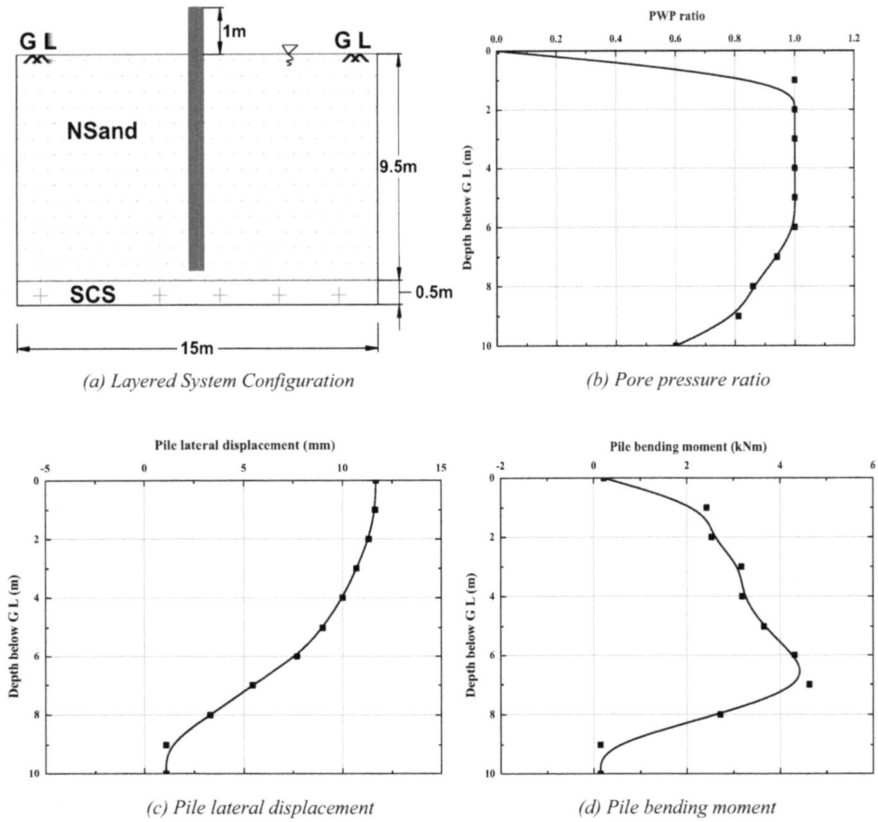

(a) Layered System Configuration *(b) Pore pressure ratio*

(c) Pile lateral displacement *(d) Pile bending moment*

Fig. 9.9 Profiles of **a** layered configuration, **b** pore pressure ratio, **c** pile lateral displacement and **d** pile bending moment for Case C in level ground condition

9.4.2 Soil-Pile Interaction in Sloping Ground

The current understanding of pile failure is based on the bending mechanism where lateral loads due to slope movements induce bending in the pile. The permanent lateral deformation is reported to be the main source of distress in piles. To simulate the effect of sloping ground on the pile response, the soil profile considered above was provided with a gentle slope of 2° with respect to the horizontal as shown in Fig. 9.10a.

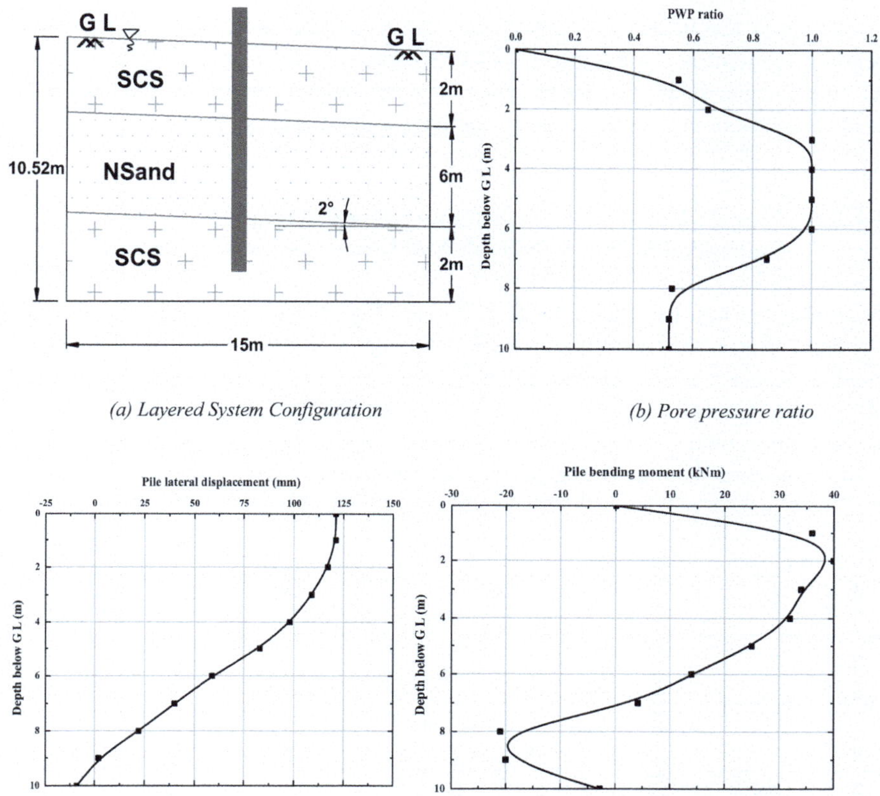

(a) Layered System Configuration

(b) Pore pressure ratio

(c) Pile lateral displacement

(d) Pile bending moment

Fig. 9.10 Profiles of **a** layered configuration, **b** pore pressure ratio, **c** pile lateral displacement and **d** pile bending moment for Case A in sloping ground condition

9.4.2.1 Case A: Non-liquefiable Crust Overlies Liquefiable Soil, and Pile is Embedded in Non-liquefiable Soil Below the Liquefiable Soil

Figure 9.10a shows the schematic of the soil-pile model. The pile is embedded at the centre of the soil block wherein the top and bottom 2 m is of slightly cemented sand and at the middle 6 m is of Nevada sand (Nsand) forming a total thickness of soil profile as 10 m on one side and by providing gentle slope of 2°, the thickness on the other side of the model is 10.52 m. When the tip of the pile is embedded sufficiently into the non-liquefiable layer, as the dynamic load is applied, the pore pressure ratio reaches a value of 1 resulting in liquefaction of sand (Nsand), thereby the lateral displacement of the pile is 121 mm at the ground level and reduces to zero at the tip of the pile.

Further at the interface of liquefiable and non-liquefiable soils, the maximum bending moment is observed to be 40 kNm and −21 kNm at the top and bottom interface, respectively. The settlement of pile tip under the absence of axial load is 1.02 mm is more in sloping ground compared to level ground.

9.4.2.2 Case B: The Top Non-liquefiable Layer is Absent, and the Pile is Embedded in Non-liquefiable Soil Below the Liquefiable Soil

When the top non-liquefiable layer is absent (Fig. 9.11a), the pore pressure ratio reaches 1 (Fig. 9.11b) resulting in the liquefaction of Nsand. Figure 9.11c shows the maximum lateral pile top displacement of 89 mm and the maximum bending moment of 22 kNm is developed at 2 m depth and has attained a negative value of 20 kNm at the bottom interface (Fig. 9.11d). The pile tip has shown vertical settlement of 1.305 mm which is more in sloping ground compared to level ground.

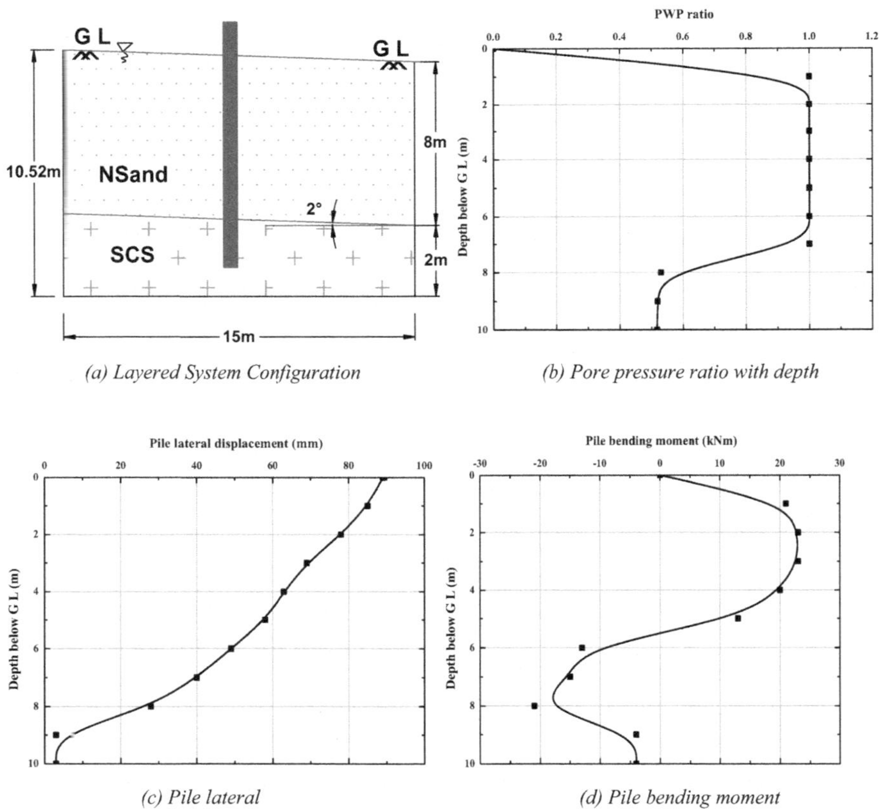

Fig. 9.11 Profiles of **a** layered configuration, **b** pore pressure ratio, **c** pile lateral displacement and **d** pile bending moment for Case B in sloping ground condition

9.4.2.3 Case C: When Thin Non-liquefiable Crust is Overlaid by Liquefied Soil, and Pile Tip Rests on Liquefied Soil

Figure 9.12a shows the schematic of a layered system in sloping ground. As a result of the pore pressure ratio reaching a value of 1, the complete liquefaction of Nsand exerts a lateral load on the pile, thereby the lateral displacement of the pile top is 53 mm and the maximum bending moment attained is 16 kNm at 7 m depth. The vertical settlement of the pile is found to be 1.4 mm which is more compared to level ground.

Figure 9.13a–c shows the comparison between the displacement values and Fig. 9.14a–c shows the comparison between bending moment values in level and sloping ground for all three cases which clearly represents that in the case of sloping ground, vulnerability is very severe in terms of increased displacement and bending moments in the pile.

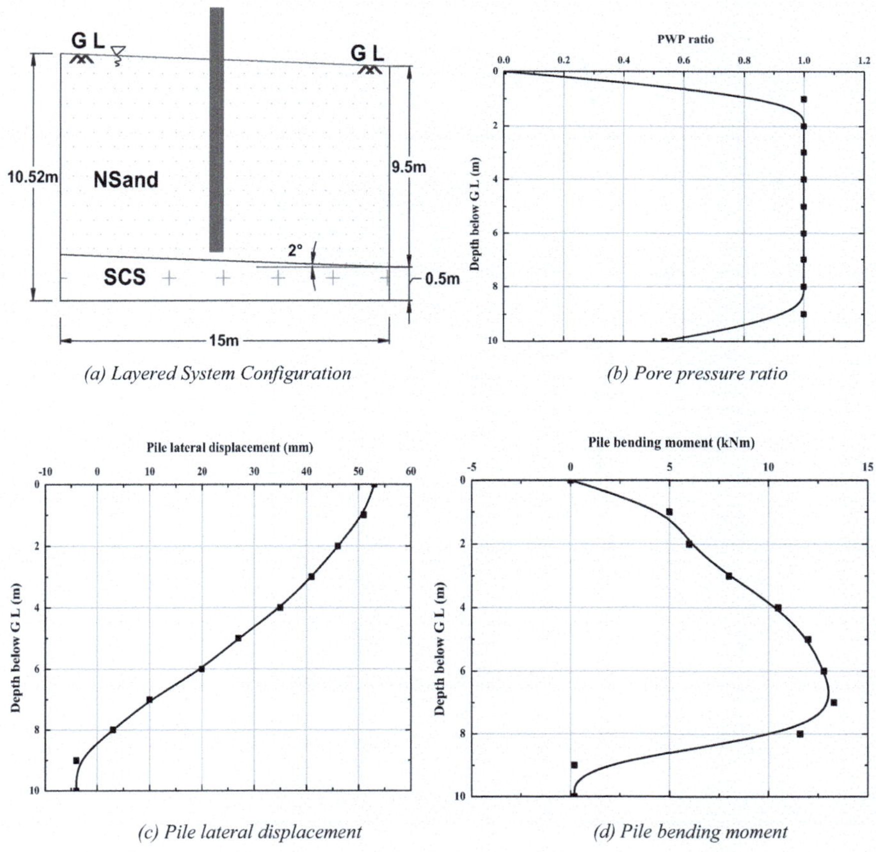

Fig. 9.12 Profiles of **a** layered configuration, **b** pore pressure ratio, **c** pile lateral displacement and **d** pile bending moment for Case C in sloping ground condition

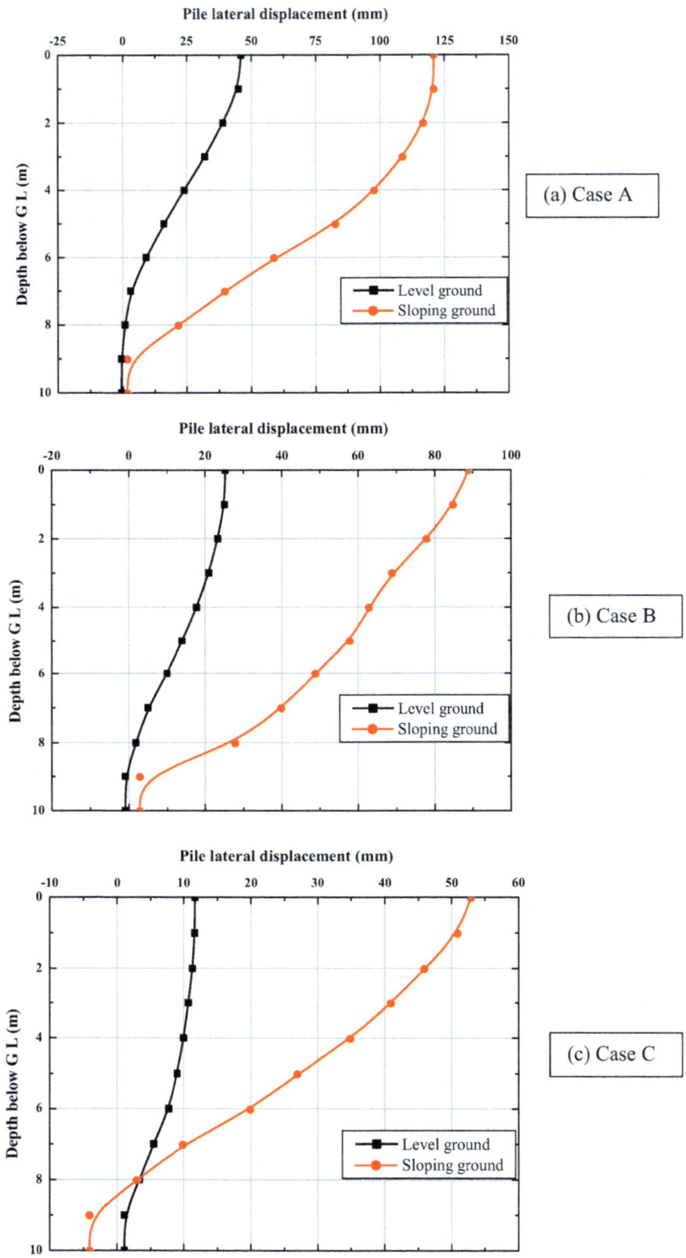

Fig. 9.13 Comparison of pile lateral displacement with depth for all three cases

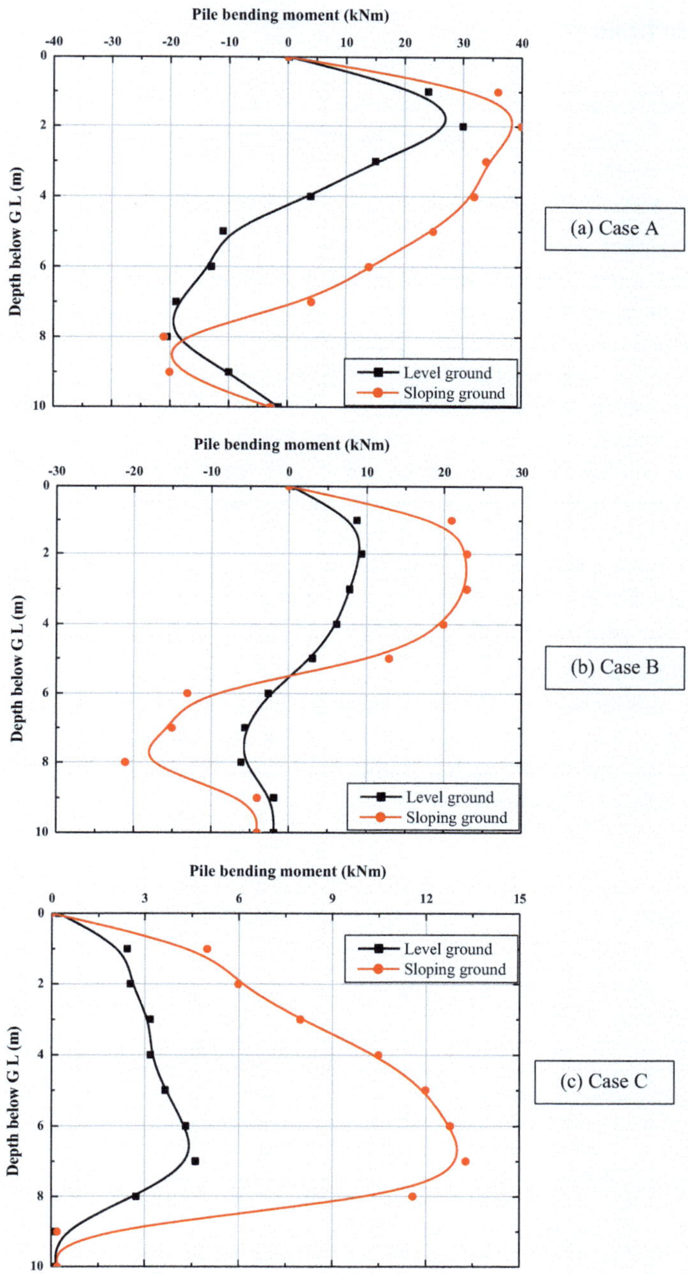

Fig. 9.14 Comparison of pile bending moment with depth for all three cases

9.5 Conclusions

The numerical analysis was performed for a pile resting in liquefiable soils using FLAC 3D. The results were presented in terms of pore pressure ratio, displacement and bending moment.

From the present study, the following conclusions are drawn:

1. The maximum values of bending moment during dynamic loading have been observed at the interface between liquefiable and non-liquefiable soils.
2. Pile displacement is maximum at the top and reduces gradually towards the bottom and tends to zero at the tip of the pile.
3. When the top non-liquefiable layer is absent, liquefaction of Nevada sand exerts lateral pressure on the pile and the pile starts to behave like a cantilever beam thereby reduction in pile displacement and bending moment sets up.
4. Pile has shown increased displacement and bending moment under sloping ground compared to level ground in the three cases discussed.
5. If the pile is not founded on the non-liquefiable hard layer below the liquefiable layer, liquefied soil can exert lateral pressure on the pile in a manner similar to a laterally loaded pile in level and sloping ground. Also, the vertical settlement can be the cause of the failure of piles.

The results of the present study are restricted to the analysis of a single pile embedded in liquefiable level and sloping ground in the absence of vertical load. However, the analysis of the pile behaviour for different ground sloping angles considering the parameters such as slenderness ratio of pile, inertial and kinematic loading deserves further study.

References

1. Abdoun T, Dobry R (2003) Evaluation of pile foundation response to lateral spreading. Soil Dyn Earthq Eng 22(9–12):1051–1058
2. Abdoun T, Dobry R, O'Rouke TD, Goh SH (2003) Pile foundation response to lateral spreads: centrifuge modelling. J Geotech Geo-Environ Eng ASCE 129(10):869–878
3. Basavanagowda GM, Gowthami P, Dinesh SV, Govindaraju L, Balareddy SM (2021) Behavior of pile group in liquefied soil deposits under earthquake loadings. In: Seismic design and performance: select proceedings of 7th ICRAGEE 2020, vol 120, p 139
4. Hamada M, O'Rourke TD (1992) Case studies of liquefaction and lifeline performance during past earthquakes. Vol 1, Japanese case studies technical rep. NCEER-92-0001, National Center for Earthquake Engineering Research, Buffalo, N.Y.
5. Mizuno H, Liba M (1982) Shaking table testing of seismic building-pile-soil interaction. In: Proceedings of the 5th Japan earthquake engineering symposium, Tokyo, Japan, pp 1713–1720
6. Mizuno H (1987) Pile damage during earthquakes in Japan (1923–1983). In: Nogami T (ed) Proceedings session on dynamic response of pile foundations. ASCE, New York, pp 53–77
7. Tokimatsu K, Asaka Y (1998) Effects of liquefaction-induced ground displacements on pile performance in the 1995 Hyogoken-Nambu Earthquake. Soils Found 2:163–178
8. Choudhury D, Chatterjee K, Kumar A, Phule RR (2014) Pile foundations during earthquakes in liquefiable soils–theory to practice. In: 15SEE, pp 327–342

9. Rao VD, Chatterjee K, Choudhury D (2013) Analysis of single pile in liquefied soil during earthquake using FLAC3D. In: Proceedings of the international conference on "State of the art of pile foundation and pile case histories" PILE–2013, pp 2–4
10. Hazzar L, Hussien MN, Karray M (2017) Influence of vertical loads on lateral response of pile foundations in sands and clays. J Rock Mech Geotech Eng 9(2):291–304
11. Bhattacharya S, Bolton M (2004) Buckling of piles during earthquake liquefaction. In: Proceedings 13th world conference on earthquake engineering, pp 1–4
12. Byrne P (1991) Acyclicshear-volume coupling and pore-pressure model for sand. In: Second international conference on recent advances in geotechnical earthquake engineering and soil dynamics, St. Louis, Missouri, pp 47–55
13. Abdoun T (1997) Centrifuge modeling of seismically induced lateral spreading of multi-layer sand deposit and its effects on pile foundations. PhD thesis, Dept. of Civil Engineering, Rensselaer Polytechnic Institute, Troy, N.Y.
14. Bowles JE (1997) Foundation analysis and design. 5th edn. McGraw-Hill, Inc
15. FLAC3D (2012) Fast Lagrangian analysis of continuum's version 5.0. Itasca Consulting Group, Minneapolis, Minnesota, U.S.A.
16. Timoshenko SP, Goodier JN (2002) Theory of elasticity, 3rd edn. Tata McGraw-Hill Education

Chapter 10
Efficacy of a Hybrid Foundation to Mitigate Liquefaction-induced Effects Under Strong Sequential Ground Motions

Ritesh Kumar and Akihiro Takahashi

Abstract A hybrid foundation is developed to mitigate the liquefaction-induced effects on a shallow foundation. The proposed hybrid foundation is a combination of the gravel drainage system and friction piles having spiral blades devised under the footing as a hybrid mitigation technique against the liquefaction-induced effects on a shallow foundation. Any structure situated on liquefiable ground may experience several moderate to strong earthquakes during its lifespan. Therefore, the effectiveness of adopted mitigation techniques under sequential ground motions is necessary to examine the long-term efficacy of their performance to mitigate the liquefaction-induced effects. A series of dynamic centrifuge experiments are carried out to investigate the effectiveness of a hybrid foundation in the liquefiable ground under strong sequential ground motions. Ground motion recorded at Hachinohe Port during the 1968 Tokachi-Oki earthquake (NS component) and design earthquake motion for highway bridges in Japan (2-I-I-3, NS component) recorded at the ground surface near New Bansuikyo Bridge, Tochigi, during the 2011 Tohoku Earthquake are used as the sequential ground motion. The second sequential ground motion is found to be diminishing the mitigation efficacy of the proposed hybrid foundation in comparison with the first sequential ground motion. The implications of strong sequential ground motions are also discussed.

Keywords Hybrid foundation · Sequential ground motion · Liquefaction mitigation

---

R. Kumar (✉)
Indian Institute of Technology Roorkee, Roorkee, India
e-mail: ritesh.kumar@eq.iitr.ac.in

A. Takahashi
Tokyo Institute of Technology, Tokyo, Japan

© Indian Society of Earthquake Technology 2023

R. S. Jakka et al. (eds.), *Earthquake Engineering and Disaster Mitigation*,
Springer Tracts in Civil Engineering, https://doi.org/10.1007/978-981-99-0081-7_10

10.1 Introduction

Soil remediation measures are a requisite for liquefaction-prone sites. Over the past few decades, extensive efforts and contributions have been made by the geotechnical earthquake engineering society to grasp the physics behind liquefaction and for the development of remedial measures for the liquefaction-induced effects. Reports presented by many researchers [1–4] described the role of liquefaction in the damage of buildings, specifically in the reclaimed land during the Tohoku 2011 earthquake. The post-earthquake settlement behavior of shallow foundations is believed to play a vital role in the sequential ground motion due to alteration in density, development of volumetric strains, and the permeability of the ground [5–9].

Soil reinforcement is intended to increase the liquefaction resistance of the ground. Soil replacement, soil densification, and bonding of soil grains with the binding agents are some means to employ reinforcement within the ground. Friction piles can be used to reduce the liquefaction-induced damage to the foundation structure system. However, many case studies show the tremendous damage to the piles and associated structures caused by excessive liquefaction [10, 11]. Liquefaction mitigation by stone column or gravel drainage piles is one of the pronounced techniques which is used to help quick dissipation of excess pore water pressure generated during the earthquake [10, 12]. In addition, the other possible benefits of gravel drains include densification of surrounding soil during installation and re-distribution of earthquake-induced or pre-existing stress [12].

The technological advancements in the experiment facilities (e.g., physical modeling using centrifuge experiments) have enabled the researchers to get a better insight into liquefaction and associated induced effects. Besides, physical modeling has been widely used to simulate geotechnical earthquake engineering-related problems and to validate finite element numerical models. In this paper, the performance of a hybrid foundation (combination of granular columns and friction piles) under strong sequential ground motions is tested for temporary structures such as buffer tank (BT) and flare stack (FS) in an industrial process plant. Friction piles in the hybrid foundation are supposed to yield frictional resistance during the earthquake and are presumed to minimize the rocking/tilting behavior of the foundation structure system. The stress of the foundation-structure system is presumed to be transferred through the ground (not through the friction piles), and hence a clear marginal spacing is kept between friction piles and footing to avoid subsidence below the footing during the shaking, whereas the granular columns in the hybrid foundation are devised to dissipate the generated excess pore water pressure.

10.2 Experimental Scheme

The development of the hybrid foundation is carried out with the help of physical modeling using the centrifuge lab facility. The performance of the hybrid foundation is investigated using two model tests, i.e., Models 5 and 6, as shown in Fig. 10.1 (all the models are shown for the sake of clarity for follow-up papers of the presented subject matter). Models 5 and 6 consist of the friction piles and gravel drainage system of different drainage capacities as individually investigated. The model ground is prepared using the Toyoura sand with a target relative density of 50% by air pluviation method using the sand hopper. The configuration of the foundation and superstructure (buffer tank and Flare stack) and the Individual performance of the different drainage systems can be found in Kumar et al. [8]. The location of different transducers for all the models is shown in Fig. 10.2. The time history of applied sequential ground motions is shown in Fig. 10.3.

10.3 Results and Discussions

The efficacy of the hybrid foundation (Models 5 and 6) is evaluated with reference to the liquefaction-induced effects on the shallow foundation during Tokachi-Oki ground motion. Time histories of EPWP at different locations are shown in Fig. 10.5 during Tohoku ground motion. Figure 10.5 depicts a similar response as observed in Fig. 10.4. It is to be noted that the presence of gravel drainage, both in the case of Models 5 and 6, could not avoid the model ground from being liquefied. However, because of rapid drainage, EPWP remains in a development state for a shorter time in the case of Model 6 in comparison with Models 1 and 5. In addition, the performance of gravel drainage seems to be diminished in terms of both generation and dissipation of EPWP during Tohoku ground motion in comparison with Tokachi-Oki ground motion.

Figure 10.6 depicts the transfer functions (TFs) during white noise, and Tokachi-Oki and Tohoku ground motions. The ratio of acceleration records at A5 to A1 in the frequency domain is used to obtain the TFs during shaking. Recorded acceleration time histories are modified using filtering and baseline correction before calculating the transfer function. Filtering is performed in the frequency domain using the band-pass Butterworth filter with corner frequencies of 0.3 Hz and 10 Hz, respectively. It is evident that the site fundamental frequency obtained for Models 1, 5, and 6 during white noise falls under the range of small-strain site fundamental frequency obtained by empirical equations [13, 14], despite the fact that the soil response is highly nonlinear. The upper and lower bounds of the fundamental frequency of the model ground (2.5–2.8 Hz) are determined by the estimated range of shear wave velocity (approximately 169–186 m/s) using empirical equations as mentioned above. This also implies that the site's natural frequency could be captured by appropriate white noise (usually a random small amplitude vibration having equal intensities at different frequencies, giving it a constant power spectral density). The site fundamental frequencies obtained during Tokachi-Oki ground motion fall in the range of 0.53–0.57 Hz for Models 1, 5, and 6. The significant drop in site fundamental frequency is because of the softening of the model ground due to shear strength mobilization during the Tokachi-Oki ground motion.

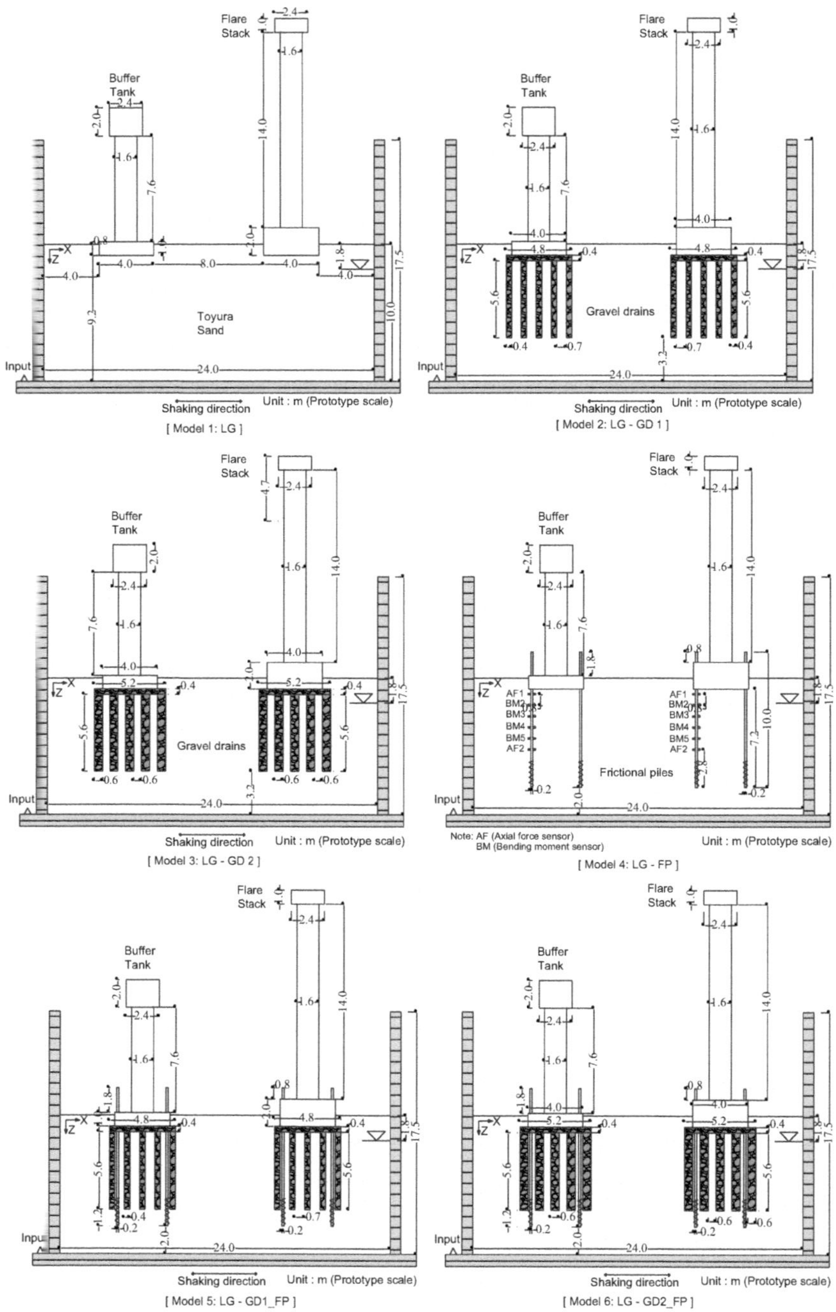

Fig. 10.1 Experimental scheme in prototype scale. Test results for Models 1, 5, and 6 are only discussed under sequential ground motions (see Fig. 10.3) in this paper (test results for all six models can be found in Kumar et al. [8])

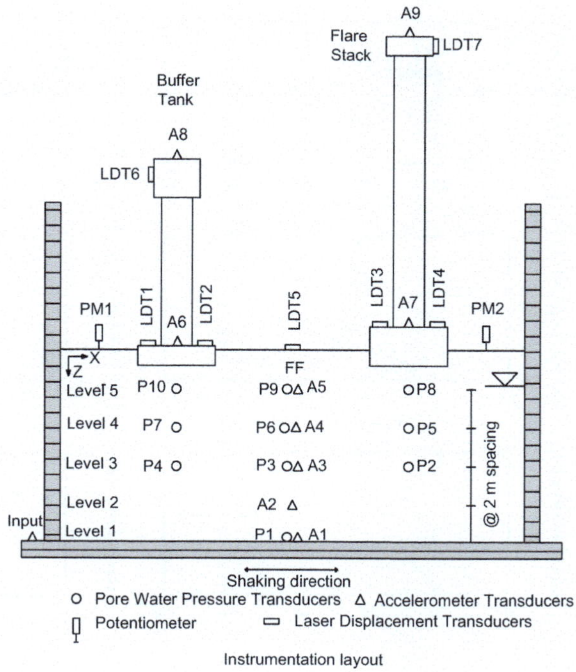

Fig. 10.2 Different transducers' layout in centrifuge model

Figure 10.7 depicts the transfer function obtained associated with the footing and superstructure's lumped mass of both BT and FS during Tokachi-Oki ground motion. It is evident that the overall seismic demand observed in the case of Model 6 is significant in comparison with Models 1 and 5 at footings, though no clear trace of magnified seismic demand at the superstructure is observed for Model 6 at a higher frequency range.

Figure 10.8 shows the structural response during Tohoku ground motion. Significant de-amplification in seismic intensity is observed at the footings of both BT and FS. This is associated with the development of liquefied state (leading to comparatively large ground softening) within the model ground during strong Tohoku ground motion. However, for superstructures, the clear trace of amplification/de-amplification is undulating in the presented frequency range.

The average settlement of BT and FS footings for different time intervals during Tokachi-Oki and Tohoku ground motions is shown in Figs. 10.9 and 10.10. The numeric values corresponding to Figs. 10.9 and 10.10 have also been tabulated separately in Table 10.1. The purpose of this illustration is to get an insight into the settlement progression of the shallow foundation. Each graph shows the average settlement of footings, i.e., the average of LDTs 1 and 2 and LDTs 3 and 4, corresponding respectively for BT and FS during different time intervals. The time intervals are selected to capture the overall settlement progression during and post-shaking events.

The effectiveness of the gravel drainage system and friction piles to minimize the settlement of both BT and FS during and after the post-shaking event in the case

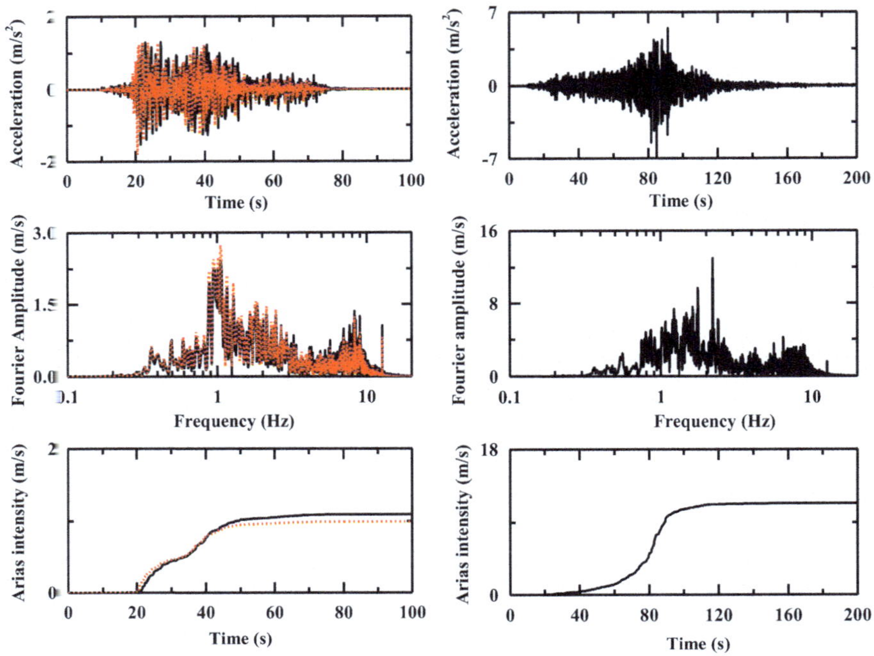

Fig. 10.3 Acceleration time histories, Fourier spectra, and Arias intensity of Tokachi-Oki ground motion (left) and Tohoku ground motion (right). Time history in red on the left (only shown for Tokachi-Oki for the shake of clarity) is the applied motion, whereas recorded motion is marked in black color

of Models 5 and 6 is apparent in comparison with Model 1, as shown in Fig. 10.9. The post-shaking settlement for both BT and FS footings is remarkably less in the case of Model 6 compared to Models 1 and 5. However, BT exhibits comparatively large settlement for BT in the case of Model 5 during the initial phase of Tokachi-Oki ground motion (i.e., 0–30 s). This might be associated with the localized higher permeability in the vicinity of BT, which steered the shear-induced deformation at the beginning of the shaking.

Figure 10.10 shows the average settlement of BT and FS footings that occurred within different time intervals during Tohoku ground motion. Surprisingly, the settlement that occurred for both BT and FS before 100 s is adversely significant in the case of Model 6 in comparison with Models 1 and 5. However, BT and FS settlement until 100 s for Model 5 is more or less the same as the one observed for Model 1. The excessive settlement for both BT and FS before 100 s of shaking indicates a sort of punching shear failure provided that the footing has a free vertical movement of the model ground. In addition, it might be possible that at the beginning of the shaking, because of the jamming of the soil between the piles and footings, the piles took some load and the vertical stress in the soil under the structure was smaller than expected. This made the soil under the structure more liquefiable, and further

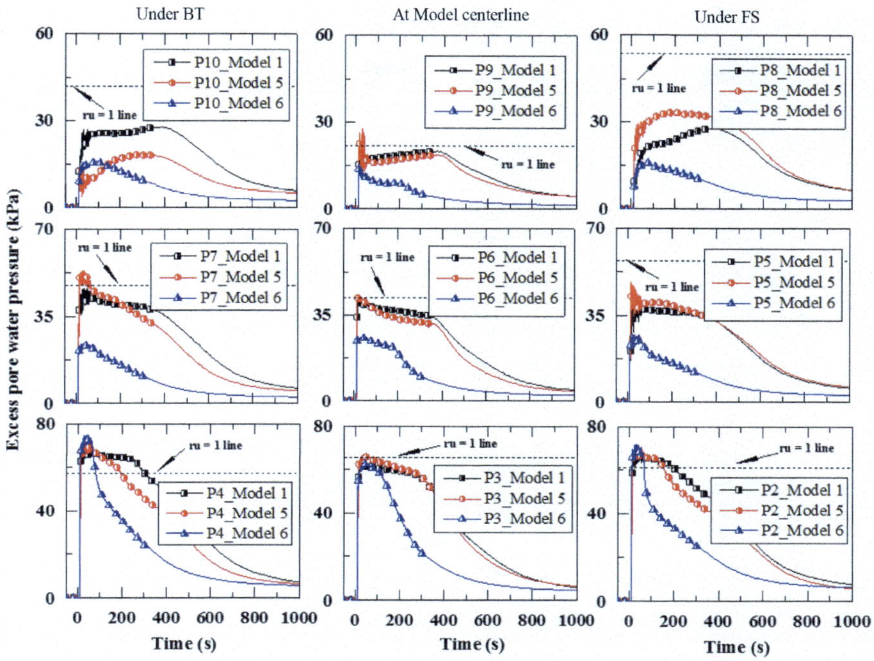

Fig. 10.4 Excess pore water pressure time histories of Models 1, 5, and 6 during Tokachi-Oki ground motion

shaking released jams of soil at the holes and allowed the footing to settle, resulting in a rather fast settlement in the case of Model 6.

10.4 Summary and Conclusions

A unique hybrid foundation has been proposed to mitigate the liquefaction-induced effects on a shallow foundation. The proposed hybrid foundation is a combination of gravel drainage and friction piles with spiral blades. It was found that a shallow foundation resting on liquefiable ground is prone to excessive settlement. Shear and volume change-induced deformation was seen to dominate the overall settlement of the shallow foundation. The presence of gravel drainage increased the dissipation rate of the generated EPWP and decreased the post-shaking settlement. However, the performance of the hybrid foundation seems to be diminished in terms of the overall settlement of both BT and FS footings during Tohoku ground motion, whereas the evolution of EPWP within the model ground indicated otherwise. The hybrid foundation is a complex arrangement of the gravel drainage system and friction piles passing through the footing with minimal clear spacing. The alteration in ground condition after the Tokachi-Oki ground motion significantly affected the performance

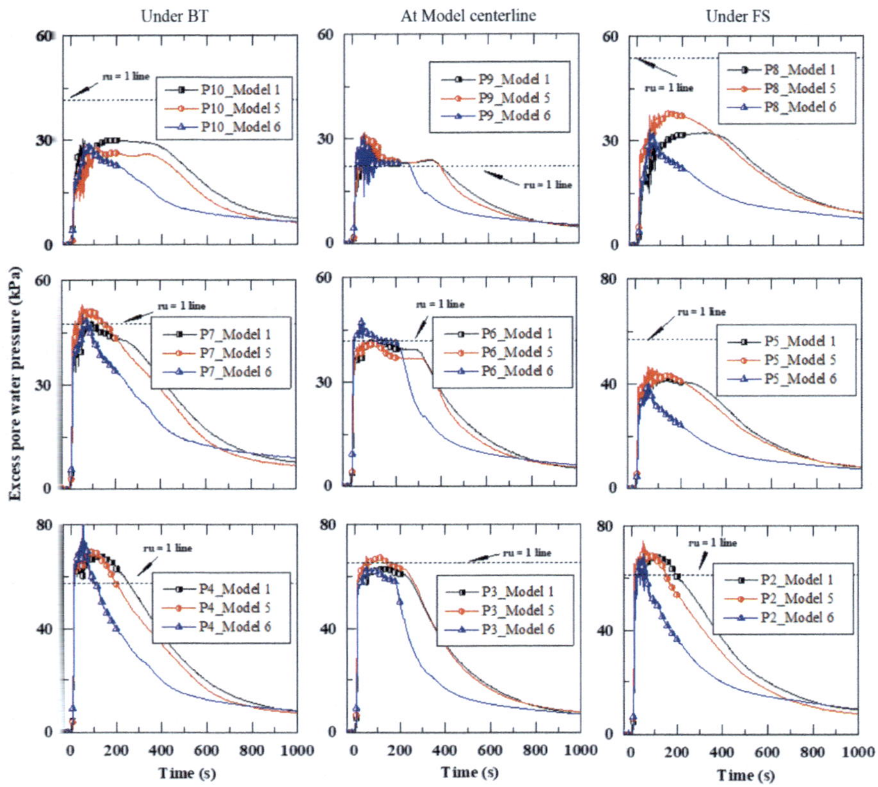

Fig. 10.5 EPWP time histories of Models 1, 5, and 6 during Tohoku ground motion

of the hybrid foundation under the Tohoku ground motion. The extent of softening of the ground due to liquefaction-induced mechanism during the seismic event could be captured qualitatively by the drop of site fundamental frequency. More substantial is the drop in the fundamental frequency, and softer is the ground during shaking, which is also related to the severity of the shaking. Settlement progression of the shallow foundation during and post-shaking is also associated with the liquefaction extent (model ground softening, to be specific). The shallow foundation resting on the liquefiable ground is found to undergo excessive settlement during Tohoku ground motion caused by during- and post-shaking-settlement mechanism.

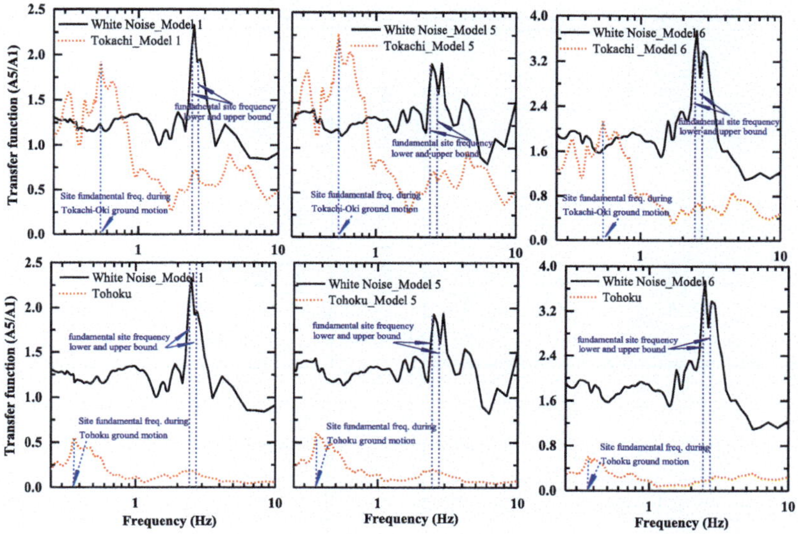

Fig. 10.6 Model ground behavior during white noise, Tokachi-Oki, and Tohoku ground motions

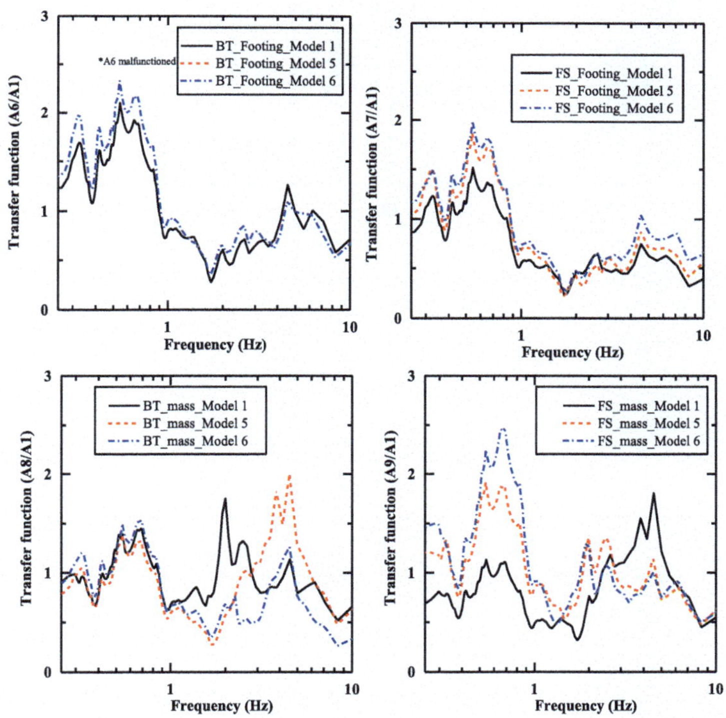

Fig. 10.7 Foundation-superstructure kinematic response during Tokachi-Oki ground motion

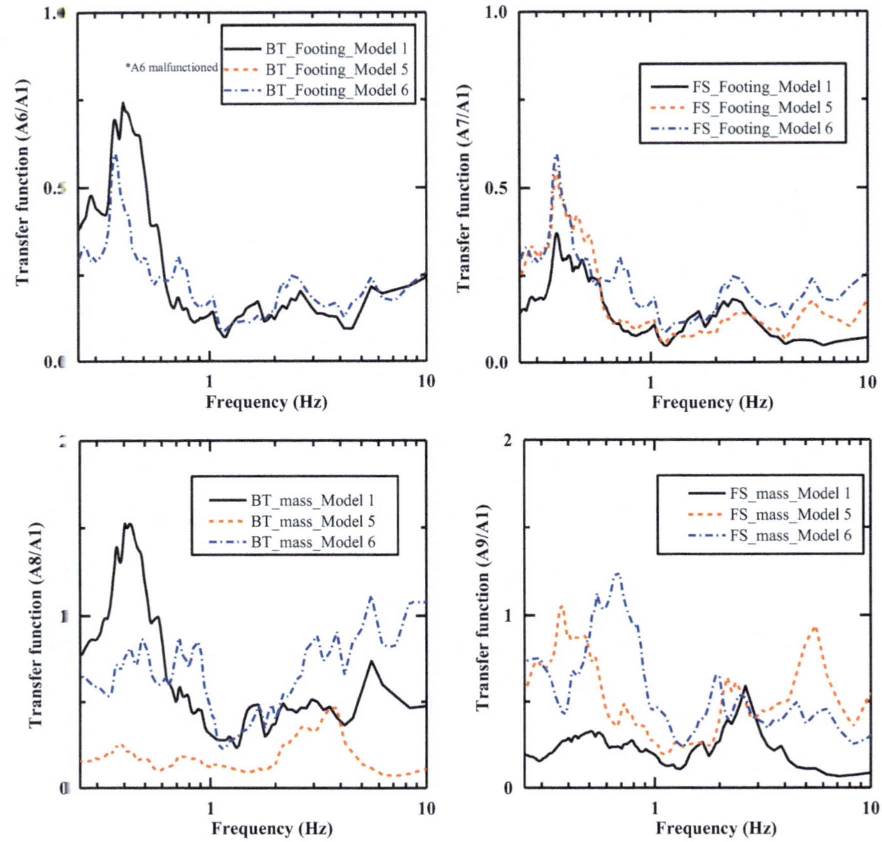

Fig. 10.8 Foundation-superstructure kinematic response during Tohoku ground motion

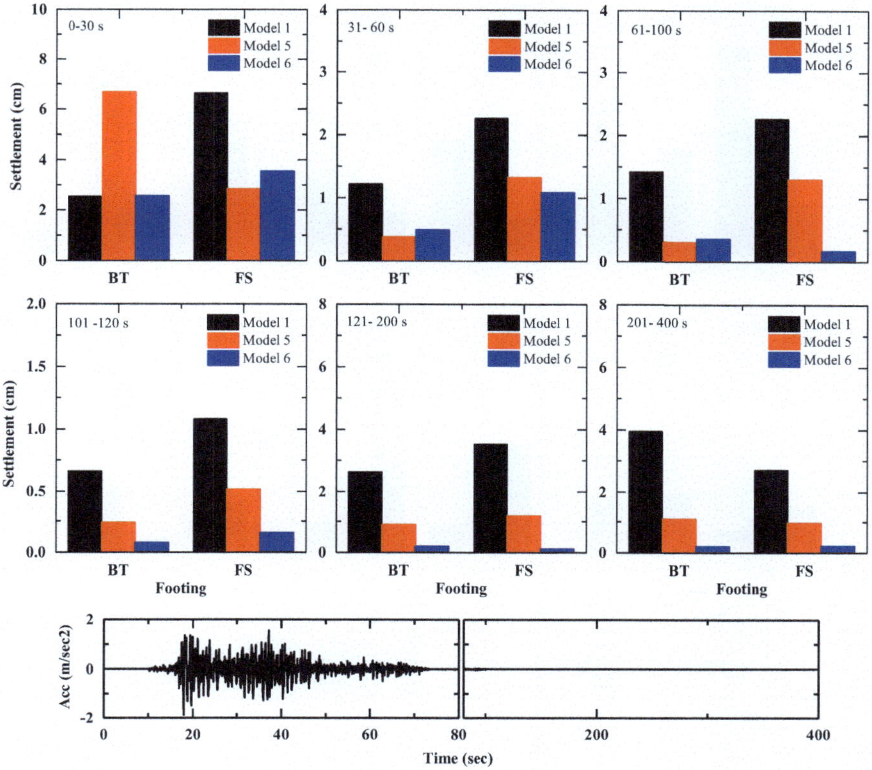

Fig. 10.9 Average settlement of BT and FS footings for different time periods during Tokachi-Oki ground motion

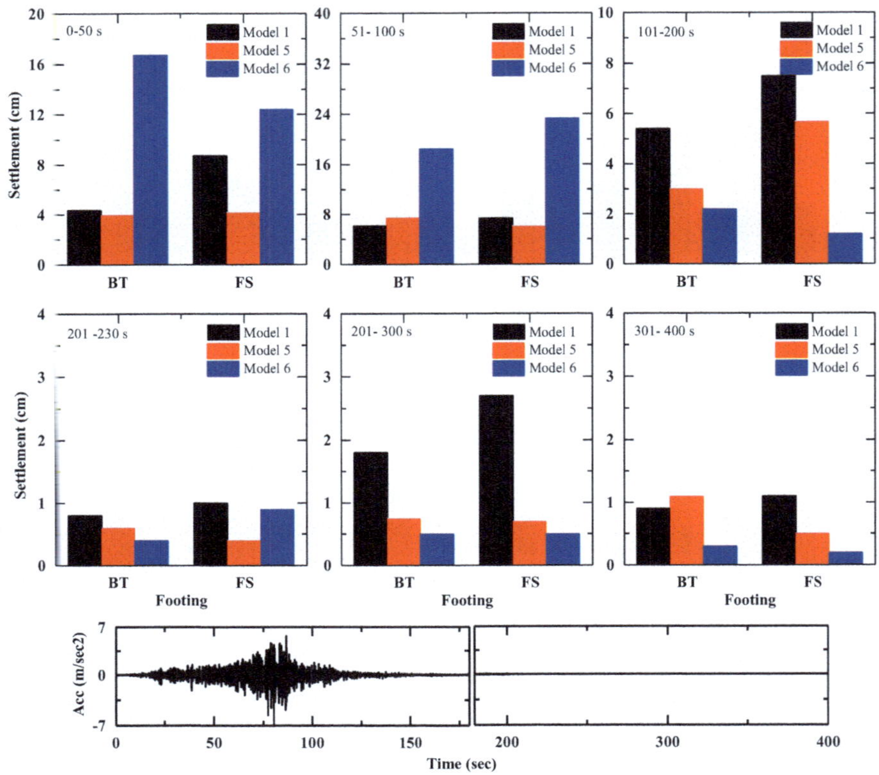

Fig. 10.10 Average settlement of BT and FS footings for different time periods during Tohoku ground motion

Table 10.1 Foundation settlement during Tokachi-Oki and Tohoku ground motions for BT and FS under different time durations

Time	Structure	Settlement for Tokachi-Oki ground motion (cm)			Settlement for Tohoku ground motion (cm)		
		Model 1	Model 5	Model 6	Model 1	Model 5	Model 6
0–30	BT	2.54	6.68	2.57	4.34	3.96	16.76
	FS	6.64	2.85	3.56	8.7	4.16	12.43
31–60	BT	1.22	0.38	0.49	6.1	7.49	18.52
	FS	2.26	1.33	1.09	7.3	6.1	23.42
61–100	BT	1.42	0.31	0.36	5.4	3.01	2.2
	FS	2.26	1.31	0.17	7.5	5.7	1.2
101–120	BT	0.66	0.24	0.08	0.8	0.6	0.4
	FS	1.08	0.52	0.16	1	0.4	0.9
121–200	BT	2.62	0.9	0.2	1.8	0.75	0.5
	FS	3.52	1.2	0.12	2.7	0.7	0.5
201–400	BT	3.94	1.11	0.2	0.9	1.1	0.3
	FS	2.7	0.99	0.23	1.1	0.5	0.2

References

1. Nakai S, Sekiguchi T (2011) Damage due to liquefaction during the 2011 Tohoku earthquake. In: Proceedings of the international symposium for CSMID, pp 1–8
2. Bhattacharya S, Hyodo M, Goda K, Tazoh T, Taylor CA (2011) Liquefaction of soil in the Tokyo Bay area from the 2011 Tohoku (Japan) earthquake. Soil Dyn Earthq Eng 31(11):1618–1628
3. Tokimatsu K, Tamura S, Suzuki H, Katsumata K (2012) Building damage associated with geotechnical problems in the 2011 Tohoku Pacific earthquake. Soils Found 52(5):956–974
4. Yamaguchi A, Mori T, Kazama M, Yoshida N (2012) Liquefaction in Tohoku district during the 2011 off the Pacific Coast of Tohoku earthquake. Soils Found 52(5): 811–829
5. Ishihara K, Yoshimine M (1992) Evaluation of settlements in sand deposits following liquefaction during earthquakes. Soils Found 32(1):173–188
6. Liu L, Dobry R (1997) Seismic response of shallow foundation on liquefiable sand. J Geotech Geoenviron Eng 123(6):557–567
7. Shamoto Y, Zhang J-M, Tokimatsu K (1998) Methods for evaluating residual post-liquefaction ground settlement and horizontal displacement. Soils Found 38:69–83
8. Kumar R, Sawaishi M, Horikoshi K, Takahashi A (2019) Centrifuge modeling of hybrid foundation to mitigate liquefaction-induced effects on shallow foundation resting on liquefiable ground. Soils Found 59(6):2083–2098
9. Zhang J-M, Wang G (2012) Large post-liquefaction deformation of sand, part I: physical mechanism, constitutive description and numerical algorithm. Acta Geotech 7(2):69–113
10. Tokimatsu K, Yoshimi Y, Ariizumi K (1990) Evaluation of liquefaction resistance of sand improved by deep vibratory compaction. Soils Found 30(3):153–158
11. Seed HB, Booker JR (197) Stabilization of potentially liquefiable sand deposits using gravel drains. J Geotech Eng Div 103(7): 757–768

12. Adalier K, Elgamal A, Meneses J, Baez JI (2003) Stone columns as liquefaction countermeasure in non-plastic silty soils. Soil Dyn Earthq Eng 23(7):571–584
13. Seed HB, Idriss IM (1971) Simplified procedure for evaluating soil liquefaction potential. J Soil Mech Found Div 97(9): 1249–1273
14. Hardin BO, Drnevich VP (1972) Shear modulus and damping in soils: design equations and curves. J Soil Mech Found Div 98(7):667–692

Chapter 11
Influence of Vertical Seismic Coefficient in Seismic Analysis of Hydro-Tunnel in Rock

Ambika Srivastav and Neelima Satyam

Abstract As a result of increased urbanization and the need for better infrastructure, the scope of tunnel projects is also expanding. The country has much potential for hydropower, and the hydropower sector has produced the most tunnels. Hydro projects have some of the country's longest tunnels. In the Himalayan areas, almost 75% of the full potential for hydropower production is concentrated. Himachal Pradesh has the most projects with a total tunnel length of 500 km, followed by Uttarakhand (160.8 km) and Jammu and Kashmir (135.14). Due to geographical and geological difficulties, tunneling in the Himalayan area is extremely difficult. Due to geological issues such as sheared rock, high water intrusion, and high geothermal gradient, long hydro-tunnels have experienced time and expense overruns. Extensive studies must be conducted to use the hydropower potential fully, the appropriate technique must be adopted, and risks must be adequately identified and managed. Geological prediction during tunneling should be standard practice to reduce geological uncertainty and prevent unanticipated hazards. To complete a project on schedule and safely accurate evolution, analysis, and interpretation of rock mass quality play a key role. This paper provides a numerical analysis of the seismic response of a circular lined tunnel running through a jointed rock mass. The effect of tunnel depth, frequency, and peak ground acceleration on the axial force produced in the tunnel liner is studied. The outcomes of the numerical computation have verified the patterns of seismic damage observed in the past.

Keywords Rockmass · Tunneling · Rock support · Seismic loading · Peak ground acceleration

A. Srivastav (✉)
Geotechnical Engineering Laboratory, Earthquake Engineering Research Centre, International Institute of Information Technology Hyderabad, Hyderabad, India
e-mail: ambika.srivastav@gmail.com

N. Satyam
Discipline of Civil Engineering, Indian Institute of Technology Indore, Simrol, Indore, India

© Indian Society of Earthquake Technology 2023

R. S. Jakka et al. (eds.), *Earthquake Engineering and Disaster Mitigation*,
Springer Tracts in Civil Engineering, https://doi.org/10.1007/978-981-99-0081-7_11

11.1 Introduction

Tunnels have been playing a vital role in the hydropower development sector in India. Several tunneling projects have also been established in India, and several projects are under development and being designed. Maximum underground construction work relates specifically to the Himalayan region for hydropower development of the country. Approximately 600 km of tunneling has been done so far, passing through various geological formations and rock types, ranging from hard massive granites to the softest sandstones, shales, and schists using both traditional and modern methods for excavation and support systems [1, 2]. These tunnels are vulnerable due to high seismic activity in the Himalayan region. According to the Institute of Earth and Environmental Science, the University of Potsdam, Germany, which analyzed more than two hundred hydropower plants in operation, under development, or in the planning phase in the seismically active Himalayas, reveals that they are unlikely to be built to survive potential earthquakes. It offers new inspiration in Himalayan Geology for the seismic development of mountain tunnels.

The present study intends to analyze the impact of vertical seismic coefficient in analyzing seismic behavior of tunnels [3, 4]. For this study, a typical section of the power tunnel situated in low rock cover (70 m), with specification diameter—8.8 m, horseshoe shaped, has been considered for analysis. A detailed study on the seismic stability of underground structures was performed by Hashash et al. [5], which revealed that the seismic evaluation of underground structures included the following steps: (a) assessment of seismic characteristics (Swathi et al. 2018, 2019 and 2020) [6–8], (b) study of ground response to seismic waves (Satyam et al. 2007) [9], and (c) seismic design of the underground structure [10]. Therefore, three distinct methodologies, i.e., static, pseudo-static, and dynamic, can be used to assess the ground response to seismic waves (Hashash 2009). Based on empirical approaches, the pseudo-static methodology is easy to determine compared to the other two approaches and has been widely used. The summation of a permanent body force with a static finite element analysis depicting seismic shaking is a pseudo-static analysis [3, 11, 12]. By adding a body force in the direction of seismic loading, it incorporates pseudo-static seismic loads. Different authors have published distinct expressions for axial force, shear force, and bending moment produced in tunnel liners in previous work on this subject [13, 14].

The Vishnugad Pipalkoti Hydroelectric Project is one of the major projects under development on the river (ROR) in the Alaknanda Basin in Uttarakhand. It is intended to build a 65 m high diversion dam near Helong village (79° 29′30″ E and 30° 30′50″ N), creating a 2.5 km long reservoir behind the dam. Pseudo-static analyses with horizontal and vertical accelerations were conducted to understand the effect of seismic force on the lining. Figure 11.1 depicts the location of the study area.

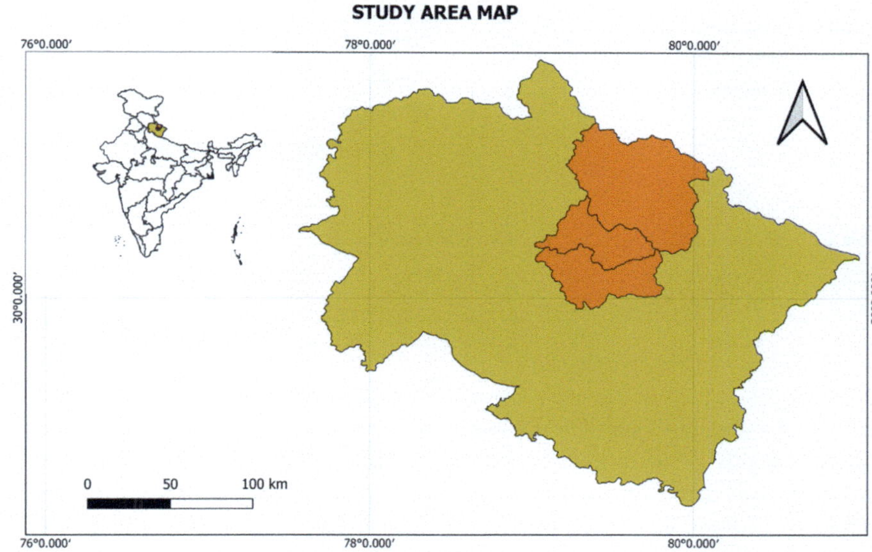

Fig. 11.1 Location map of the study area

11.2 Geology of Site

The Garhwal and Kumaun Himalayas which form the central part of the folded Himalayan belt reveal rock types of differing ages from Proterozoic to Late Tertiary phase and are grouped in four main tectonic belts, classified as the Foothill Siwalik zone, the Lesser Himalayan belt, the Central Crystalline, and Tethyan belt [15]. The project location, which forms part of the Alaknanda Valley, consists mainly of rocks belonging to the Lesser Himalayan Garhwal Group. The geological formations along the project formations are Helong formation, Gulabkoti formation, and Pipalkoti formation [16]. The rocks predicted to occur along most of the tunnel length are quartzite with biotite schists, interbedded and interbanded gray slates and dolomites/calcareous, gray thinly bedded slates with tiny calcareous interbeds, dolomite calcareous with subordinate gray slates, gray pyrite/slates, thinly bedded dolomite calcareous, gray slate/phyllite and dolomite belonging Garhwal group [17].

11.3 Geotechnical Data

For an adequate understanding of engineers and designers, quantitative evaluation of the rock mass is essential. The rock mass rating has been modified in the study area owing to complicated Himalayan structures. Intact rock properties were obtained through laboratory tests as per ISRM [18] specifications on the number of rock samples collected from the dam site and powerhouse area. Predominant rock types

Table 11.1 Structural details observed in slates [19]

Joint details	Dip/Dip direction	Spacing (cm)
FJ	28°–32°/N025°–N30°	30–50
J1	50°–58°/N190°–N180°	20–50
J2	75°–85°/N265°–N275°	80–100

Table 11.2 Geotechnical parameters of rock mass considered as [19]

Rock type	E_i Intact rock Modulus (GPa)	GSI	m_i	σ_i (MPa)	Cohesion (MPa)	Friction (degree)
Slates	11.39	20	10	101.5	0.314	45.82

found in the area are Quartzite, Dolomitic Limestones, and Slates. Since seismic activity is highly prone to tunnels constructed in weak deposits, Slates' properties have been considered for numerical modeling. Rock masses consist of a continuous rock medium and discontinuities like joints, faults, and bedding planes. For numerical modeling of the rock mass, numerical methods are typically required to evaluate continuous rock and discontinuity (Youjun et al. 2016). Two joint sets were observed on the site in addition to the foliation. Structural details observed in slates are presented in Table 11.1.

Table 11.2 shows the average rock mass characteristics utilized in this research. The rock mass behavior was described using the Mohr–Coulomb failure criterion.

11.4 Methodology

The effects of seismic stress on tunnel deformation were investigated using Phase2. Phase2 [20] is a flexible 2D elastoplastic finite element stress analysis system for underground and surface excavations. For assistance, Phase2 provides a wide choice of modeling possibilities. Shotcrete, concrete steel set structures, foundations, reinforced multi-layer liners, geotextiles, and other applications can all benefit from liner (pre-programmed) components. When a shotcrete liner is modeled, the axial force, shear force, and bending moment of each liner component may be calculated in this module. The Seismic Axial Force, which is a change in axial force on the liner caused by seismic loading during earthquakes, is applied as a factor to reflect the earthquake effect. By estimating the axial force produced in the lining, the damage analysis of the tunnel would assist us in identifying the likely collapse scenario.

11.4.1 Model Geometry and Mesh Details

The effect of seismic stress on tunnels was investigated in this study, and the support system was built to accommodate the seismic scenario. To assess the deformation surrounding the tunnel and the axial forces produced in the tunnel lining, a pseudo-static finite element analysis was used. To reduce boundary impacts, the model's overall size is 150 × 150 m. The lower boundary was fixed vertically and horizontally, while the lateral boundary was only fixed vertically. For this study, the ground level is referred to as a free surface which can move in both horizontal and vertical directions; the bottom boundary is fixed (restrain horizontal and vertical), as shown by "pin" symbols; and both sides are referred to as "rollers" (Restrain vertical and Free horizontal). The size of the finite element, which affects the model's sensitivity, is determined by the parameters set during meshing and discretization. Hence, three nodded triangles with a graded mesh make up the grid. Figure 11.2 illustrates the numerical model used in this study. The research was conducted on the assumption that the in situ stress was lithostatic and hydrostatic. The overburden depth has been considered at 70 m. Using the pseudo-static seismic loading technique, a series of numerical analytical case studies were conducted in Phase2 to evaluate the influence of the vertical seismic coefficient. The four seismic loading scenarios are applied to the tunnel model as illustrated below.

Case 1: The peak horizontal acceleration (PHA) value of 0.36 g and the peak vertical acceleration (PVA) value of 0. The impact of the vertical seismic coefficient is ignored in this case.

Case 2: The peak horizontal acceleration (PHA) value of 0.36 g and the peak vertical acceleration (PVA) value of 0.24 g. In this case, the vertical seismic coefficient adds an inertial force in the opposite direction of gravity.

Case 3: The peak horizontal acceleration (PHA) value of 0.36 g and the peak vertical acceleration (PVA) value of −0.24 g. This loading scenario is similar to Case 2, albeit the vertical seismic coefficient has a negative sign. As a result, because the inertial force, which simulates seismic loading, is in the same direction as the downward gravitational force, it is added to the self-weight.

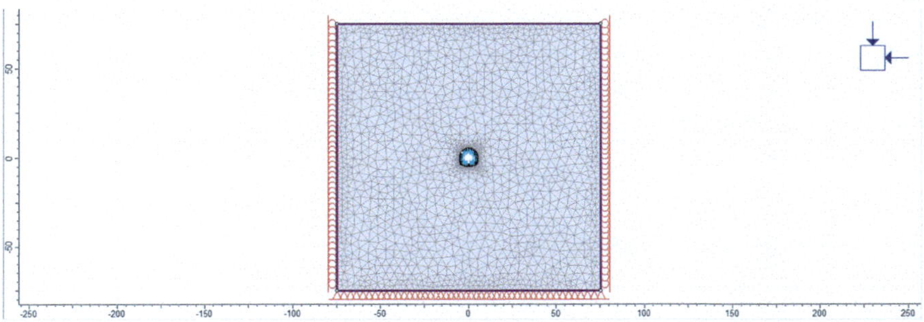

Fig. 11.2 Phase2 model of the 8.8 m diameter horseshoe shape tunnel

Table 11.3 Strength properties of shotcrete

Parameters	Value or description
E Modulus (MPa)	30,000
Poisson's ratio	0.15
Thickness (m)	0.150
Material type	Elastic
Liner type	Reinforced concrete

Case 4: The peak horizontal acceleration (PHA) value of −0.36 g and the peak vertical acceleration (PVA) value of −0.24 g. This scenario is similar to Case 3, with the exception that the horizontal seismic coefficient is negative. This scenario was created to look into the effect of the horizontal seismic coefficient's direction on the axial force on the tunnel lining.

The excavation and support installation processes were recreated in reasonable detail in this study. Stage 1 depicts the unexcavated tunnel's boundary modeled under hydrostatic conditions, Stage 2 depicts stress relaxation, Stage 3 depicts the tunnel after the shotcrete liner installation, and Stage 4 depicts seismic loading.

The excavation of the tunnel and the installation of the support are constantly delayed during Stage 2 tunnel construction. Before the support is installed, some tension is relieved, allowing the installed support to bear less of the strain. As a result measuring the amount of stress released during relaxation, or deformation, before the support is placed is required to determine the real size of the load carried by the support system. The tunnel core is replaced with an unstressed elastic material and the Young Modulus values are over-reduced for this investigation, resulting in a 70% stress relaxation [21].

Stage 3 explores the rock mass-support interaction, with the ultimate tunnel support being a reinforced concrete liner of M30 grade. After testing for cracking resistance and serviceability, the 150 mm thick liner was selected as the final reinforcement IS456 [22]. The excavated tunnel was supported by 4 m long rock bolts spaced 2.0 m apart. Table 11.3 shows the strength characteristics of the shotcrete utilized in the analysis.

The difference in axial force between seismic (Stage 4) and static loading (Stage 3) aids in determining the earthquake's effect. The effect of seismic loading during earthquakes may be ascribed to the difference in axial force on the lining, referred to as Seismic Axial Force, which can be used as a parameter to characterize the influence of earthquakes [23].

11.5 Results and Discussions

Figure 11.3 shows the major principal stress for the four cases in the static scenario as it is the same for each case. Figures 11.4, 11.5, 11.6, and 11.7 show the contour plots for major principal stress in the seismic scenario for Cases 1, 2, 3, and 4. The distribution of the major principal stress around the tunnel periphery forms an ellipse, with the highest stresses placed along the ellipse's long axis, as can be observed. The long axis of the ellipse is reflected along the vertical axis when the direction of the horizontal seismic coefficient is changed. The location of the long axis of the ellipse does not vary considerably when the vertical seismic coefficient is changed.

In Case 2, if the seismic coefficient is positive and acts in the opposite direction to the gravitational force, the magnitude of the maximum axial force decreases. When seismic coefficient is negative and operating in the same direction as the gravitational force, maximum axial force increases dramatically under a seismic scenario in Case 3. Case 3 has a 58% increase in maximum axial force compared to static loading, whereas Case 1 has a 5% increase. This demonstrates that including the vertical seismic coefficient in the equation results in a considerably greater maximum axial force during seismic loading. Ignoring the vertical seismic coefficient might lead to an underestimate of the earthquake effect. When comparing Cases 3 and 4, it is evident that the direction of the horizontal seismic coefficient does not influence the

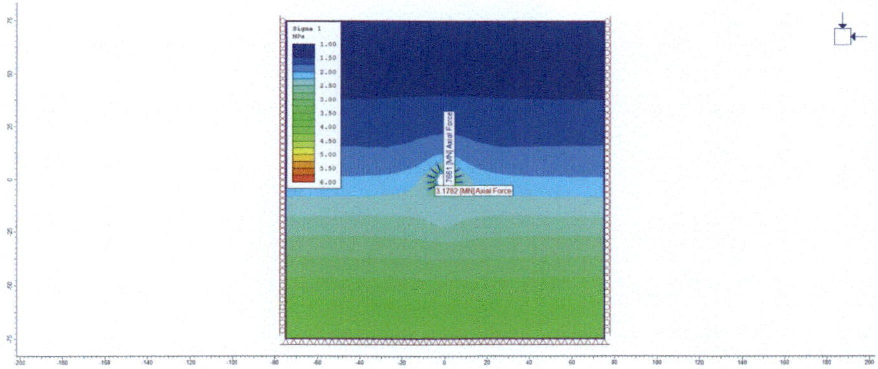

Fig. 11.3 Major principal stress plots for static scenario for Cases 1, 2, 3, and 4

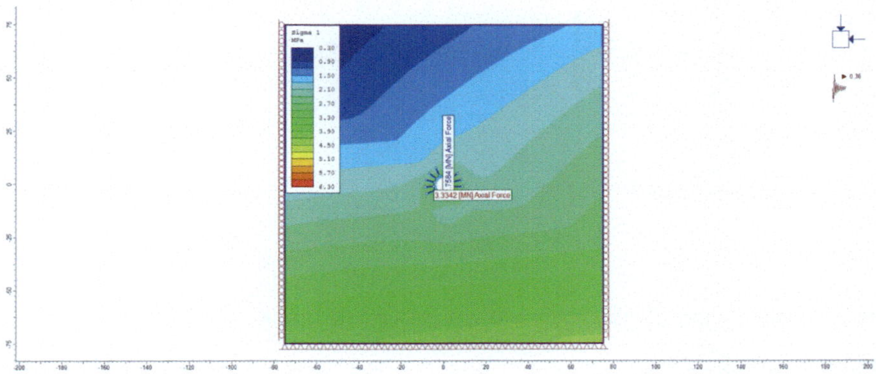

Fig. 11.4 Major principal stress plots for seismic scenario for Case 1

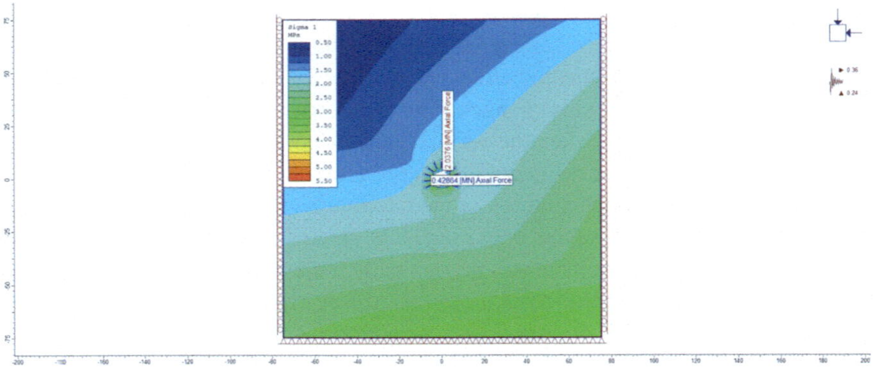

Fig. 11.5 Major principal stress plots for seismic scenario for Case 2

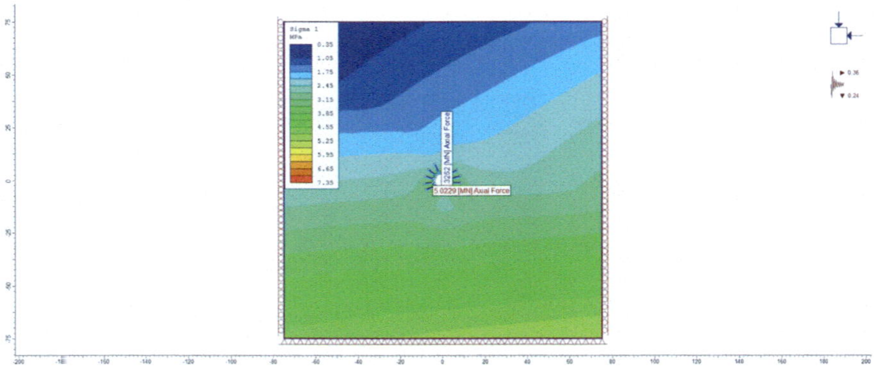

Fig. 11.6 Major principal stress plots for seismic scenario for Case 3

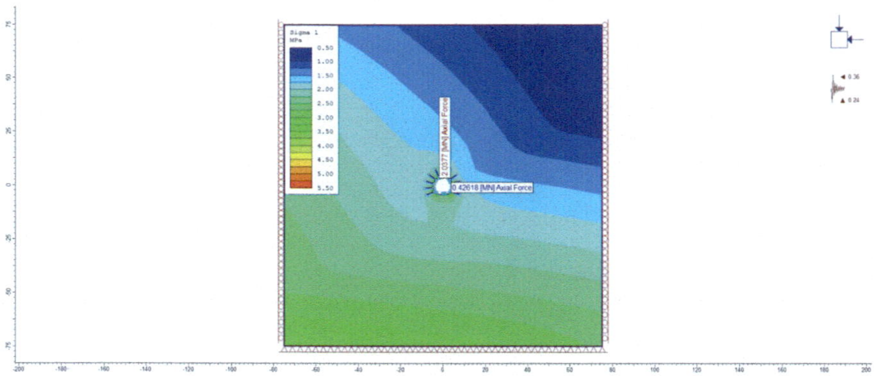

Fig. 11.7 Major principal stress plots for seismic scenario for Case 4

Table 11.4 Percent increase in the maximum axial force developed in liner

Scenario	Maximum axial force in the lining for static case (MN)	Maximum axial force in the lining for static case (MN)	Percentage increase in MAFL
Case 1	3.17	3.33	5
Case 3	3.17	5.02	58

amount of maximum axial force in the liner. This conclusion is expected, given that the vector sum of extra seismic force and the downward-acting body force governs the pseudo-static seismic method used here. The direction of the horizontal seismic coefficient in Cases 3 and 4 is perpendicular to the gravitational force, even though they are opposite each other. The percentage increase in the axial force in the lining for different cases is shown in Table 11.4.

Because of the findings mentioned earlier, all studies utilizing pseudo-static seismic loading to analyze underground structures use a vertical seismic coefficient that is always applied in a downward (negative) direction.

11.6 Conclusions

In this study, the numerical method in conjunction with the pseudo-static approach was applied to estimate the stability and preliminary support design of an 8.8 m diameter horseshoe-shaped power tunnel excavated in Uttarakhand, India, using an FE-based computer program, Phase2. The effect of earthquake loadings on tunnels that are constructed in weak rock mass has been studied. Numerical analysis was carried out in rock mass as an equivalent continuum for an 8.8 m diameter horseshoe-shaped tunnel. The results were interpreted in terms of major principal stress, minor principal stress, and total displacement around the tunnel for the static and seismic conditions. The change in the maximum principal stress state is more significant when a downward vertical seismic coefficient is applied to simulate peak vertical acceleration than when only horizontal peak acceleration is applied. The forces generated in the tunnel lining were studied for static and seismic loading conditions. The results for the axial force pattern reveal that the tunnel shows some periodicity for both static and seismic loading. There is a 58% increase in the maximum axial force developed in the tunnel lining after superimposing the seismic loads in the vertical direction. It can be concluded that incorporating the vertical seismic coefficient in the equation results in a much higher maximum axial force during seismic loading. Neglecting the vertical seismic coefficient might result in an underestimation of the earthquake effect. Hence, it is advisable to use a vertical seismic coefficient that is consistently applied downward to analyze underground structures.

References

1. Goel RK (2001) Status of tunnelling and underground construction activities and technologies in India. Tunn Undergr Space Technol 16(2):63–75
2. Panthi KK (2012) Evaluation of rock bursting phenomena in a tunnel in the Himalayas. Bull Eng Geol Env 71(4):761–769
3. Sun QQ, Dias D (2019) Assessment of stress relief during excavation on the seismic tunnel response by the pseudo-static method. Soil Dyn Earthq Eng 117:384–397
4. Zou Y, et al (2017) A pseudo-static method for seismic responses of underground frame structures subjected to increasing excitations. Tunn Undergr Space Technol 65: 106–120
5. Hashash YMA, Hook JJ, Schmidt B, Yao JI-C (2001) Seismic design and analysis of underground structures. Tunn Undergr Space Technol 16(4):247–293
6. Putti SP, Devarakonda NS (2018) Ground response analysis and liquefaction hazard assessment for Vishakhapatnam city. Innov Infrastruct Solutions 3(1): 1–14
7. Putti SP, Devarakonda NS (2020) Evaluation of site effects using HVSR microtremor measurements in Vishakhapatnam (India). Earth Syst Environ 4: 439–454
8. Putti SP, Devarakonda NS, Towhata I (2019) Estimation of ground response and local site effects for Vishakhapatnam, India. Nat Hazards 97(2): 555–578
9. Rao KS, Devarakonda NS (2007) Liquefaction studies for seismic microzonation of Delhi region. Curr Sci 646–654
10. Srivastav A, Satyam N (2020) Understanding the impact of the earthquake on circular tunnels in different rock mass: a numerical approach. Innov Infrastruc Solutions 5(1):1–9
11. Bilotta E, Lanzano G, Russo G, Santucci de Magistris F, Aiello V, Conte E, Silvestri F, Valentino M (2007) Pseudostatic and dynamic analyses of tunnels in transversal and longitudinal direction. In: Proceeding of the fourth international conference on earthquake geotechnical engineering, Thessaloniki, Greece; Paper no. 1550
12. Lee, T-H, et al (2016) Damage analysis of cut-and-cover tunnel structures under seismic loading. Bull Earthq Eng 14(2): 413–431
13. Sivarajan TK (2016) Seismic load considerations in the design of underground structures for hydropower projects in the Himalayan region. In: Recent advances in rock engineering (RARE 2016). Atlantis Press
14. Tshering T (2011) The impact of earthquakes on tunnels in different rock mass quality Q. Master's Thesis in Geosciences, Department of Geosciences, University of Oslo
15. Thakur VC, Sriram V, Mundepi AK (2000) Seismotectonics of the great 1905 Kangra earthquake meizoseismal region in Kangra-Chamba, NW Himalaya. Tectonophysics 326(3–4):289–298
16. Richards A, Argles T, Harris N et al (2005) Himalayan architecture constrained by isotopic tracers from clastic sediments. Earth Plan Sci Lett 236: 773–796. https://doi.org/10.1016/j.epsl.2005.05.034
17. Goel RK (2014) Tunnel boring machine in the Himalayan tunnels 54–67
18. ISRM (1981) Rock characterization testing and monitoring, ISRM suggested method. Int Soc Rock Mech 211
19. Lakshmanan K (2016) Geotechnical investigations of Vishnugarh Pipalkoti HYDEL project, Garhwal, India. PhD Thesis
20. Rocscience Inc (2011) Phase2 program reference manual
21. Paraskevopoulou C, Vlachopoulos N, Diederichs MS (2012) Long term tunnel behaviour and support response analysis using 2D numerical modelling methods. In: 46th US rock mechanics/geomechanics symposium. American Rock Mechanics Association
22. IS 456 (2000) Plain and reinforced concrete—code of practice [CED 2: Cement and Concrete]
23. Geng P, Mei S, Zhang J et al (2019) Study on seismic performance of shield tunnels under combined effect of axial force and bending moment in the longitudinal direction. Tunn Undergr Space Technol 91:103004

Chapter 12
Experimental Studies on the Dynamic Response of Buildings Supported on Pile and Piled Raft Foundation in Soft Clay

K. G. Subramanya, L. Govindaraju, and R. Ramesh Babu

Abstract Conventionally, seismic analysis of structures has been carried out by considering the base of the structure to be fixed without the soil-foundation-structure interaction effect. The primary aim of the present study is to understand the effect of the dynamic soil-pile-structure interaction on the seismic response of a 10-storey scaled model representing a prototype. Laboratory experiments were carried out using the Shake table facility. The response of the model structure was investigated for fixed and flexible base conditions considering two types of foundation systems, namely supported by pile groups and small piled raft foundations in soft clay. From the experimental measurements, it has been observed that the soil-pile-structure system amplifies the lateral deflections and the storey drift of the superstructure in comparison with the fixed base. Building supported on a small piled raft shows a reduction in the response of the superstructure compared to a model building supported on a pile foundation.

Keywords Soil-structure interaction · Shake table studies · Scaled model · Piled raft foundation

12.1 Introduction

In the seismic analysis and design of structures, Soil-Structure Interaction (SSI) plays an important role. Conventionally, the pile-supported structures are designed by considering the base of the structure to be fixed. The interaction between the soil, foundation and superstructure is ignored due to the complexity involved in the analysis. The failure of bridges, flyovers, towers and buildings supported on

K. G. Subramanya (✉)
Department of Civil Engineering, GSKSJTI, Bengaluru, India
e-mail: mail2subbu.kg@gmail.com

L. Govindaraju
Department of Civil Engineering, UVCE, Bengaluru, India

R. Ramesh Babu
Central Power Research Institute, Bengaluru, India

© Indian Society of Earthquake Technology 2023

R. S. Jakka et al. (eds.), *Earthquake Engineering and Disaster Mitigation*,
Springer Tracts in Civil Engineering, https://doi.org/10.1007/978-981-99-0081-7_12

pile foundations has been witnessed in past earthquakes. The forensic analysis after the failure of the Hansin expressway in the 1995 Kobe earthquake revealed the determinantal effect of Dynamic SSI [1]. Dynamic SSI is very important, especially for heavy and huge structures like bridges, dams, tall buildings and nuclear power plants resting on soft soil [2].

The incorporation of SSI in the analysis allows flexibility at the base of the structure resulting in the elongated period and high damping [3], which is usually considered to be beneficial to the structural system. In contrast to the inertial interaction, the soil-pile-structure interaction may increase the response of the superstructure due to the translation and rotation of the foundation (Kinematic Interaction). The combined effect of both interactions can lead to a determinantal response during excitations with frequencies near resonance [4]. The structures resting on soft soil is much prone to catastrophic effects as the soft soil can amplify ground motion and intensify the problem [5].

There have been several cases in which SSI has been studied experimentally. Boulanger et al. [6] carried out dynamic p-y analysis with centrifuge tests on models subjected to base excitation and found reasonable agreement. To access the effect of pile configuration, pile cap and superstructure, Durante et al. [7] carried out laboratory investigations using a shake table. Kinematic and inertial components were observed in the response to centrifuge model tests conducted by Hussien et al. [8] for various pile configurations. A series of centrifuge experiments were conducted by Martakis et al. [9] to find the effect of soil properties and structural properties on SSI. Most of the experimental investigations on SSI are shake table studies or centrifuge tests on scaled models. There are only a few cases where dynamic tests on a full-scale model were carried out, such as a large-scale gantry railway bridge by Zangeneh et al. [10]. As such experiments are often complex and expensive, researchers have studied model studies and numerical modelling strategies to analyse the effects of SSI.

Pile foundations are commonly used when the structure has to be built on soft soil. The use of composite foundation is becoming popular in recent years from the point of view of excessive and differential settlement when appropriate load bearing soil does not exist. The composite foundation consists of a raft with a few number of friction piles called a piled raft foundation [11]. The piled raft can be named as large piled raft and small piled raft based on the dimensions of the raft and piles beneath the raft. If the size of the raft is smaller than the length of the pile, it is classified as a small piled raft [12]. The primary reason for providing the piles in the case of a small piled raft foundation system is to achieve a factor of safety against the bearing failure. The present study focuses on the experimental investigations of the dynamic soil-pile-structure interaction of a 10-storey scaled-down model structure representing the prototype structure using appropriate scaling laws. A series of laboratory experiments on scaled building models have been carried out using the Shake table facility. The response of the model structure was investigated for the fixed base and flexible base condition considering two types of foundation systems, such as buildings supported on pile groups and those supported on small piled raft foundations in soft clay.

Numerical studies have been carried out to compare the response of the prototype and the scaled model structure.

12.2 Construction of Scaled Model Representing Prototype Structure

A prototype structure of moment resisting building frame having 10 storeys is considered for the study. The dynamic response of the prototype building frame was evaluated for both fixed base and flexible base conditions considering the soil-structure interaction. Two types of foundations such as pile group and piled raft foundations in soft clay were considered for the analysis in flexible base condition. Figure 12.1 shows the schematic diagram highlighting the scope of the investigation.

Earthquake loading is one of the important design considerations, because of its destructive nature. The behaviour of building frames to earthquake force is an important input parameter in analysis and design structural systems. In order to understand the realistic dynamic behaviour of the structure, one can conduct experimental investigations. It is always very difficult to analyse the prototype or full-scale model in a laboratory, hence model tests are conducted by fabricating a small-scale physical model which represents a full-scale prototype structure.

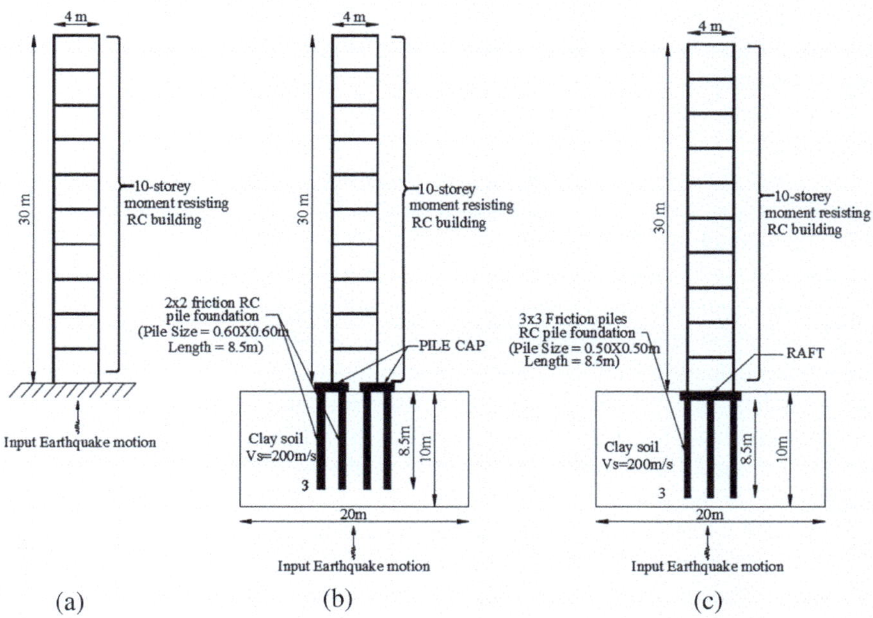

Fig. 12.1 **a** Prototype building with fixed base; **b** Prototype building supported by pile group; **c** Prototype building supported by pile raft

Table 12.1 Summary of similitude factors for earthquake response of structures [4]

Parameter	Scaling relationship
Mass density	1
Force	λ^3
Modulus	λ
Acceleration	1
Shear wave velocity	$\lambda^{1/2}$
Time	$\lambda^{1/2}$
Frequency	$\lambda^{-1/2}$
Length	λ
Stress	E
Strain	1
EI	λ^5

Adopting a suitable scaling factor for scaling the prototype structure is one of the important steps in fabricating a small-scale physical model representing the prototype. Weixing [13] discussed the similarity theory of the high-rise building model for the shaking table experiment and suggested a scaling factor of 1:25–1:50 for reinforced concrete high-rise buildings because of the weight and height of the building. Hokmabadi et al. [4] conducted a series of shake table experiments on a 5-, 10- and 15-storey scaled-down model with a 1:30 scaling factor. In the present study, a scaling factor of 1:30 is adopted considering the size of the shake table and the prototype structure for both fixed base and flexible base conditions. Several researchers have proposed scaling factors for different types of investigations. The dynamic similitude relation presented by Hokmabadi et al. [4] is adopted in the present study as tabulated in Table12.1.

Similitude analysis is a mathematical framework for relating the quantities measured in a scaled model to those measured in the actual building (prototype). The scaled models can be classified as true, adequate and distorted to the degree to which scaled models meet the requirements of similitude law. The true model meets all of the similitude criteria. The first-order (controlling) parameters of an appropriate model maintain similarity, while the second-order parameters are allowed to deviate from an adequate model and the second-order parameters are those that have a minor impact on the analysis. If both first-order and second-order parameters deviate from the similitude law, then it is a distorted model [14]. It is always very difficult to create a true model but usually; an adequate model will be sufficient to meet the requirements of the study.

The scaled models can be defined as having geometrical, kinematic or dynamic similitude similarities [15]. Geometric similarity defines a model and prototype with homologous physical dimensions. Kinematic similarity refers to a model and prototype with a homologous particle at homologous points at a homologous time. Dynamic similarity describes a condition where homologous parts of the model and prototype experience homologous net forces. For the purpose of scaling the model

Table 12.2 Properties of prototype and scaled model

Parameter	Prototype	Scaled model (1:30)
Time period	1.0 s	0.18 s
Frequency	1.0 Hz	5.5 Hz
Super structure material	Concrete	Aluminium
Young's modulus	25 GPa	69 GPa
Size of column	450 mm × 450 mm	10 mm × 3 mm
Slab thickness	150 mm	8 mm
Bay dimension	4 m	135 mm

through dynamic similitude, the methodology presented by Hokmabadi et al. [4] has been adopted in the present work.

In order to scale down the superstructure, a modal analysis of the prototype structure was carried out using a finite element approach. Based on the results of modal analysis, the structure is scaled to 1:30 for dynamic model studies. The similitude factors adopted for the study are shown in Table 12.1. The scaled model parameters are found by satisfying the equivalent stiffness in the fundamental mode direction and keeping the mass density ratio the same as in the prototype and the scaled model [4, 14]. Numerical analyses were carried out to match the similitude parameters such as natural frequency and stiffness using the trial and error method by varying the cross section of the column and the thickness of the slab. Lumped symmetrical mass was also considered at each floor level to achieve the same mass density ratio of the model as that of the prototype. The properties of the scaled model are tabulated in Table 12.2.

Similar to the scaling of the superstructure, the scaling criteria should also be applied to the substructure. By adopting geometric similarity, the model pile also should satisfy the scaling flexural rigidity. It requires a low-strength material for modelling pile, pile cap and raft, hence acrylic material was used to fabricate the substructure system. Figure 12.2 shows the schematic representation of a fabricated small-scale physical model.

12.2.1 Model Soil Container for SSI

Model tests on soil require an infinite soil medium to be modelled with a finite boundary, however, the confining effect as in continuous medium needs to be incorporated in the soil model. This is achieved by confining the soil in a model container. Many researchers have used laminar container, which generally contains a stack of laminae supported on bearings.

Initially, researchers conducted experiments using rigid soil containers. The drawback of the rigid soil container is that the waves get reflected from the walls of the

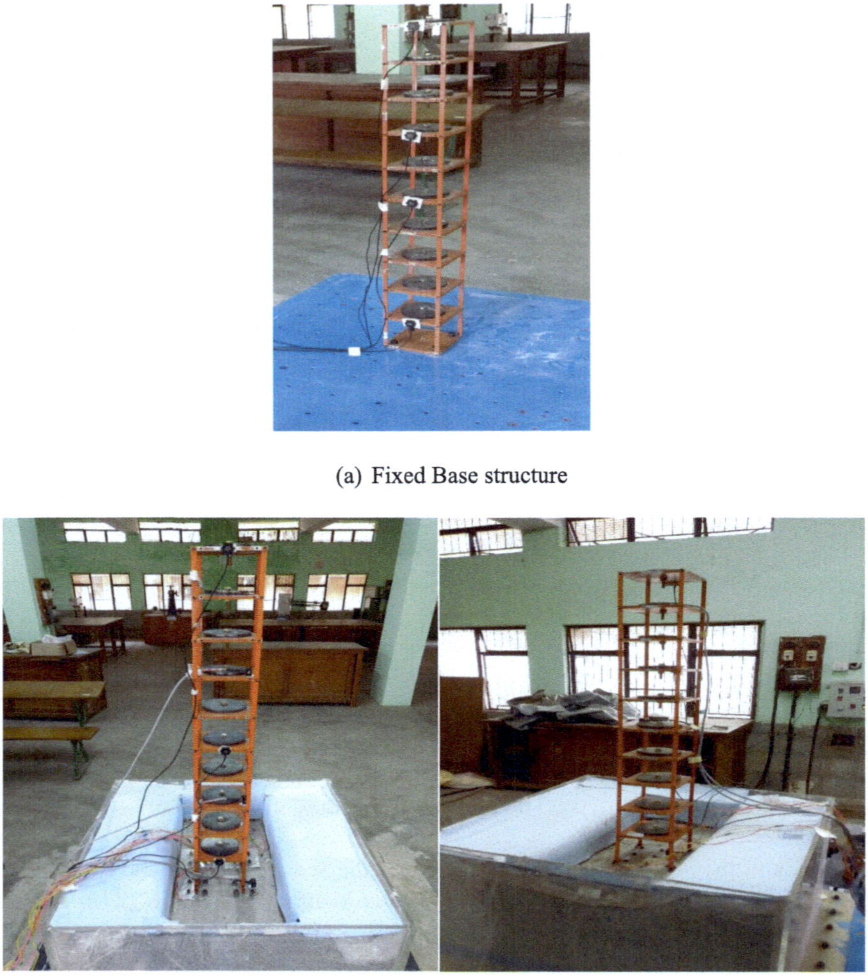

(a) Fixed Base structure

(b) Flexible Base Structure- Pile group (c) Flexible Base Structure- Piled Raft

Fig. 12.2 Experimental setup of scaled models: **a** fixed base; **b** flexible base supported by pile group; **c** flexible base supported by piled raft

container due to the artificial boundary. Numerical studies have reported that zones up to 1.5–2 times the height of the container close to walls are affected by artificial boundaries and to minimise the artificial boundary effect, the ratio of the length to the height of the container is more than 4 [16]. To overcome the drawbacks of the rigid container, several researchers have studied and proposed alternatives or modifications to the rigid container. Lombardi et al. [16] used an absorbing boundary to a rigid wall container to limit the reflection of waves from the boundary. The flexible boundary is introduced by glueing a soft material like a sponge along the end walls of the container.

In the present study, a rigid soil container is used for soil-pile-structure interaction studies. The size of the tank is so designed that the length of the tank is five times the width of the model and also a flexible boundary is introduced by glueing a very low-density sponge along the end walls of the container in order to minimise the boundary effect.

12.2.2 Preparation of Soil Mix

Based on the scaling relationships, the model soil must have a shear-wave velocity of 36 m/s to satisfy the required dynamic similitude between the prototype soil of shear wave velocity of 200 m/s. Meymand et al. [17] worked on a synthetic clay mix which can offer the required bearing capacity underneath the structural model and can also meet the requirement of the shear-wave velocity. To find a suitable soil mix for conducting laboratory experiments, an artificial synthetic clay mix was designed using kaolinite clay, bentonite, Class F fly ash, lime and water by Hokmabadi et al. [4]. The same mix proportions were used with trials and the density of 1480 kg/m^3 was found. The density and the undrained shear strength of the clay mix were in good agreement with the values reported by the author. Further, the same clay mix with the proportion of 60% kaolinite clay, 20% bentonite, 20% Class F fly ash and lime with 100% water was used for the synthetic clay mix.

12.3 Analysis of Prototype Structure and Scaled Model

The dynamic analysis of the scaled model was carried out experimentally and numerical analysis of the prototype structure was carried out using the finite element approach. The scaled response of the model was compared with the response of the prototype structure.

12.3.1 Experimental Investigation

Shake table tests were conducted on scaled models that represent a prototype to assess the dynamic behaviour. All tests were carried out on a 1.0 m × 1.0 m shake table; the details of the facility are outlined in Table 12.3. Experimental investigations were conducted on the model structure under different base conditions. Firstly, the models were tested under fixed base conditions and subjected to harmonic loading at different frequencies. The dynamic properties like fundamental frequency and modal damping

Table 12.3 Details of shaking table

Component	Discerption of property	
Shake table	Size of table	1.0 m × 1.0 m
	Frequency range	1–25 Hz
	Max displacement of table	±10 mm
	Frequency control	0.05 Hz (±5%)
	Motion	Simple harmonic (unidirectional)

were calculated. The response of the structure in terms of lateral displacement and acceleration was recorded with the help of Data Acquisition System (DAQ).

The primary aim of the study was to find the effect of the soil and foundation system on the dynamic behaviour of the framed structure. Two types of foundations, namely pile and piled raft foundations, were designed for the prototype structure. Model pile and pile raft systems were fabricated by applying dynamic similitude relations using acrylic material. Synthetic clay in a rigid soil container with absorbing boundaries was used to represent the soil medium. The model supported on these foundation systems in synthetic clay was tested to understand the behaviour of the soil-foundation-structure system. Dynamic tests were performed to find the dynamic properties of the soil-foundation-structure system and its response to dynamic excitation. The details of the experimental program are outlined in Table 12.4. The response of the model under different base conditions is compared.

12.3.2 Numerical Analysis

In the present research, numerical modelling and analysis of prototype structure with and without soil-structure interaction analysis were performed by finite element method using ETABS and SAP2000 FEA software [18] and are compared with the scaled experimental response. SAP2000 was used to model the soil-structure system and solve equations for complex geometries and boundary conditions. The width of the soil mass considered was five time the width of the model in order to minimise the boundary effect [16]. Structural elements beam and columns were modelled as a one-dimensional frame element, and slabs were modelled as a two-dimensional shell element. In addition to the modelling of the superstructure, the substructure and soil are modelled as a continuum model [19, 20]. The pile cap, pile and soil are modelled with an 8-noded three-dimensional solid element. The periphery of the soil boundary was constrained in translation degrees, and the base of the soil mass is fixed assuming that the soil below is hard strata [21]. Experimentally computed damping ratio for fixed and flexible systems was updated in the numerical model for dynamic analysis. A scaled harmonic motion was generated based on the captured base motion of the experimental model for different base conditions. The numerical models were

Table 12.4 Detailed experimental program

S. no	Experiment type	Test type	Peak acceleration (input) (m/s^2)	Frequency (Hz)	Objective of the experiment
1	Fixed base	Free vibration	–	–	To determine the fundamental time period and frequency
2	Fixed base	Forced vibration	0.74	2–10 Hz	Response of fixed base for harmonic loading
3	Flexible base condition with pile (2X2) group in soft clay	Free vibration	–	–	To determine the fundamental time period and frequency of the system
4	Flexible base condition with pile group in soft clay	Forced vibration	0.74	2–10 Hz	Response of soil-structure system for harmonic loading
5	Flexible base condition with piled raft in soft clay	Free vibration	–	–	To determine the fundamental time period and frequency
6	Flexible base condition with piled raft in soft clay group in soft clay	Forced vibration	0.74	2–10 Hz	Response of soil-structure system for harmonic loading

analysed for the scaled dynamic excitation, and the response of the scaled model and prototype are compared. Numerical model of the prototype is presented in Fig. 12.3.

12.4 Results and Discussion

In the first stage of dynamic analysis of the prototype and models were carried out for fixed base condition to find the dynamic response. Flexible base condition was considered in the second stage of analysis with the building resting on a pile foundation and piled raft foundation.

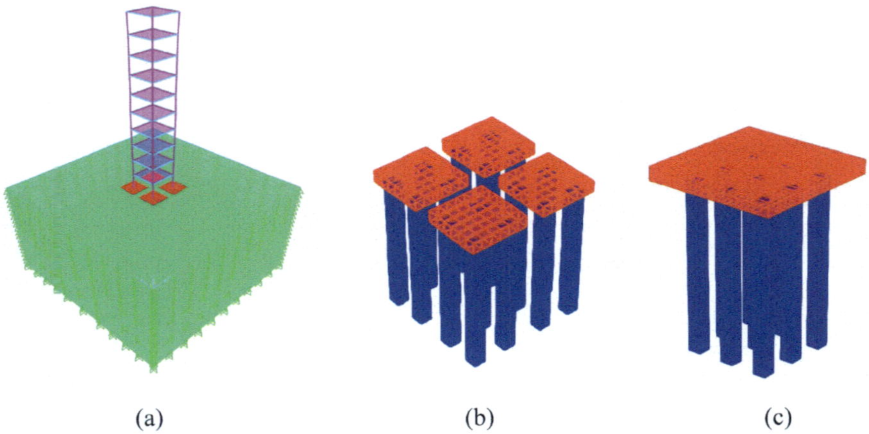

Fig. 12.3 Three-dimensional view of numerical model **a** prototype flexible base; **b** pile group; **c** piled raft

12.4.1 Fixed Base Condition

Free vibration was performed initially to estimate the fundamental frequency of the fixed base structure and the first fundamental frequency of the scaled model was found to be 6.1 Hz. The model was subjected to motion with a sine sweep, and frequency is varied from 2 to 10 Hz with an increment of 0.25 Hz. The response of the model was recorded with the help of DAQ. It is found that the response of the model has reached a maximum at the resonant frequency. The variation of top-storey displacement for harmonic base excitation for different frequencies is presented in Fig. 12.4.

Fig. 12.4 Frequency versus top-storey displacement

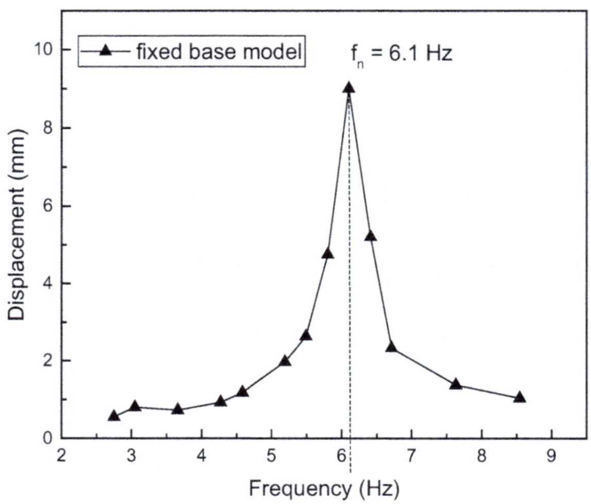

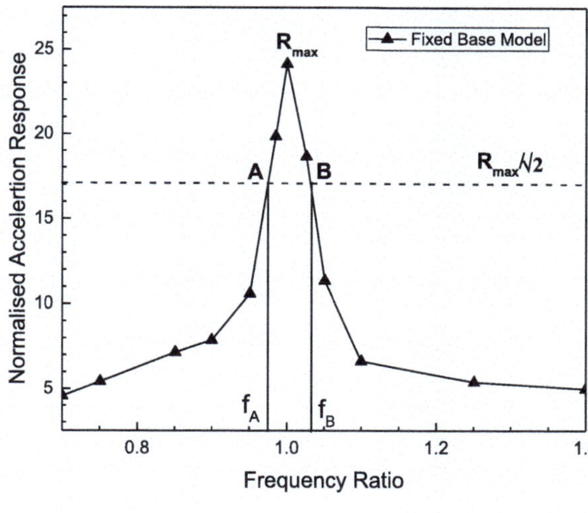

Fig. 12.5 Top-storey displacement verses frequency for fixed base model

The damping value of the model is calculated by the half power bandwidth method considering the frequency response curve. Normalised response of the model was calculated by taking the ratio of top-storey acceleration to base acceleration plotted against the frequency ratio [22]. On the response curve at harmonic forced vibrations, points A and B are identified corresponding to the maximum response multiplied by $1/\sqrt{2}$. The damping ratio is given by Eq. 12.1:

$$\zeta = \frac{f_B - f_A}{2f_r} \tag{12.1}$$

where $f_A and f_B$ are corresponding frequencies of points A and B. f_r is frequency ratio corresponding to maximum response. The damping ratio was calculated from Fig. 12.5 using the half power bandwidth method and was found to be 2.75% for the fixed base model.

Harmonic force is generated by elongating the timestep according to the resonant frequency of the prototype structure. An amplitude of 0.74 m/s^2 acceleration was recorded at the base of shake table at resonance. The harmonic excitation of the model and the corresponding scaled harmonic load for prototype respectively are presented in Fig. 12.6. The Fourier amplitude spectrum of the harmonic loading is presented in Fig. 12.7. It can be observed that the frequency of the prototype is nearly $\sqrt{30}$ times that of the prototype. Numerical modelling and dynamic analysis were carried out using finite element software adopting the computed damping ratio. The analysis was carried out by exciting the prototype structure with the generated harmonic motion. The displacements at different storey levels of both the scaled model and prototype are shown graphically in Fig. 12.8.

From Fig. 12.8, it can be seen that the deviation in the response of the prototype and the scaled model are well within the acceptable limits [22]. The inter-storey

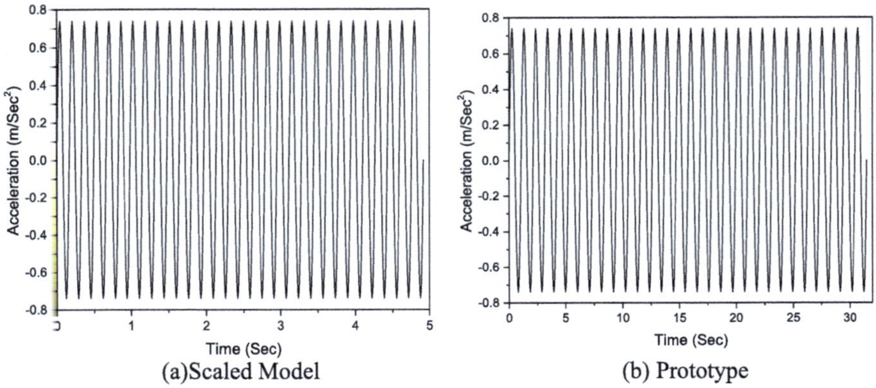

Fig. 12.6 Harmonic base excitation for **a** scaled model; **b** prototype

drifts were computed using Eq. 12.2 for both model and prototype structures and are illustrated in Fig. 12.9.

$$D_{(i,i+1)} = \frac{d_{i+1} - d_i}{h} \tag{12.2}$$

where $D_{(i,\,i+1)}$ is inter-storey drift between the (i) and $(i + 1)$ floor; d_{i+1} is deflection at the $(i + 1)$ floor; d_i is deflection at the (i) floor; h is distance between the floors under consideration.

12.4.2 Flexible Base Condition

Flexible base analysis was the second stage of experimental investigation to study the effect of the SSI on the dynamic response of the structural system. Acrylic material was used to fabricate the model piles, pile cap and raft. The pile cap and the piles were connected with screws and intern the pile cap was connected to the column. The strain gauges were placed near the pile head, mid-length and near the free end. The stain gauges were sealed with a waterproof cover to ensure the workability of the strain gauges. The entire model was installed carefully ensuring all the connections.

After connecting all of the components of the system, the accelerometers were placed on the model structures at different floor levels and in addition to that an accelerometer was placed at the pile cap and raft. Similar events were performed to estimate the fundamental frequency and the response of the flexible base models (pile group and piled raft). The models were subjected to motion with sine sweep, and frequency is varied from 2 to 10 Hz with an increment of 0.25 Hz. The response of the model was recorded with the help of DAQ. It was found that the response of the models reached maximum at a resonant frequency of 5.3 Hz and 4.2 Hz respectively

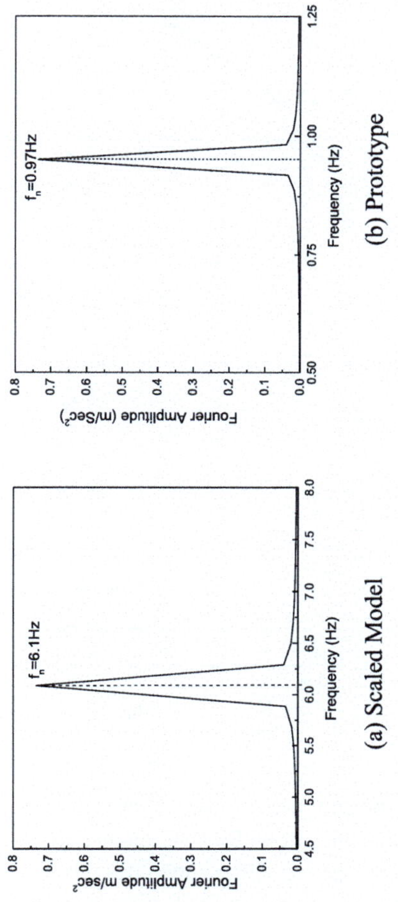

Fig. 12.7 Fourier amplitude spectrum

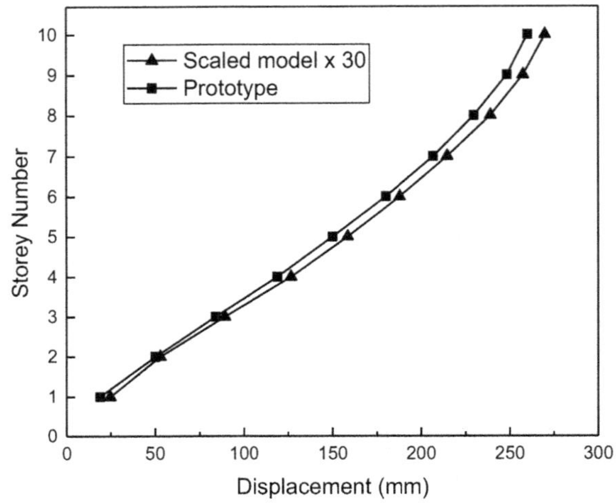

Fig. 12.8 Variation of displacement along height of the structure

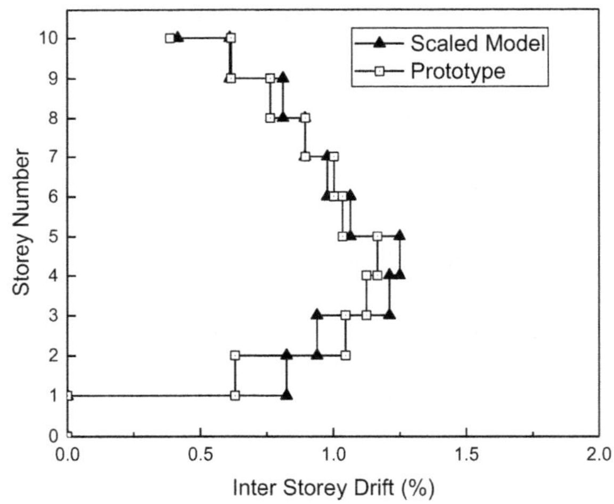

Fig. 12.9 Variation storey drift along height of the structure

for flexible base supported on the pile and piled raft foundations. Thus, it shows that the soil-pile-structure system alters the fundamental dynamic property of the structure thereby altering the response of the superstructure to the dynamic loading. It can be observed from Table 12.5 that the damping has increased for the flexible base condition. The piled raft system shows higher damping than that of the model supported on a pile foundation.

The storey response of the models with varied frequencies for different base conditions is shown in Fig. 12.10. It is observed that the response of the flexible base models has shown an increase in the response to base excitation in comparison with the fixed base model. The structure supported on piled raft shows a decrease in the response compared to the structure supported on a pile foundation because of increased damping and stiffness.

Table 12.5 Damping ratio for flexible base models

Model type	Frequency response curve—half power bandwidth Method	Damping ratio
Scaled model supported on pile group in soft clay	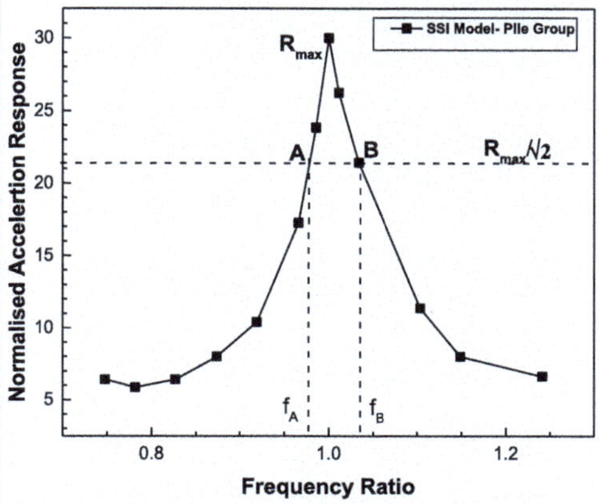	$\zeta = \frac{f_B - f_A}{2}$ $\zeta = 3\%$
Scaled model supported on piled raft in soft clay	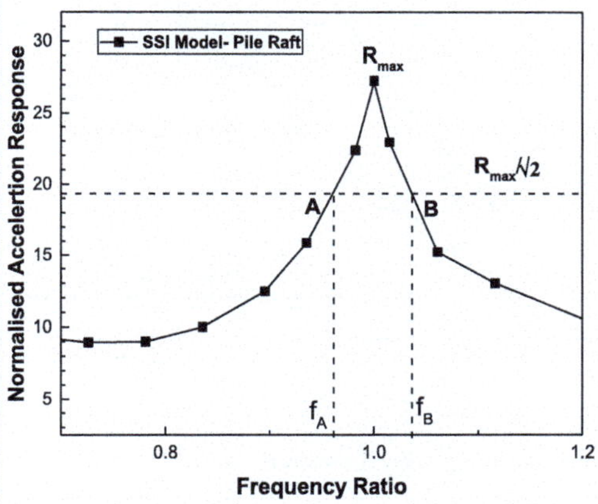	$\zeta = 3.3\%$

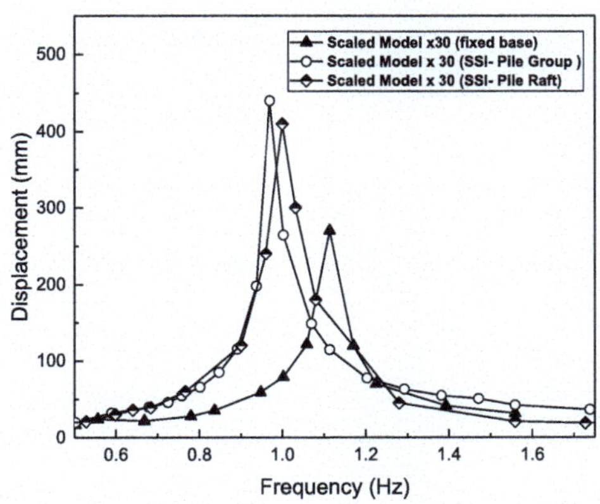

Fig. 12.10 Top-storey displacement verses frequency of scaled models

Displacements at different floor levels are presented in Fig. 12.12 for both the scaled model and prototype structure supported on the pile foundation system and piled raft system, respectively. The response of the model and prototype are in good correlation. The flexible base models show an increase in top-storey displacement in comparison with the fixed base condition. The model supported on the pile group shows higher displacement than the model supported on the piled raft as depicted in Fig. 12.11. The model supported on pile foundation and piled raft shows an increase in top-storey displacement of about 63% and 48% respectively compared to the fixed base model.

The inter-storey drifts of flexible base models were computed using Eq. 12.2 and are presented in Fig. 12.13. The storey drift value has increased with the flexible

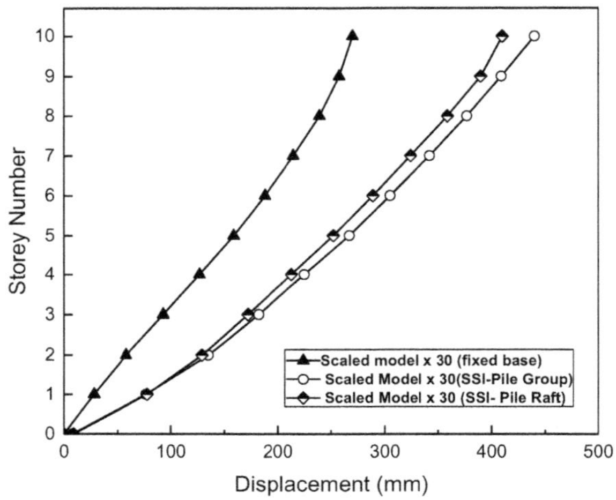

Fig. 12.11 Storey displacement (mm) for scaled models

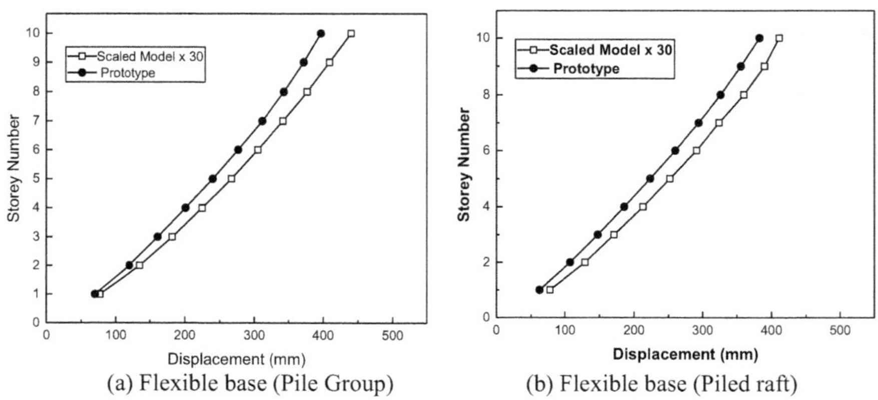

Fig. 12.12 Variation of displacement along height of the structure

base condition compared to the fixed base condition. The maximum storey drift for the fixed base was found to be near the mid-height of the structure, whereas the SSI model shows a maximum drift near the first floor because of the increased flexibility at the base. The same behaviour has also been reported by earlier studies [4, 23, 24]. The piled raft case shows lower drift values compared to the model supported on a pile foundation.

The stresses in the pile were measured by means of a strain gauge. The arrangement of the strain gauge is depicted in Figs. 12.14 and 12.15. The bending moments were calculated based on the bending strain measured along the length of the pile. It can be observed from Fig. 12.16 that the exterior piles exhibit a higher value of bending moments compared to interior piles. The bending moment is more near the pile head. The same pattern is observed in piles beneath the raft also. Bending strains in piled raft case is less compared to pile group condition. The bending stress is within the permissible limit.

The increase in the lateral response of the flexible base is also dependent on the rocking foundation. The total lateral displacement of the superstructure in SSI has rocking and distortion components [4]. The effect of the rocking of the raft and the lateral displacement is depicted in Fig. 12.17. The maximum rocking angle of the piled raft foundation was found to be 0.087°. Thus, the displacements due to rocking and inertia were found to be 54.6 mm and 355.4 mm, respectively. The model structure supported on piled raft foundation exhibits an average of 13.3% of the maximum lateral response due to the rocking component, while 86.6% was due to the distortion component.

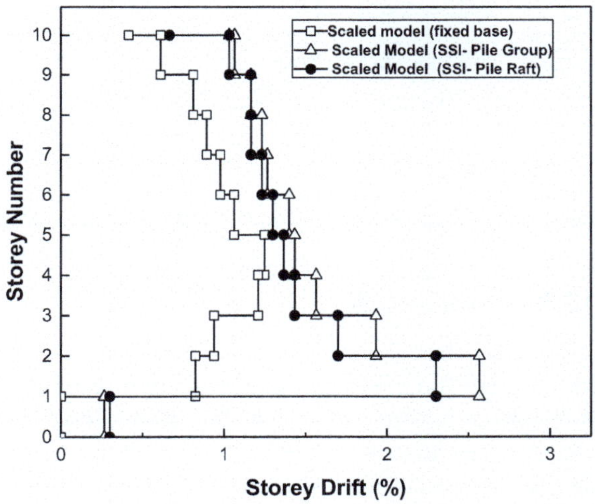

Fig. 12.13 Variation storey drift along height of the structure

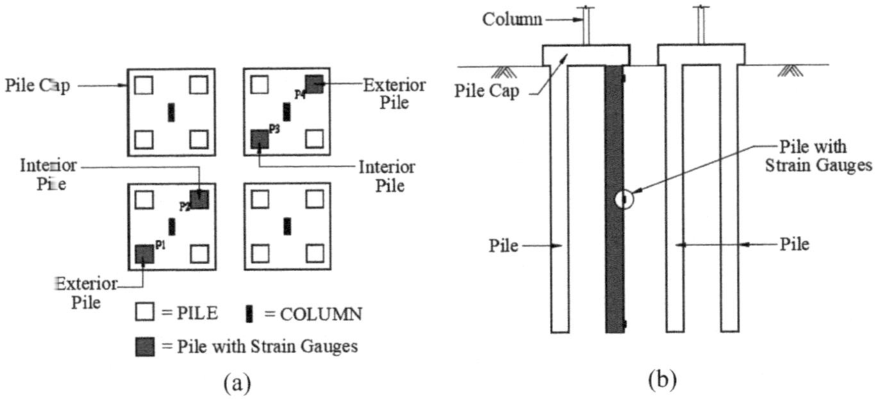

Fig. 12.14 Schematic representation of model pile and pile cap: **a** Plan; **b** Sectional elevation

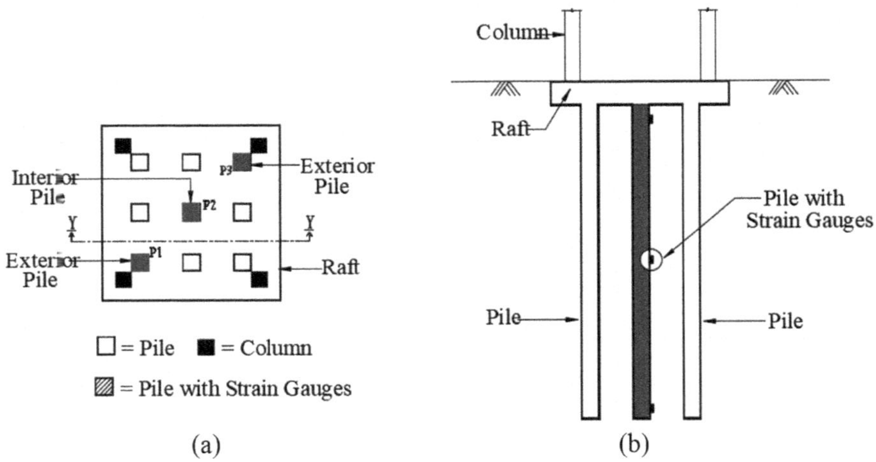

Fig. 12.15 Schematic representation of model piled raft: **a** Plan; **b** Sectional elevation

12.5 Conclusion

In this study, an effort has been made to investigate the dynamic behaviour of the soil-foundation-structure system. Shake table tests were conducted on a scaled model which represents a prototype. Tests were conducted with the fixed base condition and flexible base condition considering pile foundation and piled raft foundation in soft clay. It is observed that the dynamic properties and the response of the superstructure were altered by considering the foundation and the subsoil. The lateral deflection of the building was amplified in the case of a building with a flexible base in comparison with the fixed base indicating the significant role of foundation and subsoil on the response of the superstructure.

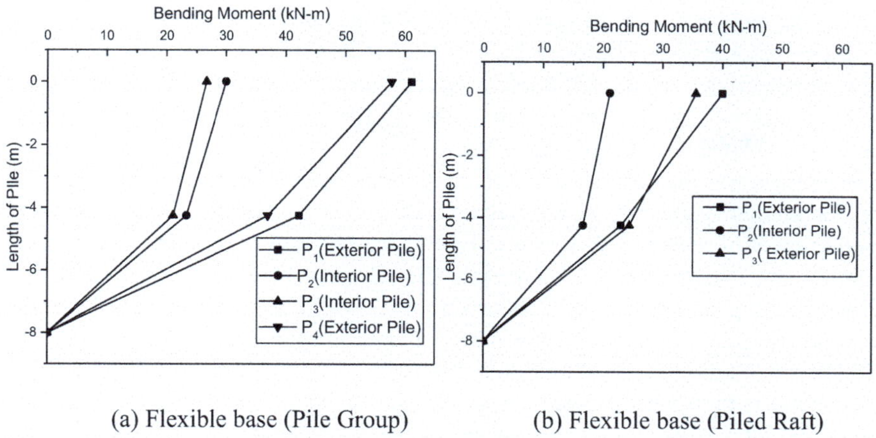

(a) Flexible base (Pile Group) (b) Flexible base (Piled Raft)

Fig. 12.16 Variation of bending moment along the length of pile at resonance

Fig. 12.17 Rocking and inertial components of displacement in piled raft system

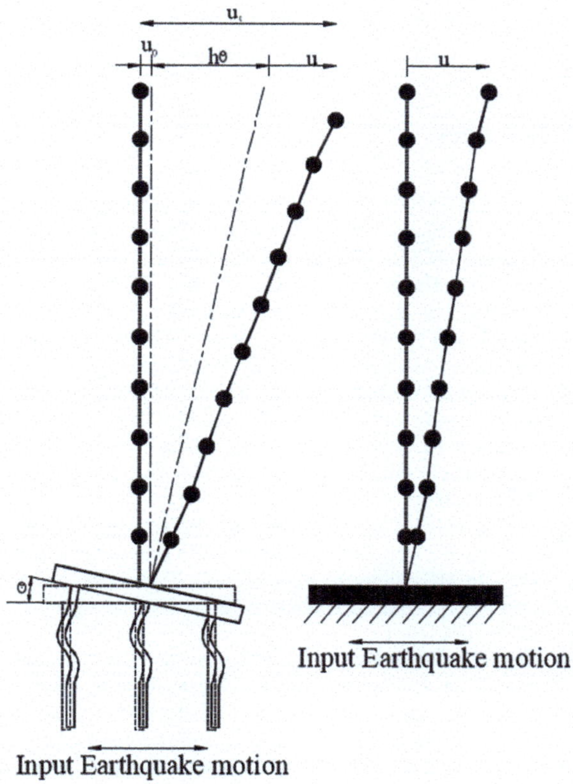

The damping ratio has increased for the SSI model compared to the fixed base model on account of flexibility of the soil. The model supported on piled raft system shows higher damping and reduction in the response to dynamic loading compared to the model supported on the pile foundation. The piled raft system shows better resistance to lateral load compared to the pile group system. The increase in lateral displacement was due to kinematic and inertial interaction. The response of the piled raft foundation system shows the influence of the rocking of the raft on the lateral displacement of the framed structure. Due to the increase in the inertial force, the bending strain in the pile has increased. The pile group foundation system shows higher bending strains in piles compared to piles with a raft. It is observed that the bending strains are more in exterior piles than interior piles.

The increased response of the framed structure supported on the soil-foundation system shows higher drift values compared to fixed base conditions. The increased flexibility at the base of the structure in the SSI model resulted in the increase of storey drift on lower floors. The increase in the storey drift can change the performance level of the building. It is more realistic to consider the soil-foundation-structure system in the seismic design of buildings resting on soft soil.

References

1. Gazetas G, Mylonakis G (1998) Seismic soilstructure interaction: new evidence and emerging issues. In: State of the art paper, geotechnical earthquake engineering and soil dynamics geo-institute ASCE conference, vol II, pp 1116–1174
2. Gatmiri B, Haeri SM (1996) Guidelines of SSI analysis and its effects on dynamic response of structures. In: Natural Disaster Defense Research Center, First Edition (in Farsi)
3. FEMA 440 (2005) Improvement of nonlinear static seismic analysis procedures. Federal Emergency Management Agency, Washington, DC, U.S.A
4. Hokmabadi AS, Fatahi B, Samali B (2015) Physical modeling of seismic soil-pile-structure interaction for buildings on soft soils. Int J Geomech 15(2):04014046
5. Meymand PJ, Riemer M, Seed RB (2000) Large scale shaking table tests of seismic soil-pile interaction in soft clay. In: Proceeding of the 12th world congress on earthquake engineering, Paper no. 0915
6. Boulanger RW, Curras CJ, Kutter BL, Wilson DW, Abghari A (1999) Seismic soil-pile structure interaction experiments and analyses. J Geotech Geoenviron Eng 125(9):750–759
7. Durante MG, Sarno L, Mylonakis G, Taylor CA, Simonelli AL (2015) Soil-pile-structure interaction: experimental outcomes from shaking table tests. Earthq Eng Struct Dynam 45(7):1041–1061
8. Hussien MN, Tobita T, Iai S, Karray M (2016) Soil-pile-structure kinematic and inertial interaction observed in geotechnical centrifuge experiments. Soil Dyn Earthq Eng 75–84
9. Martakis P, Taeseri D, Chatzi E, Laue J (2017) A centrifuge-based experimental verification of soil-structure interaction effects. Soil Dyn Earthq Eng 103:1–14
10. Zangeneh A, Svedholm C, Andersson A, Pacoste C, Karoumi R (2018) Identification of soil-structure interaction effect in a portal frame railway bridge through full-scale dynamic testing. Eng Struct 159:299–309
11. Nakai S, Kato H, Ishida R, Mano H, Nagata M (2004) Load bearing mechanism of piled raft foundation during earthquake. In: Proceedings third UJNR workshop on soil-structure interaction, Menlo Park, California, USA

12. Mali S, Singh B (2019) 3D numerical modelling of large piled-raft foundation on clayey soils for different loadings and pile-raft configurations. Studia Geotechnica et Mechanica 1–17
13. Weixing SHI (2000) Shaking table experimental study of reinforced concrete highrise building. In: Proceedings of the 12th world congress on earthquake engineering, Paper no. 0218
14. Harris HG, Sabnis G (2000) Structural modelling and experimental techniques, 2nd edn. CRC Press, Boca Raton
15. Sulaeman A (2010) The use of lightweight piles for deep foundation on soft soils. PhD thesis in Civil Engineering, University of Tun Hussein Onn, Malaysia
16. Lombardi D, Bhattacharya S, Scarpa F, Bianchi M (2015) Dynamic response of a geotechnical rigid model container with absorbing boundaries. Soil Dyn Earthq Eng 69:46–56
17. Maymand P, Reimer M, Seed R (2000) Large scale shaking table tests of seismic soil-pile interaction in soft clay. In: Proceedings 12th world conference earthquake engineering New Zealand, vol 5, no 0915
18. SAP, Advanced 14.2.4. Computers and Structures Inc., University Avenue Berkeley, California, United State (2000)
19. Pulikanti S, Ramancharla PK (2014) SSI analysis of framed structure supported on pile foundations-with and without interface elements. Front Geotech Eng (FGE) 1(3)
20. Hirave V, Kalyanshetti M (2018) Seismic response of steel braced building frame considering soil structure interaction (SSI): an experimental study. J Inst Eng (India) Ser A 99(1): 113–122
21. Tabatabaiefar HR, Fatahi B, Samali B (2012) An empirical relationship to determine lateral seismic response of mid-rise building frames under influence of soil-structure interaction. Struct Des Tall Spec Build 23: 526–548
22. Cruciat R, Ghindea C (2012) Experimental determination of dynamic chracteristics of structures. Math Model Civil Eng 4:51–59
23. Tahghighi H, Mohammadi A (2020) Numerical evaluation of soil–structure interaction effects on the seismic performance and vulnerability of reinforced concrete buildings. Int J Geomech 20(6):04020072
24. Kabtamu HG, Peng G, Chen D (2018) Dynamic analysis of soil structure interaction effect on multi story RC frame. Open J Civil Eng 8(04):426

Chapter 13
Seismic Performance of a Masonry Infilled Reinforced Concrete Frame Building Designed as per Indian Codes

M. Sharma, Y. Singh, and H. Burton

Abstract In the Indian subcontinent and many other parts of the world, it is a common practice to use masonry infills as partitions in framed buildings. Masonry infills are widely known to interact with the surrounding frames and change the dynamic characteristics and seismic performance of the frame buildings. However, the current Indian design codes do not adequately cover the design of frame buildings with masonry infills. As a result, the designers continue to design frame buildings while ignoring the infills. It is assumed that the infills added to the frame buildings will increase the strength and stiffness of frame buildings and hence enhance the seismic performance. This article reviews provisions of different national codes regarding the design of masonry infilled frames. Further, the effectiveness of the current design procedure being followed in the Indian design industry is evaluated. Performance-based seismic engineering framework is used to quantify the performance of frame designed using the mentioned design procedure. Subsequently, the performance is also compared with equivalent bare frame and fully infilled frame.

Keywords Performance estimation · Masonry infilled reinforced concrete frames · Multiple stripe analysis

13.1 Introduction

The introduction of masonry infills complicates the behavior of reinforced concrete frames due to frame-infill interaction. The addition of infills significantly increases

M. Sharma (✉) · Y. Singh
Deparment of Earthquake Engineering, IIT Roorkee, Roorkee, India
e-mail: msharma@eq.iitr.ac.in

Y. Singh
e-mail: yogendra.singh@eq.iitr.ac.in

H. Burton
Department of Civil Engineering, University of California, Los Angeles, CA, USA
e-mail: hvburton@ucla.edu

© Indian Society of Earthquake Technology 2023

R. S. Jakka et al. (eds.), *Earthquake Engineering and Disaster Mitigation*,
Springer Tracts in Civil Engineering, https://doi.org/10.1007/978-981-99-0081-7_13

the strength of the frame. However, increase in strength is not equal to a direct summation of strength of the infills and bare frame [30]. Addition of infills also introduces multiple failure modes due to frame-infill interaction and possible introduction of vertical or horizontal irregularities. Therefore, it is difficult to say whether the addition of infills improves or degrades the performance of infilled frames. There are multiple studies that suggest that introduction of infills improve the performance of frames [7, 21, 24, 26]. On the other hand, there are few studies and reconnaissance reports that seem to suggest otherwise [9, 10, 20, 28]. The predominant failure pattern observed in the reconnaissance reports is failure of adjacent concrete members by stronger infills. A possible explanation for poor performance of infilled frames could be that due to their higher stiffness, infilled frames are more severely loaded during an earthquake.

The current practice in the Indian design industry is to ignore the strength and stiffness of infills during the design process. It is assumed that this practice is conservative as infills only add to the lateral resistance of the frame. However, few reconnaissance reports have identified disregard for infills during the design process as one of the reason for large damage sustained by reinforced concrete frames [22]. The predominant failure mode reported by Li et al. [22] was simultaneous occurrence of column ductile flexural yielding and infill brittle cracking due to frame-infill interaction. Designers usually lack the knowledge that introduction of infills also introduces multiple failure modes and possible irregularities in the infilled frames.

Haldar et al. [16] identify four failure modes for infilled panels, i.e., sliding shear failure, diagonal tension, diagonal compression and corner crushing. Stavridis and Shing [30] based on the experiments done by Mehrabi et al. [26] identify three failure modes. They combine both diagonal tension and compression under a single mode named diagonal cracking. Sliding shear failure tends to happen in ductile frames with weak infill wherein sliding takes place within the masonry bed joints accompanied by flexural yielding in columns. Diagonal cracking tends to happen in nonductile frames with strong infills and eventually leads to the brittle shear failure of the adjacent column (Fig. 13.1). Corner crushing of infills usually takes place in case of strong frames with relatively weak infills and is often accompanied by flexural yielding in columns.

Additionally, the introduction of infills also introduces possible vertical or horizontal irregularities in the structure. Often, infills are omitted in the ground story as it is being used as space for vehicle parking or commercial establishments. This leads to vertical irregularity in such buildings. Open ground story buildings are known to perform significantly worse than their fully infilled counterparts. Asymmetric planar placements of infills in a building lead to torsion. Buildings with dominant torsional modes are known to perform poorly than comparable symmetric buildings. Further, buildings with partially infilled panels produce enhanced shear demand in the adjacent columns which could lead to shear failure of adjacent columns (Fig. 13.2). As the strength and stiffness of infills are ignored during the design of infilled frames, these issues are left unnoticed.

Damage to infilled frames during an earthquake is most severe in infills and adjacent columns. Relatively fewer instances of damage to the beams are observed. Haldar

Fig. 13.1 Diagonal cracking in an infill panel and shear failure in adjacent column

et al. [16] attribute this observation to the construction sequence being followed during the construction of infilled frames. Usually, the frames are constructed first, and they are filled with infills at a later stage (Fig. 13.3). This allows for a small gap between the top of the infill panel and the bottom of the beam. The small gap permits movement between the infill and beam and causes beams to sustain lesser damage during earthquakes.

This study aims to compare the results from two design approaches mentioned in the Indian codes and quantify the performance of the resulting frames. Initially, the frame is designed while ignoring the strength and stiffness of the infills but considering their weight. Thereafter, the frame is designed while modeling infills as compression-only diagonal struts. Bending moment, axial force diagrams and reinforcement requirement are compared in both cases. Thereafter, a performance-based seismic engineering framework is used to quantify the performance of the resulting frame. Eight sets of 40 ground motions representing 8 return periods are selected to match the site-specific spectrum for a site located in Dehradun. The collapse performance of the frame is quantified using the fragility curves obtained from non-linear Multiple Stripe Analysis (MSA). The performance of the frame is compared with the performance of a bare frame and a fully infilled frame.

Fig. 13.2 Shear failure of columns due to partial infills

13.2 Code Provisions

When it comes to the design of reinforced concrete infilled frames, most building codes do not provide working details. IS code (IS 1893: Part 1, 2016) mandates the use of a lower time period for infilled frames as compared to bare frames. IS code also mandates the modeling of infills in case the infill plan density increases beyond 20%. However, such a high value of infill plan density is rarely reached in modern reinforced concrete frames. Additionally, the code states that the vertical irregularity must be checked while considering the strength and stiffness of infill walls. IS code also provides an empirical formula to calculate the equivalent width of the diagonal strut, w_{ds}. The formula originally proposed by Stafford Smith and Carter [29] has since been reproduced in multiple codes [1, 6, 13].

The formula is given in Eq. 13.1:

$$w_{ds} = 0.175\alpha_h^{-0.4} L_{ds} \tag{13.1}$$

$$\alpha_h = h\left[\sqrt[4]{\frac{E_m t \sin 2\theta}{4E_f I_c h}}\right]$$

Fig. 13.3 Construction sequence being followed during construction of infilled frame building

where

L_{ds} Diagonal length of the infill panel
E_m Modulus of elasticity of masonry
E_f Modulus of elasticity of RC frame
t Thickness of the infill panel
θ Angle of the diagonal strut with the horizontal
I_c Moment of Inertia of the adjoining column
h Height of the infill panel

However, this expression was removed from subsequent versions of ASCE 41 [2, 3] and was replaced by an expression of total in-plane stiffness of uncracked infill frame based on the work of Stavridis et al. [31]. The frame is assumed as an I-shape cantilever column with concrete columns acting as flanges and masonry walls acting as the web of the column. The total stiffness of the frame is estimated as the resultant stiffness of this cantilever column.

Eurocode 8 (2004) is based on the philosophy to prevent any adverse effect of the infills while not profiting explicitly from the beneficial ones to reduce the seismic action effects in structural members [12]. Eurocode 8 (2004) states that the shear capacity of the adjacent reinforced concrete column needs to be checked for the shear force generated by the infill diagonal struts. But it doesn't provide details on modeling the equivalent diagonal strut. Fardis [12] recommends using Eq. 13.1 to calculate the equivalent width of diagonal struts. A set of empirical equations are developed by Huang et al. [18] through calibration with experimental data of 113 infilled frames using multi-variate regression analysis. These equations have been used in the later portion of this paper to define the backbone curve of diagonal struts for non-linear analysis.

13.3 Design Procedure

The design of a representative infilled reinforced concrete frame is illustrated using the two approaches suggested by the Indian code. Firstly, the strength and stiffness of the infills are ignored, whereas their weight is included in the design. This approach represents the current practice being followed by structural designers in India. Secondly, the infills are included in the design and are modeled as compression-only diagonal struts. The equivalent width of the diagonal strut is obtained from the formula provided in the IS code [6]. Both approaches use the reduced infilled frame period of vibration given in Eq. 13.2 [6] and have the same design base shear. Popular design software ETABS is used for the design. The reinforcement requirements are compared for both methods to comment on the effect of ignoring infill strength and stiffness on the design process of infilled reinforced concrete frames:

$$T_a = \frac{0.09h}{\sqrt{d}} \quad (13.2)$$

The representative frame is a fully infilled 8-story 3-bay reinforced concrete frame assumed to be located in central Dehradun, a city in northern India. The frame is assumed to be infilled with fair quality infills with a compressive strength of 4.13 MPa. The elevation of the frame along with the plan of the building is shown in Fig. 13.4. The frame is designed and detailed as per the relevant Indian codes [6, 8]. A zone factor of 0.24 (Zone IV), applicable to Dehradun, is considered in the design. The soil is assumed to be medium or stiff soil. A response reduction factor of 5 and an importance factor of 1 are used during the design process. All design parameters remain exactly the same during both approaches.

In the first approach, the frame is designed while ignoring the infill strength and stiffness but considering infill weight whereas in the second approach, the frame is designed by modeling the infills as compression-only diagonal struts using the modeling recommendations of IS codes [6]. The bending moment and axial force diagrams, for the seismic (response spectrum) load case, obtained using both approaches are shown in Figs. 13.5 and 13.6, respectively. While the axial force diagram obtained is close for both approaches, there is a large difference in the

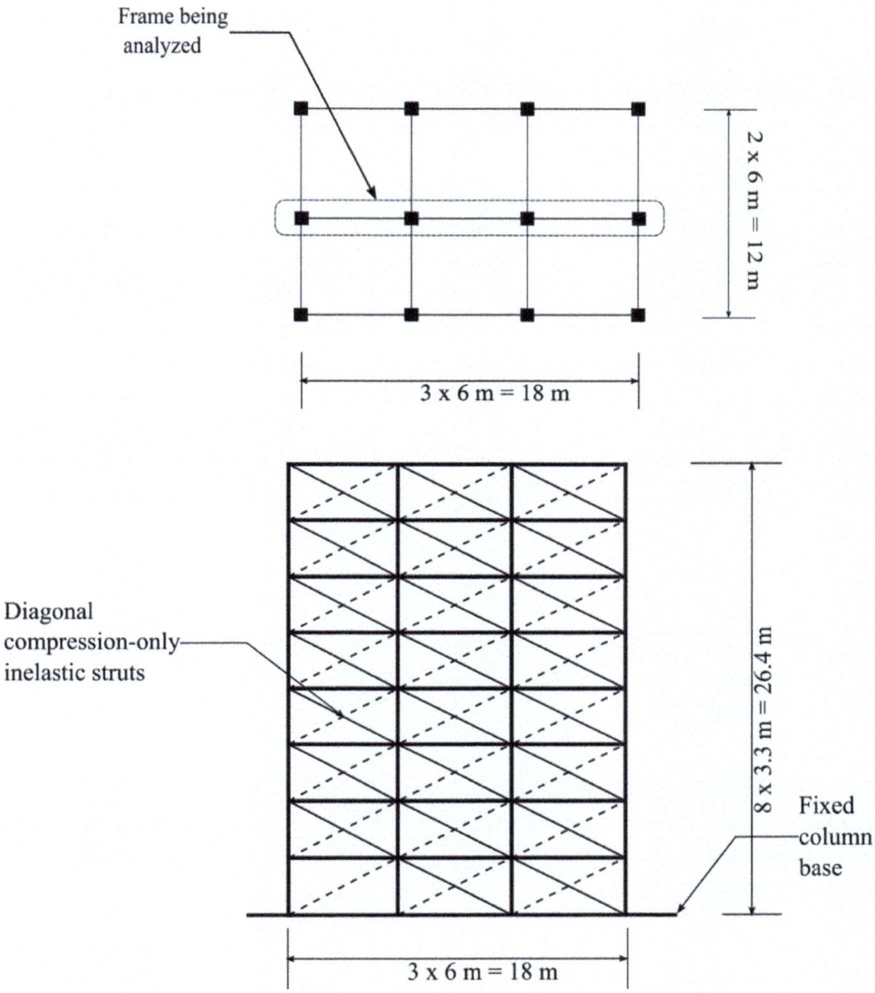

Fig. 13.4 Plan of the building and elevation of the frame being analyzed

bending moment diagram. The bending moment is greatly reduced when infills are considered in the model for analysis. The comparison of the reinforcement requirements obtained from ETABs for both approaches is given in Table 13.1. It can be seen that while including infills as diagonal struts in the model, the longitudinal reinforcement requirements in the frame members are drastically decreased. This makes sense as the action of the moment frame changes to that of a braced frame when infills are modeled. The reinforcement requirement is coherent with the bending moment demands in the frame members.

The reinforcement requirement is higher while ignoring infills. Therefore, it can be concluded that ignoring the stiffness of infills in the case of infilled frames designed

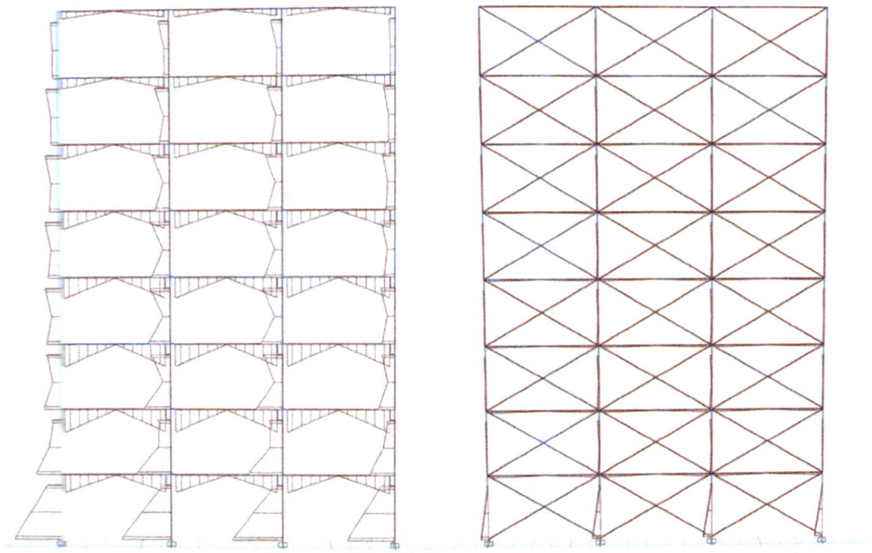

(a) (b)

Fig. 13.5 Bending moment diagram for the seismic load case: **a** without modeling of infills; **b** with explicit modeling of infills

using the code-specified reduction in the period results in a conservative design as compared to the case when the stiffness of the infills is explicitly modeled. However, as recommended by the code it is always a good practice to check the design using both approaches.

13.4 Performance Assessment

In this section, we quantify and compare the performance of two frames, with and without considering the strength and stiffness of the infills. Both frames are designed for the code-specified reduced design period and mass of infill. Therefore, the design base shear remains the same for both frames. The performance is also compared with a third frame, which is bare (without any infills), and designed using the period specified for bare frame buildings by the code [6]. The first frame represents the current practice being followed in the design industry and is the same as the frame designed 'without modeling of infills' in Table 13.1. The 8-story frame is named '8SBFI'. The second frame named '8SIF' is the same as 8SBFI, except that it includes infills modeled as non-linear compression-only diagonal struts. The third frame is an 8-story bare frame, named '8SBF', and is designed using the design period specified for the bare frame.

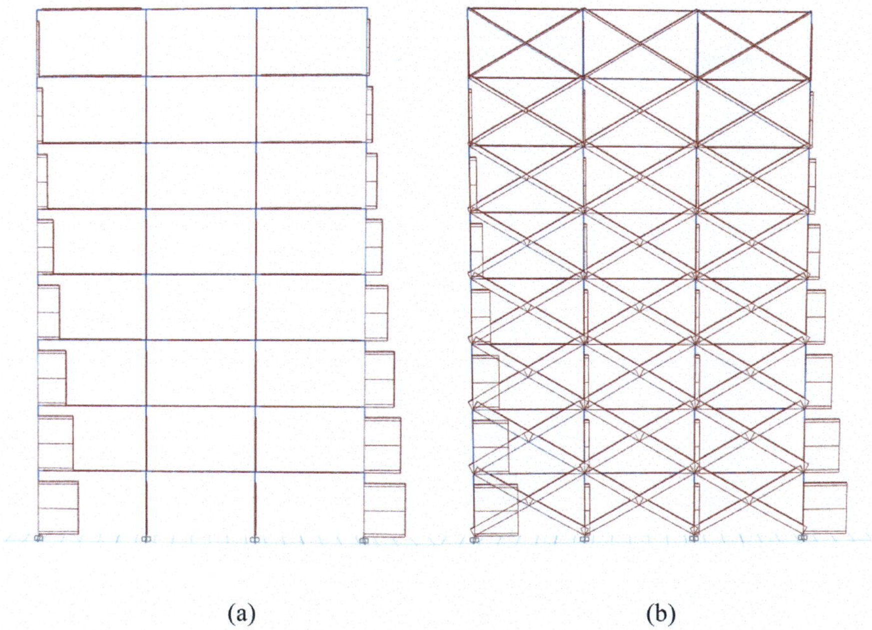

Fig. 13.6 Axial force diagram for the seismic load case: **a** without modeling of infills; **b** with explicit modeling of infills

Non-linear structural models for the three frames are developed in *OpenSees*. Columns and beams are modeled as elastic elements with lumped plastic hinges at the ends. Both flexural and shear hinges are assigned at the column ends. Flexural hinges use the peak-oriented Ibarra–Medina–Krawinkler model [19], whereas the shear hinges use the Elwood shear failure model [11]. Backbone parameters for all flexural hinges are calculated from the predictive equations developed by Haselton et al. [17]. Backbone parameters for column shear hinges are calculated from equations developed by Elwood [11].

Masonry infills are modeled as compression-only struts using a non-linear truss element. Pinching4 material, also known as Lowes–Mitra–Altoontash [23] material in *OpenSees*, is used to simulate the masonry infills. Backbone parameters for the infill struts are obtained from the empirical equations developed by Huang et al. [18]. The model for an intermediate infill frame is shown in Fig. 13.7.

MSA is performed for all frames using eight stripes selected at return periods of 500, 1000, 1500, 2000, 2500, 3500, 4500 and 5700 years. A suite of 40 unidirectional ground motions is selected for each stripe. The number of collapse cases are noted at each stripe, and the maximum likelihood method [4] is then used to fit a fragility curve to the observed fractions of collapse cases at different intensity levels. Collapse is defined as the occurrence of any one of the following two conditions. Firstly, if the maximum inter-story drift ratio at any story exceeds 0.1, this drift limit is consistent with the collapse drift limit used in previous studies [5, 15, 32]. Secondly, if the

Table 13.1 Comparison of the reinforcement requirements obtained while ignoring infills versus while modeling the infills in elastic analysis for design

Analysis performed	Columns				Beams			
	Story levels	Longitudinal reinforcement (%)	Dimension (mm)	Shear reinforcement	Story levels	Longitudinal reinforcement (%)	Dimension (mm)	Shear reinforcement
Without modeling of infills	Base to 1st	3.76	550 × 550	4L—T10 @ 100 mm	1st	1.31	300 × 550	4L—T8 @ 100 mm
	2nd to 3rd	1.81			2nd to 5th	1.55		
	4th to 5th	0.87			6th to 7th	1.11		
	6th to 8th	1.32			8th	0.52		
With explicit modeling of Infills	Base to 1st	0.80	550 × 550	4L—T10 @ 100 mm	1st	0.42	300 × 550	4L—T8 @ 100 mm
	2nd to 3rd	0.80			2nd to 5th	0.42		
	4th to 5th	0.80			6th to 7th	0.39		
	6th to 8th	0.80			8th	0.22		

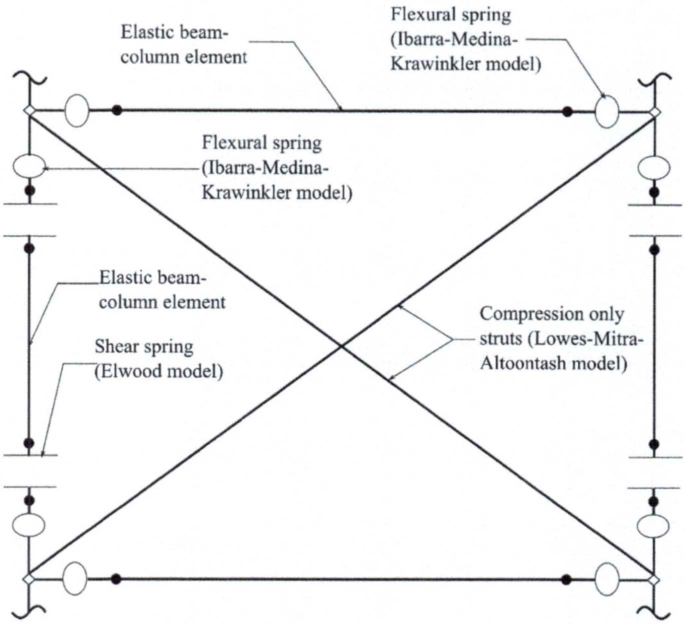

Fig. 13.7 Detailed OpenSees model of one of the infill bays

shear strength of any column reduces to zero, it leads to loss of its axial load carrying capability.

All three frames follow a ductile failure pattern. Initial yielding of lower story struts (if present) is followed by flexure failure yielding of lower story beams, followed by flexural yielding of lower story columns and finally collapse (if any) due to shear failure of lower story columns. Fragility curves in terms of normalized intensity measure ($\overline{IM} = \mathrm{Sa}(T_1)/\mathrm{Sa}(T_1)_{MCE}$) for the reference frames are shown in Fig. 13.8. The collapse performance results are summarized in Table 13.2.

It can be observed that 8SBFI performs better than both competing frames. At lower intensities performance of 8SIF and 8SBFI is very close to each other, however, at higher intensities, 8SBFI performs considerably better. This is because 8SBFI while having the same frame design as 8SIF has significantly lesser seismic demands at MCE (0.5 g compared to 1.22 g). 8SBF performs poorly with the probability of collapse at MCE at an unacceptable 30% [14]. It can be concluded that, in this case, the bare frame, 8SBFI designed with a design period of infilled frame, while ignoring the infills but considering their weights, performs better than the fully infilled frame, 8SIF, and significantly better that the bare frame, 8SBF. This observation further enforces the finding of the previous section that in the case of an infilled frame designed with the infilled frame design period, the current design industry approach to ignore the infills while considering their weight is safe.

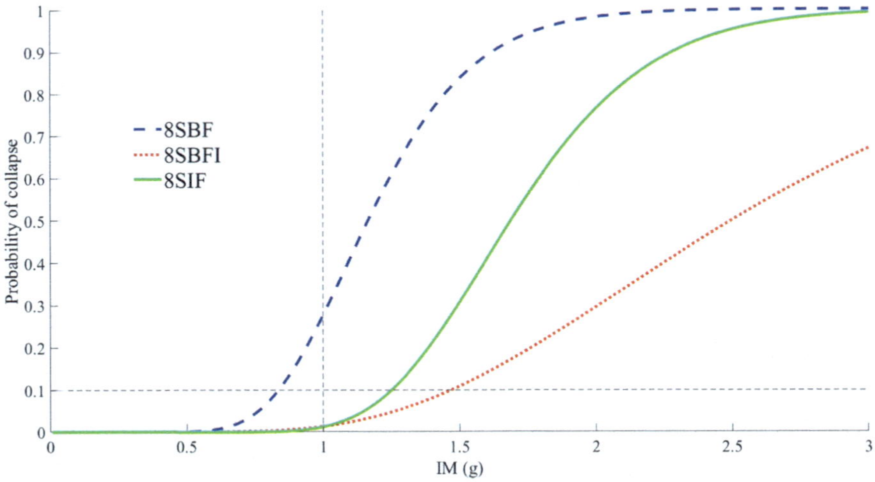

Fig. 13.8 Normalized fragility curves for the three frames

Table 13.2 Comparison of collapse performance of the three frames using MSA

Frame ID	$SaT_{1,MCE}$ (g)	Median collapse capacity (g)	Record-to-Record VARIABILITY	P (collapse) at MCE (%)
8SBFI	0.50	1.26	0.42	1.38
8SIF	1.22	2.00	0.23	1.68
8SBF	0.35	0.41	0.26	29.60

13.5 Conclusion

Design clauses, governing the design of masonry infilled frames, of Indian and other commonly referred national codes have been briefly reviewed. Designs resulting from two design approaches suggested in the Indian code have been compared. It is found that the bending moment and the reinforcement requirement are significantly higher when the strength and stiffness of infills are ignored during design, compared to when the stiffness is included by means of compression-only diagonal struts. Hence, the current industry practice to ignore the strength and stiffness of the infills during design has been found to be reasonably conservative, in this case.

Subsequently, the performance-based seismic engineering framework has been used to compare the performance of the resulting frame (8SBFI) with the corresponding fully infilled frames (8SIF), and a bare frame (8SBF). Probability of collapse at MCE, estimated from multiple stripe analysis, has been used as the metric for the comparison. It has been found that the performance of 8SBFI is superior to both the bare frame and the fully infilled frame. The probability of collapse of 8SBF is more than 20 times that of 8SBFI. This result indicates the large effect that the control over the design period based on empirical relations in the code can have on the seismic performance of buildings. The performance of 8SBFI is much closer to 8SIF. The probability of collapse of 8SIF is 21% more than the probability of collapse of 8SBFI. Both the frames have the same design period, base shear and design of reinforced concrete members. The only difference is the presence of infills, simulated as compression-only non-linear struts, in 8SIF. The results hint that the presence of infills unaccompanied by any other design change has a negative effect on seismic performance.

Acknowledgements The first author is supported by the fellowship granted by Ministry of Education (MoE), Government of India. The authors gratefully acknowledge the financial support provided by the Ministry of Human Resource Development (MHRD), Government of India.

References

1. ASCE (2007) Seismic evaluation and retrofit of existing buildings. 41-06. American Society of Civil Engineers, Reston, VA
2. ASCE (2014) Seismic evaluation and retrofit of existing buildings. 41-13. American Society of Civil Engineers, Reston, VA
3. ASCE (2017) Seismic evaluation and retrofit of existing buildings. 41-17. American Society of Civil Engineers, Reston, VA
4. Baker JW (2015) Efficient analytical fragility function fitting using dynamic structural analysis. 21
5. Burton HV, Sharma M (2017) Quantifying the reduction in collapse safety of main shock-damaged reinforced concrete frames with infills. Earthq Spectra 33(1):25–44. https://doi.org/10.1193/121015eqs179m
6. Criteria for earthquake resistant design of structures—Part 1 : General provisions and buildings. IS 1893: Part 1: 2016. Bureau of Indian Standards (2016)
7. Dolšek M, Fajfar P (2008) The effect of masonry infills on the seismic response of a four-storey reinforced concrete frame—a deterministic assessment. Eng Struct 30(7):1991–2001. https://doi.org/10.1016/j.engstruct.2008.01.001
8. Ductile design and detailing of reinforced concrete structures subjected to seismic forces—Code of practice. IS 13920: 2016. Bureau of Indian Standards (2016)
9. Dymiotis C, Kappos AJ, Chryssanthopoulos MK (2001) Seismic reliability of masonry-infilled RC Frames. 10
10. Eberhard MO, Baldridge S, Marshall J, et al (2014) Haiti earthquake 2010: USGS/EERI Advance Reconnaissance Team Report
11. Elwood KJ (2004) Modelling failures in existing reinforced concrete columns. Can J Civ Eng 31(5:)846–859
12. Fardis MN (2009) Seismic design, assessment and retrofitting of concrete buildings: based on EN-Eurocode 8, vol 8. Springer, Berlin
13. FEMA 356 (2000) Prestandard and commentary for the seismic rehabilitation of buildings, p 519
14. FEMA P695 (2009) Quantification of building seismic performance factors. Prepared by ATC for FEMA
15. Galanis PH, Moehle JP (2015) Development of collapse indicators for risk assessment of older-type reinforced concrete buildings. Earthq Spectra 31(4):1991–2006. https://doi.org/10.1193/080613EQS225M
16. Haldar P, Singh Y, Paul DK (2013) Identification of seismic failure modes of URM infilled RC frame buildings. Eng Fail Anal 33:97–118. https://doi.org/10.1016/j.engfailanal.2013.04.017

17. Haselton CB, Liel AB, Taylor-Lange SC, et al (2016) Calibration of model to simulate response of reinforced concrete beam-columns to collapse. ACI Struct J 113(6). https://doi.org/10.14359/51689245
18. Huang H, Burton HV, Sattar S (2020) Development and utilization of a database of infilled frame experiments for numerical modeling. J Struct Eng 146(6):04020079. https://doi.org/10.1061/(ASCE)ST.1943-541X.0002608
19. Ibarra LF, Medina RA, Krawinkler H (2005) Hysteretic models that incorporate strength and stiffness deterioration. Earthquake Eng Struct Dynam 34(12):1489–1511. https://doi.org/10.1002/eqe.495
20. Kam WY, Pampanin S, Elwood K (2011) Seismic performance of reinforced concrete buildings in the 22 February Christchurch (Lyttelton) earthquake. Bull N Z Soc Earthq Eng 44(4):239–278. https://doi.org/10.5459/bnzsee.44.4.239-278
21. Lee H-S, Woo S-W (2002) Effect of masonry infills on seismic performance of a 3-storey R/C frame with non-seismic detailing. Earthq Eng Struct Dynam 31(2):353–378. https://doi.org/10.1002/eqe.112
22. Li B, Wang Z, Mosalam KM, et al (2008) Wenchuan earthquake field reconnaissance on reinforced concrete framed buildings with and without masonry infill walls, 8
23. Lowes LN, Mitra N, Altoontash A (2003) A beam-column joint model for simulating the earthquake response of reinforced concrete frames. University of California, Berkeley
24. Madan A, Hashmi AK (2008) Analytical prediction of the seismic performance of masonry infilled reinforced concrete frames subjected to near-field earthquakes. J Struct Eng 134(9):1569–1581. https://doi.org/10.1061/(ASCE)0733-9445(2008)134:9(1569)
25. Maffei J, Comartin CD, Kehoe B, et al (2000) Evaluation of earthquake-damaged concrete and masonry wall buildings. http://journals.sagepub.com/doi/10.1193/1.1586111. Accessed Sept 10, 2021
26. Mehrabi AB, Benson Shing P, Schuller MP et al (1996) Experimental evaluation of masonry-infilled RC frames. J Struct Eng 122(3):228–237. https://doi.org/10.1061/(ASCE)0733-9445(1996)122:3(228)
27. Open System for Earthquake Engineering Simulation (OpenSees) (2000) Pacific Earthquake Engineering Research Center, Univ. of California, Berkeley
28. Sattar S, Eeri M and Liel AB (2015) Seismic performance of nonductile reinforced concrete frames with masonry infill walls: II. Collapse assessment, 31
29. Stafford Smith B, Carter C (1969) A method of analysis for infilled frames. Proc Inst Civ Eng 44(1):31–48. https://doi.org/10.1680/iicep.1969.7290
30. Stavridis A, Shing P (2010) Finite-element modeling of nonlinear behavior of masonry-infilled RC frames, 12
31. Stavridis A, Tempestti JM, Bose S (2017) Updating the ASCE 41 provisions for infilled RC frames, 8
32. Surana M, Singh Y, Lang DH (2018) Fragility analysis of hillside buildings designed for modern seismic design codes. Struct Design Tall Spec Build 27(14):e1500. https://doi.org/10.1002/tal.1500

The manufacturer's authorised representative in the EU is Springer Nature Customer Service Centre GmbH, Europaplatz 3, 69115 Heidelberg, Germany. If you have any concerns regarding our products, please contact ProductSafety@springernature.com

Printed and bound by CPI Group (UK) Ltd, Croydon, CR0 4YY
30/09/2026
02225304-0006